Emerging Pathogens

Emerging Pathogens

The Archaeology, Ecology, and Evolution of Infectious Disease

EDITED BY

Charles L. Greenblatt

Department of Parasitology
Hebrew University-Hadassah Medical School
Kuvin Centre for the Study of Infectious and Tropical Diseases
Jerusalem, Israel

Mark Spigelman

Department of Medical Microbiology
University College of London
London, UK
and
Hebrew University-Hadassah Medical School
Kuvin Centre for the Study of Infectious and Tropical Diseases
Jerusalem, Israel

OXFORD
UNIVERSITY PRESS

This book has been printed digitally and produced in a standard specification
in order to ensure its continuing availability

OXFORD
UNIVERSITY PRESS

Great Clarendon Street, Oxford OX2 6DP

Oxford University Press is a department of the University of Oxford.
It furthers the University's objective of excellence in research, scholarship,
and education by publishing worldwide in

Oxford New York

Auckland Cape Town Dar es Salaam Hong Kong Karachi
Kuala Lumpur Madrid Melbourne Mexico City Nairobi
New Delhi Shanghai Taipei Toronto
With offices in
Argentina Austria Brazil Chile Czech Republic France Greece
Guatemala Hungary Italy Japan South Korea Poland Portugal
Singapore Switzerland Thailand Turkey Ukraine Vietnam

Oxford is a registered trade mark of Oxford University Press
in the UK and in certain other countries

Published in the United States
by Oxford University Press Inc., New York

© Oxford University Press, 2003

The moral rights of the author have been asserted

Database right Oxford University Press (maker)

Reprinted 2006

ISBN 0-19-850901-4

Preface

Why this book and what it attempts to do

Motivation for this volume was based on a number of factors. Perhaps foremost was the sheer lack of material on pathogens presented in the ancient DNA (aDNA) meetings and literature. 'Ancient DNA' is a relatively new field in which the tools of molecular biology are being applied to archaeological materials. At the 'Ancient DNA III' meeting held in 1995 at Oxford, only three presentations in more than 50 dealt with infectious diseases. In the Research Reports of Ancient Biomolecules Initiative (Ancient Biomolecules Initiative 1996), one report dealt with halophilic bacteria, but none with pathogenic forms. It seemed to me and to a number of others that ancient pathogens seemed too good a target to be neglected by the molecular biologists. Powerful new tools have the potential to detect ancient biomolecules that are either the DNA or the chemical signatures of pathogens. Very recently Martin Jones (2001) has written enthusiastically of how these techniques are changing archaeology, and certainly this is true for the study of ancient pathogens.

The polymerase chain reaction (PCR) was one such tool (Mullis and Faloona 1987). It allows the retrieval of minute amounts of DNA and its amplification from biological samples. A few copies of intact DNA present in samples where the majority of the molecules are damaged and degraded provide enough targets for PCR. The special genius of PCR and its inventors (Mullis and Faloona 1987) was to utilize an enzyme from a thermophilic bacterium, which is resistant to the elevated temperatures required for strand separation. Therefore no additional enzyme is required for each cycle and by avoiding this the reaction is possible for both

technical and financial reasons. Successful and specific amplification is dependent on a number of factors including the uniqueness of the sequences at the chosen target site, the damage incurred to the DNA in millennia of storage, the accumulation of inhibitors, etc. Reading out the DNA sequences is somewhat akin to PCR, but uses the basic letters of the genetic code—labelled either radioactively or with fluorescent dyes.

Since some pathogens combine the properties of multiple copies, in some cases an appropriate stable envelope, and an essentially simple DNA, they provide the best opportunities for successful amplification by PCR. However, at various times during the first years of aDNA research there was a certain unease. Artifacts caused by contamination and the failure to reproduce the work of other laboratories produced an occasional sense of dispair. In our own laboratory we would interface modern work with the aDNA studies just to make certain that students would survive the frustration of aDNA work. At least the modern robust DNA yielded results! I personally was delighted when three published works raised my confidence in the entire endeavor. These were the studies of Krings *et al.* (1997, 1999) on Neandertal man and Taubenberger *et al.* (1997) on the influenza virus of the great pandemic of 1918. In the first study, several independent teams successfully amplified and cloned a portion of the mitochondrial genome. The work was elegantly done and the results independently confirmed by the different teams. The retrieval of DNA from such an ancient specimen (30 000 years old) was an important demonstration of the power of the techniques of molecular biology. The second major contribution is described in this volume by Taubenberger and Reid (Chapter 16). The influenza work, although performed on material that some of you would call comparatively modern (a little over 80 years), was also a *tour de force*. Fixed paraffin sections, from soldiers who died in the 1918 flu pandemic, were extracted for their viral RNA. Frozen material from an Eskimo who died in the same pandemic also provided viral sequences. This kind of medical archaeology is the subject of this book.

The problems cited above have not gone away, but as we understand the limits and potential of these tools they are being utilized in addressing questions regarding ancient emerging diseases. There is no question that techniques will improve and today's methods in a few years will seem primitive, inefficient, and lacking in sensitivity.

Those organisms that enter the bloodstream, offer another advantage because they lodged in bones or teeth, where the DNA seems best preserved. One of the first goals of this new field was to join the 'classical' palaeopathologists, with their wonderful skeletal materials, in confirming the diagnoses based essentially on morphology. Beyond this beginning we felt intuitively that if we knew the nature of ancient pathogens and the host responses over evolutionary history we would be in a better position to meet the challenge of the emerging diseases.

It seemed that a book that focuses on ancient pathogens might launch a critical look at the subject so as to bridge a gap that separates today's infectious disease specialists and the palaeopathologists who had described pathology in skeletal and mummified remains. Such a linkage between the people looking at old diseased bones and those worrying over patients sick with infectious diseases, when complemented by the views of the anthropologists, medical ecologists, and molecular biologists might just help to fill the missing gap. To this end, a symposium was held in Jerusalem in 1997, which led to the publication of a monograph (Greenblatt 1998). This work suffered from attempting to do too much, but I believe it succeeded in stimulating interest because of the pressing need of the emerging diseases. The material was incomplete, but the hope was that it would be complemented by future studies dealing with the survival of ancient signature biomolecules of microbes, the evolution of pathogenesis, the influence of disease on the evolution and structure of populations, and many other subjects that will aid us in understanding emerging diseases. This volume is a second step in that direction.

The plan for the book

We can now begin to build the paleoepidemiology that we dreamed of in 1997. In the five years that have passed since that Jerusalem symposium, the field has changed so much that significant results can be presented. What has happened? First of all the human genome project has been 'completed' with profound implications for our understanding of ourselves and all organisms. For the specific field that this book treats, the studies of the ancient Neandertals (Krings *et al.* 1997, 1999) mentioned above and several diseases have removed some of the doubts about the validity of aDNA research. They have defined the time range of these endeavors and at the same time have fixed new standards (Cooper and Poinar 2000). Furthermore, appropriate tools for study are available or are being developed and great improvement has occurred in the methods of aDNA extraction. Only a few of these techniques have been widely applied, but there is certainty that they will be. We have found suitable diseases where collections of remains exist. In two examples in this book—influenza and tuberculosis—there is a complete archive so that details of the disease in the victim are known (Taubenberger and Reid, Chapter 16) and it is possible to build pedigrees and annotate the infected family members (Spigelman and Donoghue, Chapter 15). In 1998, Herrmann and Hummel (Greenblatt 1998) listed 10 publications in which authors reported the retrieval of pathogen DNA. The same authors in this volume list 42 examples with many cases of confirmation. Although it has been difficult to convince others to repeat one's work, this is becoming more commonplace. With sufficient support for this research, it may be possible to establish reference laboratories to verify results.

This volume is divided into four sections: (I) setting the stage for infection, (II) disease with and of humans, (III) past and ancient diseases studied by new techniques, and (IV) a look at what happens next.

Part I concentrates on evolution and ecology. It speculates on the nature of the beginnings of infection, examines the role of earth history on change in transmission of disease agents, cites some surprising associations of protozoa and bacteria, explores the borderline between symbiosis and parasitism in bacterial symbionts that survive in amber entrapped bees, and describes the evolution of vector-borne disease.

Part II presents different aspects of human disease from the viewpoints of co-evolution, the role of primate and human behavior on disease, the evidence and its proper recording in the skeletal record, and the differential diagnosis of the recorded findings as disease entities. It then asks how the evolution of parasite virulence relates to host behavior.

Part III presents the state of the art in new technologies, which will help us to define paleoepidemiology in meaningful evolutionary and molecular terms. It discusses evidence from the survival of important biomolecules that can be used in describing ancient disease, emphasizes the importance of molecular diagnosis in confirming morphological signs of disease, presents the problems of extrapolating from grave to population (or from plague pit to population), and then examines the molecular bases of enteric diseases, tuberculosis, and influenza.

Part IV is a brief 'virtual' discussion of the problems of ancient pathogen research and the promise it holds for developing new strategies in emerging disease control.

The different authors and our readers will be coming at this subject from a number of different tangents. The chapters vary in detail, but they ask important questions and attempt to clarify the future directions of the research.

Although I am anxious for the reader to read to the end, I do not think I will be revealing too much if I tell something of the punch line here at the very beginning. The essence of the various chapters are rooted in the realm of evolutionary biology. It is this editor's view that the study of evolutionary biology may in the end give humans an edge on our adversaries, the pathogenic microbes. Paul Ewald (2000) most certainly ascribes to this view, while my other contributors show varying levels of agreement and disagreement. However, all of us feel that a central lesson is that political will must back up the scientific effort. This does not say that in the end our days are not numbered, just like the dinosaur and perhaps unlike the microbe. To repeat Richard Krause's words: 'and it is likely they will be here 2 billion years after we depart'. We have the opportunity to scan the big picture in a way the microbe cannot. Obviously the microbe has been fashioned by thousands or even millions of trials and errors, but while the microbe has been 'doing' evolution, we have been studying its multiple paths and processes for little over 140 years and I believe we are beginning to discern its logic.

Jerusalem   C.L.G.
June 2002

Acknowledgements

I am most grateful to the Center for Emerging Diseases for their generous support. The Center has understood from the very first that good research is often slow to mature and complicated applications and reports do not guarantee results. Ruth Rossing assisted me in all the things I'm incapable of doing so this volume could see the light of day. Mark Spigelman, my associate-editor, helped me focus on and understand the paleopathology literature, read and correct the more difficult chapters, and has been my constant colleague in the efforts to reconstruct the diseases of the past. His wife, Rachel, deserves our special thanks. Ilan Rosenshine patiently guided me in comprehending bacterial virulence. Several editors of Oxford University Press are to be praised for sticking with me in this endeavor, but special thanks to Ian Sherman, Chief Science Editor. Finally bless my wife Jo-Anne for having the patience of an angel.

C.L.G.

Contents

Contributors

Yann Ardagna, Department of Biological Anthropology, Research Unit 6578, CNRS, University of La Méditerranée, Faculty of Medicine of Marseilles, France.

Gila Kahila Bar-Gal, Laboratory of Genomic Diversity, NCI–FCRDC, Frederick, MD 21702-1201, USA.
e-mail: gilak@ncifcrf.gov

Jake Baum, Department of Infectious and Tropical Disease, London School of Hygiene and Tropical Medicine, Keppel Street, London WC1E 7HT, UK. e-mail: jakebaum@pobox.com

William C. Black IV, Department of Microbiology, Immunology and Pathology, Colorado State University, Fort Collins, CO 80523, USA.
e-mail: william.black@colostate.edu

D. Brian, University of Queensland, Brisbane, Australia.

Raúl J. Cano, Environmental Biotechnology Institute, California Polytechnic State University, San Luis Obispo, CA 93407, USA.
e-mail: rcano@calpoly.edu

Mark Nathan Cohen, Department of Anthropology, SUNY, Plattsburgh, NY 12901, USA. e-mail: mark.cohen@plattsburgh.edu

Gillian Crane-Kramer, Department of Anthropology, SUNY, Plattsburgh, NY 12901, USA.

Helen D. Donoghue, Department of Medical Microbiology, University College London, London, UK. e-mail: h.donoghue@ucl.ac.uk

Olivier Dutour, Department of Biological Anthropology, Research Unit 6578, CNRS, University of La Méditerranée, Faculty of Medicine of Marseilles, France.
e-mail: olivier.dutour@medecine.univ-mrs.fr

Paul W. Ewald, Department of Biology, Amherst College, Amherst, MA 01002-5000, USA.
e-mail: pwewald@amherst.edu

H.-D. Görtz, Department of Zoology, Biological Institute, University of Stuttgart, D-70550 Stuttgart, Germany.
e-mail: goertz@po.uni-stuttgart.de

Charles L. Greenblatt, Kuvin Centre for the Study of Infectious and Tropical Diseases, Department of Parasitology, Hebrew University–Hadassah Medical School, Jerusalem, Israel.
e-mail:greenbl@cc.huji.ac.il

Bernd Herrmann, Institute of Anthropology, Georg August University, Göttingen, Bürgerstrasse 50, D-37073 Göttingen, Germany.
e-mail: bherrma@gwdg.de

Susanne Hummel, Institute of Anthropology, Georg August University, Göttingen, Bürgerstrasse 50, D-37073 Göttingen, Germany.

James B. Kaper, Department of Microbiology and Immunology, University of Maryland School of Medicine, 685 W. Baltimore Street, MD 21201, USA.

Myron M. Levine, Department of Medicine, University of Maryland School of Medicine, 22 South Greene Street, MD 21201, USA.

Marta Maczel, Department of Biological Anthropology, Research Unit 6578, CNRS, University of La Méditerranée, Faculty of Medicine of Marseilles, France.

Larry D. Martin, Professor of Ecology and Evolutionary Biology; Curator of Vertebrate Paleontology, Museum of Natural History and Biodiversity Center, University of Kansas, Lawrence, KS 66045, USA.
e-mail: ldmartin@falcon.cc.ukans.edu

C. D. Matheson, Lakehead University, Thunder Bay, Canada.
e-mail: cmatheso@mail.lakeheadu.ca

R. Michel, Wiedhoehe 2, D-56581 Melsbach, Germany.
e-mail: rmichel@amoeba-net.de

James P. Nataro, Center for Vaccine Development, Departments of Pediatrics, University of Maryland School of Medicine, 22 South Greene Street, MD 21201, USA.
e-mail: jnataro@medicine.umaryland.edu

Ann H. Reid, Division of Molecular Pathology, Department of Cellular Pathology and Genetics, Armed Forces Institute of Pathology, 1413 Research Blvd, Building 101, Room 1057D, Rockville, MD 20850-3125, USA.

Bruce Rothschild, The Arthritis Center of Northeast Ohio, 5701 Market Street, Youngstown, OH 44512, University of Kansas Museum of Natural History, Carnegie Museum of Natural History, Northeast Ohio Universities College of Medicine and University of Akron, USA.
e-mail: bmr@neoucom.edu

Michel Signoli, Department of Biological Anthropology, Research Unit 6578, CNRS, University of La Méditerranée, Faculty of Medicine of Marseilles, France.

Mark Spigelman, Department of Medical Microbiology, University College London, London, UK and Kuvin Center for the Study of Tropical Diseases at the Hebrew University Medical School, Jerusalem, Israel.
e-mail: marks@md.huji.ac.il

O. Colin Stine, Department of Epidemiology, University of Maryland School of Medicine, MD, USA.

Jeffery K. Taubenberger, Division of Molecular Pathology, Department of Cellular Pathology and Genetics, Armed Forces Institute of Pathology, 1413 Research Blvd, Building 101, Room 1057D, Rockville, MD 20850-3125, USA.
e-mail: taubenbe@afip.osd.mil

Douglas H. Ubelaker, Department of Anthropology, National Museum of Natural History, Smithsonian Institution, Washington DC 20560, USA.
e-mail: ubelaker.doug@nmnh.si.edu

Part I

CHAPTER 1

An overview: how infection began and became disease

Charles L. Greenblatt

Nurs'd by warm sun-beams in primeval caves, Organic life began beneath the waves Hence without parent by spontaneous birth Rise the first specks of animated earth. (Erasmus Darwin, The temple of Nature, 1803)

1.1 Introduction and overview

When professionals are in panic, the layman has every right to be concerned. The events of September 11th 2001, followed by the successful use of the anthrax bacillus as a weapon of bio-terrorism struck fear into the hearts of many, including the public health professionals to whom we entrust our care. Those dramatic events can be seen against a background of a general dispair concerning the capabilities of physicians to handle infectious diseases. The inability to cope with anthrax was just a special case of modern science's failure to deal with one of our great natural enemies, the microbe. Today, many physicians are as stirred up by antibiotic resistance as the mother whose child does not respond to treatment. When the town council of a posh New York City suburb is up in arms over West Nile fever, the general practitioner and local public health official are as distraught as the ordinary citizen. Thus the whole medical establishment is seriously occupied with what the next plague has in store for us—whether unleashed by bio-terrorists or by mother nature (Garrett 1994).

Physician and layperson flock to buy germicidal soaps that promise a germ-free existence (Kolata 2001). Yet the nervy germs resist our best efforts to cure the diseases they cause, and even seem to promise greater problems in the future (Levy 1998). The struggle has its truly global dimensions. A new direction in the groping for understanding has been in the direction of the ecology of infection. It has become clear that the context of a disease, whether infectious or not, can hardly be ignored. Poverty breeds tuberculosis, dysfunctional societies breed poverty, social chaos breeds dysfunctional societies, etc. The society, the humidity, the warming of the globe, the very dirt, the rodents, the ectoparsites of the rodents are all interwoven so as to create an ecological context of disease.

Richard Levins (1995) has put it succinctly:

From this perspective, the current pandemic of cholera can be recognized as being possibly 'caused' by plankton blooms increasing the populations of Vibrio cholerae, international shipping transporting them in coastal ballast water, the dismantling of social services in Latin America, and the reluctance of governments to acknowledge outbreaks that might affect the tourist trade, among other factors. The plankton blooms can be related to eutrophication of coastal waters owing to erosion, agricultural fertilizers and urban sewage as well as the warming of the seas, and the dismantling of social services can be related to the budgetary crises resulting from Third World debt and the World Bank insistence on progress through impoverishment.

This volume deals 'somewhat' with all of the above. We will speak of changes in our planet, water-borne disease, human society, and creepy-crawly things. Yet the volume's chief objective is to look back at the history of infection using new tools. It is the authors' opinions that this retrospection is not simply an academic exercise. In an analogous way Charles Redman (1999) has argued that archaeological sites contain many insights into the modern human impact on global ecology. According to Redman,

> Those concerned with today's environmental problems have not focused on experiences in the past because of a belief that the crisis is a product of modern technology and a world brimming with billions of people. Both of these factors are crucial to the problems we face today. There is little question that environmental problems were more regional or local in the past than the global threats we perceive today. As real as this viewpoint is, it underestimates the impact environmental crises had on people in antiquity.

In the same way, ancient diseases (not always 'regional or local') give us case histories without jet planes, bulldozers, or weaponized anthrax, which can serve us well in understanding how diseases emerge and evolve. The agents of disease may have fine honed some of their devices in the last moments of time in which they had lived with civilized man and reacted with devastation, but surely they have acquired much more in the billions of years of their evolution. Therefore, much of what you will read here is from the viewpoint of evolutionary biology, drawing examples from both the past history of our planet, but not ignoring our written history, with the profound effects of recent multiple medical interventions. It draws upon the tools of physical anthropology, archaeology, ecology, and molecular biology in attempting to decipher the history of infectious diseases.

1.2 The new emerging diseases affect us all

Fifty years ago most physicians would have been fairly confident that the present decade would be free of plagues and serious infectious diseases. Now, nearly all of us—physician, patient, public health worker, epidemiologist, molecular biologist, and layman—see no end of surprises from the micro-dimension of life we call 'germs'. This view of our condition has been forced upon us by two unavoidable categories of news events—the threat of what are called 'emerging diseases', whether by the force of nature or the hand of man—and stories of antibiotic resistance.

The first has brought to our awareness through the impact of both bio-terrorism and what seems to be entirely new emergent diseases or the recurrence of diseases that we thought had disappeared or at least were thought to have been on the wane. In fact as I wrote the words above about the media's awareness of infectious diseases, I remembered that Richard Krause, the greatest of the 'historians' of the emerging infectious diseases, had already raised the same image (Krause 1998). From his vantage point as former director of the National Institute of Allergy and Infectious Diseases during this very tumultuous period for infectious disease experts, he wrote: 'I began this introduction the first week of January 1997 by browsing in the newspapers for recent accounts of infectious disease outbreaks. The journalists had been busy. I read about brucellosis in Wyoming bison, rabies in raccoons, … And arguments have arisen concerning the use of cattle feed … derived from the carcasses of livestock. The issue here concerns … the agent of "mad cows disease".' Since Krause's surfing of the headlines, the journalists have become even busier. It is impossible to miss the accounts of infectious diseases as their names are splashed across newspaper headlines. Anthrax, AIDS, mad cows' disease, West Nile fever, and Ebola have entered our common language, eliciting panic in some communities equal to what the fear of plague must have meant to our ancestors (Garrett 1994; Allen 2000). Sometimes the diseases are caused by well recognized pathogens, somewhat altered ('weaponized'), which exploit a new niche provided by changes in our environment due to human inventiveness. Anthrax delivered through the mail has become a household word. Another, more prosaic, is 'toxic shock syndrome' where the wonderful convenience of the tampon became a deadly threat (Krause 1994). Devised to reduce a woman's monthly inconvenience, it provided a rich culture medium for *Staphylococcus aureus* and as a result killed many.

Some of these diseases may be truly new, but most have been lurking in unexplored niches, awaiting an ecological signal to 'come forth and multiply'. Krause agrees with Levins (1995) quoted earlier that diseases emerge due to changes in human behavior, social organization, growth of crowded cities, changes in agriculture and medical interventions. He especially concurs on the role of travel or migration—which in the case of cholera was not human as much as the shipment of the microbe in the ballast of ships. Yet he adds a further aspect: microbes have 'remarkable genetic versatility that enables them to develop new pathogenic vigor, escape population immunity by acquiring new antigens, and develop antibiotic resistance' (Krause 1994).

In long evolutionary history, symbiosis, parasitism, and virulence have deep roots. The causative agents' survival or extinction must have depended on their ability or lack thereof to jump from one species to another, or to persist in the environment for long periods awaiting the right opportunity to infect a new host, or exploit unusual systems of dispersal. The ability to adapt may have been due to a random mutation, or even better, to a gift of DNA from some other member of the bacterial community. When this occurs, along with any of the ecological disruptions noted above, the diseases grab our attention and we dub them as 'emerging or re-emerging diseases'. AIDS probably existed as a monkey disease until some cardinal event gave it the chance to leap across the species barrier to infect humans. Once there, a boatload of infected Haitians moved it across the waters. Another possiblity, not widely accepted, suggested by Hooper (1999) is that human vaccine trials of polio, grown on monkey tissue culture cells, was the culprit. Cholera, on the other hand, seemed quite content in a 'viable but non-culturable' form in plankton off the coast of Ecuador. Then, shipped in the ballast of oceangoing steamers and awakened by some ecological stimulus it found new susceptible human hosts in Bangladesh. The bacteria *Legionella* likewise began as the parasite of a protozoal host until provided with the unusual opportunity to exploit air-conditioning towers where the protozoa and bacterium were able to multiply in the proximity of elderly immuno-compromised hosts and the world then became acquainted with a new infectious danger—Legionnaires' disease.

The second matter evident to us all is the appearance of bacteria that are now resistant to the doctors' best drugs. Closest to home is the child or grandchild with a stubborn middle ear effusion. Medicine's only answer is a mechanical drain, since a miracle drug is no longer available. The common causative agents of such infections have long ago become antibiotic resistant. At least for the child, the drain is usually a sufficient remedy. A less common experience, but more troubling to the physician, is that there is hardly a major hospital that does not have a patient on the critical list because they are infected with a life-threatening antibiotic resistant organism—one of which have been dubbed 'flesh eating'. The infectious disease specialists speak of the 'end of the antibiotic era'. In the lifetime of many of us, it was not always so. As an optimistic young physician, I was almost certain when I treated a middle ear infection that it stayed cured. In fact when I think back to those simpler days I remember that there were hints of antibiotic resistance when we discovered that the delivery room seemed to be rife with bladder infections due to a common gut bacterium that was misbehaving. It was not long before our simple guidelines for treating bacterial infections became more complex as laboratory tests and treatment results told us that the bacteria were rejecting our best efforts.

Our tinkering with the living and non-living world has not only been through medical intervention. Man's biblical 'dominion over the fish of the sea and over the birds of heaven; and every living thing that moveth upon the earth' is in fact an illusion. Simply by domesticating animal and plant we have opened up multiple Pandora's boxes of troubles, exposing ourselves to countless new diseases. And of course the diseases of our animal friends were caused by the microbes predating their presence in our midst and our dominion over them. Some may have been around when dinosaurs roamed the earth and human interventions were nowhere to be found.

1.3 The 'germs' that are against us

The repertoire of causative agents of infectious diseases is astonishing and bewildering. They vary from simple proteins to more complex viruses containing their own DNA or RNA, to bacteria with varying complexities of metabolic machinery, to protozoa whose cellular nuclei are like ours and to macroscopic worms. Almost anything can invade something else given a certain limitation in size, e.g. a worm does not invade a single-cell amoeba, but he does sit happily in the lumen of our intestine. These agents can be astonishingly subtle or aggressive and brutish, but what is certain is that the complexity of activities and responses of these so-called simple constructs, is built upon associations and adaptations over billions of years. However, for all the diversity in the parasitic world, they and we are connected through the common system of biological inheritance—the genomic language.

1.3.1 How did infection begin?

Can the study of ancient DNA technologies help us in looking back at ancient infectious disease scenarios? In one way it can. If diseases occurred within the time frame of our civilized life, say 15 000 or possibly somewhat beyond to 50 000 years, with luck we should be able to decipher some of the DNA messages of ancient pathogens and their hosts. Matheson and Brian (Chapter 11) will take a special look at the durability of DNA and other biomolecules that can help us to track the pathogens. To penetrate further back we will have to rely on genetic history as it is written in our genes and in those of all the germs around us. The latter method can sometimes tell remarkable things about our common past, but the message is edited to be compatible with our general genomic structure. For example, what was originally a symbiotic organism may lose its integrity and pass its genes to the nucleus, where they may be integrated into the totality of the cell's nuclear genes. Yet hopefully decipherable 'signatures' are left of their past.

The next few pages are highly speculative, since we shall never recapture the earliest infection. Although a parasitic infection may be defined as 'one life form within another doing its host damage', it may have begun even before 'forms' were put together, i.e. in the organic soup in a reducing atmosphere. Formally, these interactions may not have been 'in'fections, but they certainly could have resulted in neutral, beneficial, or detrimental associations. Interactions of single protein or RNA molecules with one another—either as protein–protein,

RNA–RNA, or protein–RNA in an enzymatic fashion may have been the first practice exercises for infection. An even earlier possibility has been postulated by Carleton Gajdusek (2001), the 1976 Nobel laureate for his discovery of the cause of kuru, the first disease for which a protein was implicated as the causative entity. Gajdusek's classical study pinpointed cannabalism as the method of transmission of the disease among natives of Papua New Guinea. Gajdusek cites the materials sciences literature to suggest that not even two organic molecules are required. In kuru, as in related diseases cited below, a protein catalyzes a conformational change in a normal protein. We now call these 'prion' diseases. In Gajdusek's view organic molecules could possibly interact with inorganic surfaces, such as ceramics, to direct molecular molding. He has written, 'Such templates of high specificity are chemically coded surfaces transcribed at a molecular level into inorganic nuclei with precise structural, orientational and positional properties' (Gajdusek 2001).

With organic molecules, Kauffman (1993) has suggested that these non-cellular events may have been the structural basis for the first primitive 'proto-organisms'. There is much discussion on whether DNA, RNA, or even proteins came first in the origin of life. RNA has a certain advantage because it can efficiently catalyze replication, excision, and ligation reactions. Proteins as enzymes are catalytic. Kauffman (1993) suggests that a proto-organism could have been composed of protein and a parasitic replicating RNA. Initially concerned with simply reproducing itself, at a later stage structural or conformational change in one molecule, either tolerated or becoming 'structurally embarrassing' could be the beginning of an infection. If purely at the nucleic acid level such transformations would be retrotransposons, selfish DNA replicating itself wherever it found a niche in the empty genome. As DNA or RNA entities they foreshadowed the viruses of a later day. These rearrangements may not have had profound deleterious repercussions for another billion years or so until the eukaryotic cell arose. Even then they may have enhanced fitness. In yeast, prions are found to act as catalytic proteins capable of altering cellular functions. True and Lindquest (2000) have suggested that prion proteins allow the expression of silent genes and thereby increase the diversity of yeast progeny. In times of stress this can be construed as beneficial. We know the prion today as the cause of scrapies and 'mad cows' disease', so that in complex organisms, where prions target the brain, the effects are catastrophic.

After self-replicating entities arose and before the eukaryotic yeast and long before humans, in that several billion years, Woese (2000) has pointed out that it may be impossible to draw cellular lineages since lateral DNA transfers were so promiscuous. Acquisition of large blocs of genetic information seemed commonplace, either directly or by the gulping up of other cells. However it occurred, more complex membrane-enclosed systems began to develop. During this period a wealth of creatures, including Archaea and bacteria, arose. The former were organisms that tended to find their niche in extreme environments and were also dubbed 'extremophiles'. Their environments were hot springs, salt lakes, and environments of high methane and sulfuric acid concentration. Hydrogen production or utilization seems a cornerstone of their metabolism, since oxygen was not yet in abundance and was in fact toxic. Their DNA is not packaged in a nuclear membrane, and their RNA was different enough to fall outside the ordinary bacterial grouping.

Let me deal with another of the new creatures that lies in the borderland between the inorganic and organic worlds. This is a minute form called the nanobacterium, which can be found nearly everywhere, and which some have argued just is not big enough (60 millimicrons) to contain 'the stuff of life'. A geologist named Robert Folk is considered their discoverer. He noted the presence of suspicious objects by scanning electron microscopy in travertine marble and then on metal surfaces. Often his tone when writing of his little critters is one of disbelief, but they are being taken more seriously day by day. Recently they have been implicated in human disease in the kidney or gall bladder in association with stone formation (Folk 1993; Hamilton 2000; Kajander and Çiftçioglu 2001). These mini-bacteria may not only be living in rocks, but may also be making them. Moreover, similar forms have been identified in a famous

meteorite from Mars. Together with the presence of aromatic compounds their appearance has been taken as evidence of life on that distant planet. In spite of the fanfare of the United States Space Agency, and like the nanobacteria on earth, this has not been accepted without controversy. Of course, if nanobacteria were passengers deeply buried in a meteorite, one need not postulate life beginning on earth, but coming from elsewhere in space.

The nanobacteria, in spite of their size, seem to be bacteria from their DNA signatures (Kajander and Çiftçioglu 2001). Their relatives are generally much larger cells, at the micron level. Some biologists believe that the more complex eukaryotic cells, i.e. those with nuclei enclosed in membranes, were built up by a coalition of two—and perhaps later more—of these prokaryotic units. Martin and Müller (1998) have hypothesized that the partnership was of a hydrogen producer, and an oxygen utilizer. Packaged together, the micro-aerobic bacterium and the anaerobic Archaea were able to form a stable symbiotic relationship. Margulis speculated on these events in a seminal book published in 1981, 'Symbiosis in Cell Evolution'. Although the idea was not entirely new, her work stated the concept in a clear and convincing way. Recently, Margulis et al. (2000) have criticized Woese, mentioned above in relationship to the Archaea, for weighing solely the molecular evidence and not considering the whole of taxonomic and morphological data in separating the Archaea as a third 'kingdom' from the eubacteria and eukaryotes. She agrees with Martin and Müller in adopting the union of the eubacterium and archaea (or in her terms archaebacteria) as the basis for the eukaryotic cell. Furthermore, she derives from the union of these forms, the eukaryotic nucleus. Most importantly she notes that extant bacterial 'consortiums' exist incorporating the two forms. The increase in oxygen atmosphere created a selective pressure on the establishment of this symbiotic relationship. The stage was set for infection as we know it.

1.3.2 Infection, symbiosis, and disease

The infections of cells probably included a range of intruders (from naked DNA to bacteria) resulting in a range of outcomes from benign to detrimental.

Remnants of certain of these association are found in protists (unicellular eukaryotes) as subcellular particles, the mitochondrion and chloroplast. When these contained DNA, with successful adaptation, they remain as entities in typical animal and plant cells and have held on to part of their information systems. However, much of their genetic equipment was transferred to the nucleus. Early eukaryotic cells, such as trypanosomes, *Leishmania*, and *Giardia* cause important parasitic diseases and possess certain of these organelles. These include the hydrogenosome and the glycosome/peroxisome, which do not contain DNA, but seem to be remnants of earlier invasions. So our parasites have parasites! These ancient associations probably produced today's eukaryotic cell as we know it, but Görtz and Michel (Chapter 3, this volume) on examining ciliates in aqueous environments as hosts of bacteria, find totally new associations of this sort. They suggest that there are multiple opportunities for these to form in sites such as sewage plants, and for the possible release of the bacteria into air-conditioning systems. At least one disease (Legionnaires' disease) has emerged in just this fashion. In these on-going evolutionary experiments in symbiosis and parasitism an especially worrisome aspect is that genomic studies show that some of the bacteria found in protists are related to known pathogens.

The infectious disease experts cite the components of an infection in terms of adhesion of parasite to host, colonization, internalization following signalling between parasite and host cell, spread of the parasite, toxin production, etc., etc. The number of mechanisms devoted to successful infection are astonishing. Early toxin precursors and host resistance and recognition factors should be identifiable by tracking their genes in extant forms found in different kingdoms and diverse forms. Razin *et al.* (1998) have recently examined one of the simplest of the bacterial groups, the mycoplasmas. When one does a simple accounting of the components of their metabolism, one finds that a certain percentage of their genome is devoted to replication, another segment to transport, another to housekeeping metabolism, etc. Then a certain proportion is devoted to the business of infection, so genes are found for attachment organelles, and repetitive

units that may be related to antigenic variation. In the authors' view, there is no 'free lunch'. As more of the entire genomes of bacteria are elucidated it is already possible in some cases to compare the entire genome of a non-pathogenic species with that of its nearest relative, which is pathogenic (see Natero *et al.*, this volume). When this type of genomic analysis is performed on the bacteria that cause enteric infections, a number of integrated systems are found—the so-called pathogenicity islands. These may involve some 20% of the genome of an organism such as virulent forms of *Escherichia coli*. In this case, garden-variety non-virulent *E. coli*, the main inhabitant of the human gut, can be compared with virulent forms called ETEC or EPEC for 'enterotoxigenic or enteropathogenic *E. coli*'. The elements that differentiate the virulent from the non-virulent form consist of grouped additional genes in the virulent organism, which may number 1300. Among this multitude of virulence products of these genes are outer membrane proteins, adhesins, iron transport systems, toxins, and serum resistance factors. Of special importance is the type III secretion system, which shares elements with other secretion mechanisms (which in ETEC and EPEC secrete hemolysins or adhesion molecules), but is distinct by being set in action by close contact with host cells. The 'syringe' mechanism itself may be coded by some 40 genes. The effector molecules of the bacteria move to its surface and there they are delivered to the host and alter its function so as to enable bacterial survival. These effector molecules may change the host cell's engulfment apparatus or its ability to destroy the bacteria. Host 'experiences' have honed and refined today's pathogens.

Hacker and Carniel (2001) have attempted to generalize these concepts by differentiating the indispensible core genome from the 'alternative' genome that allows cellular operations conferring ecological, saprophytic, symbiotic, or pathogenic fitness through 'islands' for each of these functions. This then places the performance of a potential pathogen into the mainstream of evolution, and evolutionary natural selection determines the occupancy of a particular niche. Our understanding of evolutionary biology is essential in confronting microbial selection and survival.

What is often forgotten is how the parasite survives when hosts are not available, and its systems of dispersal in the search for new hosts. Two of our chapters will deal with these issues. Raúl Cano (Chapter 4, this volume) takes us into a world where the sap of certain trees hardens as amber and accidentally traps a wonderland of organisms. Of these, bees and termites are common. Cano and Burecki drilled into the belly of such a bee dated to $25-40 \times 10^6$ years ago, and were able to culture symbiont bacilli. The system developed for spore formation of these bacilli in amber allows them to survive for millions of years. If the bacterium can wait long enough, certainly new opportunities for infection can arise. Amber spewed up from the depths of the Baltic Sea or mined in deep Dominican earth give the enclosed bacteria new opportunities to establish themselves. Cano also points out another disturbing matter. The symbiotic bacillus of the bee is helpful, even necessary for bee survival, but genetically it is very close to harmful relatives, some bearing toxins. Evidence is presented indicating that interconversion of these forms is probable in nature. One such harmful relative is the anthrax bacillus.

Persistence is only one aspect of transmission. Another important component is the role of vectors. William Black (Chapter 5, this volume) presents the evolution of vector blood feeding to show that many of the habits of blood feeding were pre-adaptations to plant feeding. The bee was successful feeding on flowers, but using similar mouthparts, Dipteran relatives were capable of finding open sores on vertebrate hosts or on their carrion as suitable food sources. As plant evolution exploded with exuberance, insect vectors went along with matched exuberance. Out of that diversity, mouthparts of certain insects, adapted to gaining a plant meal, were sometimes pre-adapted to taking a meal from an animal host as well. Wherever a feeding opportunity occurred, there went the insects.

Having finished with this highly speculative discourse on the beginnings of infection, we can then turn to the human experience. We now have: (i) any number of infectious agents with a wide range of relatives with potential for causing damage, (ii) the recognition that perturbations in ecology open the door to emergence, (iii) that there are certain pre-adaptations in both the agents and the vectors to promote infection given the opportunity, and (iv) traces of these natural experiments may be found in the phenotype and genotype of extant organisms.

1.4 Infection and the human lineage

The appearance of disease in the last several hundred million years may have been relatively unexciting for *Homo sapiens* up to the time of settlement or sedentism, but for the pathogen the times were momentus. Martin (Chapter 2) perceived the tribulations in 'earth history' as the prerequisites of disease. He contrasted the times of plenty of 50 000 000 years ago when tropical vectors inhabited the face of the earth without restriction—to the later times of great limitation in the face of glaciers and the earth's cooling. Host mass extinctions pressed parasites with the need to adapt so that they themselves would not become extinct. During mass extinctions dependence on vector transmission, medium-sized hosts, and durability might have been the characteristics of surviving pathogens. A major need for adaptation occurred with the ice age, when ectothermic hosts were replaced by endothermic ones. Martin ponders whether the appearance of mammals and birds with their warm body temperatures stressed pathogens during the ice age. The instabilities in ecology, in host populations, in technologies of food, water, and transport, and simply in the playfulness of DNA have enriched the repertoire of parasite and host.

Mark Cohen and Gillian Crane-Kramer (Chapter 7) stress primate and human behavioral change as it relates to pathogens, arguing that the intensity of infectious diseases has increased with the advent of civilization. They follow the transitions of infective diseases from the primates to humans. As anthropologists they trek after primates as they evolve to early humans and begin to live in the crowded conditions that characterize civilized creatures. Certain specific causal agents adapt to these changed human conditions. The diseases that troubled our ancestors in their arboreal surroundings differed from those in the savanna and were even further from the disease spectrum that struck them as they congregated around the community fireplace. As populations grew and sedentism became prominent, parasite loads increased and nutritional status worsened. Plant domestication further changed diet so profoundly that human disease susceptibility was altered. Recording these changes becomes an exact science under the watchful eye of Douglas Ubelaker (Chapter 8). In his work one finds confirmation that there is an increased disease load as civilization evolves in Ecuador. Bruce Rothschild (Chapter 9) bridges the gap between diseases of prehistory and those recorded in historical texts. He has applied his clinical expertise in human bone and joint diseases to their evolution from ancient animal diseases, with a special emphasis on spondyloarthritis, syphilis, and tuberculosis.

Paul Ewald (Chapter 10) was among the first to grasp the notion that diseases do not always tend to reach a happy equilibrium with their hosts in a 'no great loss no great gain' scenario. In fact a number of diseases seem to go for broke and even gain in virulence. Ewald argues that pathogens with different modes of transmission tend to have differing levels of harmfulness. Understanding these tendencies can aid in understanding pathogenesis and disease emergence. He places stress on the predictability of virulent associations and the wide range of these implications. Vector borne pathogens tend to be more harmful than parasites that 'sit and wait' for hosts to come to them. These latter pathogens, on the other hand, are more durable and persist in the environment. Sexually transmitted pathogens must be less destructive of their hosts while awaiting another sexual encounter. Recently, Ewald has been telling us to explore the chronic diseases, even mental illness, for further evidence of ancient pathogen activities.

Animal and plant domestication can be taken as a prime example of ecological change that brought about an expansion of human disease. Any disease agent accidentally or purposely in place in the animal host now had a free tramp to its next human victim. Human behavioral change is reflected in animal behavioral change. Special focus will be made on zoonotic diseases, since many emerging diseases have done so from animal reservoirs. Throughout the chapters of Cohen and Krane-Kramer, Rothschild, and Martin, diseases passing from flocks to their human herders will be prominent. Jake Baum and Gila Kahila Bar-Gal attempt to understand whether there is a direct co-evolution in these relationships.

In turning to constructing the field of paleopathology-based molecular biology, perhaps the first question is what remains of the microbes, their products, or host reactive substances that can serve to identify these ancient pathogens? This challenge is taken up in the chapter on taphonomy or post-mortem change by Carney Matheson and Debbie Brian (Chapter 11). In their study as well as that of Mark Spigelman and Helen Donoghue (Chapter 15), the question is asked as to whether the mechanisms used by bacteria in stationary phase to stave off starvation are also at work in preserving prokaryotic aDNA in the host post-mortem state so that it can be retrieved by aDNA extraction techniques. They then probe the wide range of molecular signatures and determine which of these survive post-mortem change.

Herrmann and Hummel (Chapter 12) survey the technical advances that may allow us to find answers to the questions that modern disease specialists seek by citing the range of diseases so far detected by new techniques. We then deal with four diseases where specific molecular biological information may be exceedingly important in deciphering disease emergence. The European continent from the medieval period until today was examined by Olivier Dutour and his colleagues (Chapter 13) with respect to plague and tuberculosis. The numerous victims of the plague, struck down with no apparent bias of age or gender, supply a cross section of ancient populations. Dutour *et al.* give special attention to reconstructing the populations of antiquity. A sharp focus is brought to bear on enteric diseases by James Nataro and his colleagues (Chapter 14), who challenge those interested in past diseases to elucidate the serotypes of ancient epidemics of cholera if we are to be prepared for vaccine development for the future. Although retrieving ancient cholera DNA has not been successful using the techniques of molecular biology, much can be learned about the evolution of enteric pathogens from the comparative studies of extant organisms. Mark Spigelman and Helen Donoghue (Chapter 15) create the beginnings of paleo-epidemiology of tuberculosis in 18th-century Hungary by retrieving the DNA of the tubercle bacillus from a collection of mummies whose life data are recorded. This 'medical archaeology' gives us insight into life and death in the face of the constantly present tubercle bacillus, at a time when pasteurization and antibiotic therapy did not exist. Before and after these great selective pressures, one can now squeeze out of the tubercle bacillus its evolutionary story.

Influenza has also begun to yield up its secrets. Jeffery Taubenberger and Anne Reid (Chapter 16) tell how they retrieved the RNA of the virus from the formalin-fixed paraffin blocks of soldiers who died in the great 1918 pandemic. Samples from Eskimos buried in the permafrost also contributed to this group's major effort to try to decipher the genetic change in the virus of 'Spanish' influenza. We are beginning to understand what made this virus perhaps the greatest killer in a single pandemic in the entire recorded history of infectious disease (Crosby 1989; Kolata 1999).

We are far from completing the saga of man and the microbes. In this volume we will barely touch the macroparasites and host responses. Some new directions are discerned and a few generalizations are beginning to emerge. The genomes of host and parasite both conserve shared themes, but depart from one another in many aspects. Yet our existences are intertwined with fine tuned dependencies and communications. We must learn to understand these interactions.

CHAPTER 2

Earth history, disease, and the evolution of primates

Larry D. Martin

We, too, have come mostly from caverns in the rocks. Our flesh is linked to these great bones that we recover. (Loren Eiseley, Why did they go? In The Innocent Assassins, 1973)

2.1 Introduction

Recent theoretical work on the population ecology of emergent diseases provides a model for understanding disease in the context of earth history. As world climates change, ecologies are modified as are the parameters governing the origin and spread of new categories of sickness. Human disease is the end product of numerous evolutionary modifications and can best be understood within a historical context.

It is through understanding this historical context that we may isolate and modify our own cultural contribution to emerging diseases. The extinction at the end of the Cretaceous provides a baseline for the discussion of disease origin, and the climatic and evolutionary events that follow suggest a probable sequence of emergence. Eventually, humans through domestication formed symbiotic relationships with plants and animals that opened up new avenues for the emergence and spread of disease. In spite of the advances of modern medicine, the potential disease load of any human population is probably much greater now than at any time in the past.

The discovery of avascular necrosis in a group of extinct, giant, marine lizards (mosasaurs) was seminal to the modern resurgence of paleopathology. Subsequent publication in the prestigious journal *Science* (Rothschild and Martin 1987) led to articles in *Discovery Magazine, Science News,* and a lengthy article in the journal *American Scientists* (Martin and Rothschild 1989). New interest was generated in the applications of paleopathology to the discovery of behavior in extinct animals and the origin and dispersal of disease. Such applications only became meaningful as diagnosis improved through the development of population and distributional statistics. This is largely the result of a Herculean attempt to analyze all available collections of certain vertebrates by Dr Rothschild. Another factor becoming increasingly important is the improving ability to extract and sequence early DNA. This area has been shadowed by problems of contamination and analysis (Austin *et al.* 1997). Almost all of the early results are suspect to a greater or lesser degree. At the present time any DNA claims over 100 000 years old need to be approached with caution (Poinar *et al.* 1996). That said, DNA work has proved very useful when applied to Holocene samples, and recently (Rothschild *et al.* 2001) tuberculosis has been confirmed in 17 000 year old *Bison* using gene sequencing.

The fossil record is most useful when it provides distributional data in time and space. The resultant pattern provides insight into why disease evolves and how it spreads. These are pragmatic questions of interest to humanity as a whole. Our success in using this data is dependent on its reliability and density. As in all sciences, reliability can only be measured through independent confirmation of results and the application of independent tests. Truth is not only elusive in science; in a very real sense, it cannot be demonstrated. What we accept as true is the convergence on a common answer of two or more independent lines of argument. Diagnosis of fossil

13

disease needs to be tested by more than one line of evidence and, whenever possible, confirmed in many individuals where epidemiology may itself become an additional and independent test. This is especially true for a historical science like paleopathology.

Historical sciences (paleontology, archaeology) are sometimes excluded from 'science' because they are not experimental. This is a misunderstanding of how experimental science works. An experiment is designed to provide an answer to a question. The acquisition of this answer usually results in another question requiring an additional experiment. Some of the alternative explanations for each experiment are excluded by controls. Knowledge grows incrementally. In 'historical' sciences, the experiments have already occurred and we are confronted with answers. It is our job to find the right question for a particular result, and our manipulation of how we frame that question provides controls on the experiment. In many ways, the differences between experimental and historical sciences are not profound. This is recognized in physics, where astronomy is embraced, although the light of stars may be more ancient and the actual star more inaccessible than the fossil progenitor of any organism. It is also important that independent replication of results remains an important criteria for scientific acceptance of observations. Some maintain that every event is historically unique, and therefore, historical sciences cannot rely on replication! This is also based on an inadequate understanding of scientific observations. To some extent, all observations are unique and rooted in the past. By the time we comprehend something, it is already history and the exact sequence of conditions to some degree lost. Replication simply requires that we control our observational parameters and can reasonably predict what we will observe in future observations. Without replication, science is useless. This is one of the reasons that the reports of isolated occurrences of pathological expression that once dominated paleopathology led to few interesting hypotheses.

2.2 Population biology of disease

The formulation of interesting questions results largely from speculations based on previous observations. The lack of a theoretical framework for many interesting problems with living pathogens has probably impeded paleopathology more than any inadequacy of the fossil record. In contrast, recent attempts towards a theoretical framework for emerging diseases provided new opportunities for paleopathology (Ewald 1998). Much of the new theoretical work was presented at a conference in Israel sponsored by the Center for the Study of Emerging Diseases and published in the book *Digging for Pathogens* (Ewald 1998). One of the most interesting aspects of this conference was the application of population theory to the evolution of pathogen virulence by Paul Ewald. In his 1998 study, virulence (measured in the number of deaths per 10 000 infected) was viewed as an evolutionary trait. He pointed out that virulence is very variable among strains of pathogens. To a large extent, the negative effect on the host is accidental; it results from toxins and physiological interference, resulting from the rapid expansion of the pathogen colony. In other words, it is largely a byproduct of reproductive rate. Many have suggested that pathogens should evolve in such a way as to protect the host. Ewald answers this question in terms of pathogen dispersal capabilities. In so doing, he harks back to selection models that look at the success of demes more than that of individuals. Especially pertinent is a work by Van Valen (1971), where dispersal is viewed as a kind of 'group selection'. In such a model, dispersal evolves because the original range occupied by the deme is only a temporarily suitable habitat. In the long view of geological time, this would be true of any contiguous area occupied by a species and non-dispersers are always doomed to extinction at some time scale. The converse (founders of surviving populations must have had the necessary genes for dispersal) tells us that all populations contain a potential for dispersal. Conflicting with the group need for dispersal is the population genetics within the individual deme where dispersers are removed as completely as if they were dead. Within each population, there must be selection against dispersal (Van Valen 1971). An example of this may be the evolution of flightlessness in island birds where flighted members are lost to the population when they attempt to disperse. Vertebrate hosts are islands to their pathogens, and there must be some selection against dispersal. We might also expect the development of symbiotic relationships, as harming the host would not seem to be in the pathogen's long-term interest, and that it would be better if both could benefit. Why isn't this model more prevalent? It has to do with the stability of the island. If a suitable environment is maintained for many generations, selection against dispersal will be more marked as will features that maintain this stability. If change resulting in local extinction is rapid, populations are less separated in time from their founders (dispersers whose genes have little vested interest in the fate of their ancestral population). The time interval on islands is long enough that many flightless birds evolve but transitory enough that most of those lineages are now extinct.

Vertebrate hosts may cease to be available to their pathogens through death or acquired immunity. The latter greatly decreases the time that an individual host is available and increases the stakes for the pathogen. The pathogen must reproduce and disperse rapidly, conditions that increase the likelihood that it will be harmful to the host. If the host becomes immobilized or dead, the opportunities for dispersal are greatly reduced. The same is true if potential new hosts are already infected and immune to further colonization. There is thus a balance between dangerously rapid reproduction of the pathogen, which through death or disability may reduce the time span available for a dispersal opportunity, and a slower, less virulent reproductive strategy that increases that time interval, but loses potential hosts because they have already been occupied by competing colonies resulting in immunity. If many new hosts are available and can be easily accessed, rapid dispersal may take precedence over host survivability with an increase in virulence. If suitable new hosts are rarely met, then virulence is reduced and the time span for searching increased. If hosts are plentiful but rapidly removed due to death or immunity, an initial highly virulent interval might develop with a decrease in virulence as the suitable hosts are used up. This is a pattern observed in some highly contagious diseases where epidemics seem to burn themselves out and leave a residual with reduced virulence (bubonic plague, swine flu, etc.). In such a model, the development of specific immunity

may itself have contributed to increased virulence by increasing the potential costs of delaying dispersal.

If we combine this model of disease evolution with what we know about earth history, we can formulate a number of interesting hypotheses about the evolution of pathogens and explain some of the factors that lead to emergent diseases. We begin by creating a baseline characterizing the probable pathogen load borne by our earliest primate ancestors and then follow the history of primates as they adapt to changing environments.

2.3 A time line for the emergence of human disease

Conditions at the end of the Cretaceous help to establish a baseline. Catastrophic extinction of many of the most important animals and plants completely reorganized communities and must have been accompanied by severe extinction of pathogens. The dominant vertebrate organisms became extinct throughout the world 65 000 000 years ago. Most of these were probably ectotherms whose body temperatures fluctuated between night and day. The world changed to one where the dominant species (mammals and birds) had constantly elevated body temperatures. Microbes sensitive to elevated temperatures might be pressed close to their tolerance levels in such hosts. We do not know when fevers first appear as part of the arsenal to fight infection, but their significance may have increased with the relative abundance of endothermic hosts.

It is important to realize that during most of the last 65 000 000 years global temperatures were much warmer than they are now. For a rather long period, there was little effective latitudinal climatic zonation, with tropical conditions extending almost to the poles. The absence of high latitude climatic barriers to potential vectors permitted tropical diseases to be cosmopolitan. Pathogens that evolved their life cycles under these conditions must now be confined largely to the tropics or to the artificially tropical environments created culturally by humans. Land connections between Asia and North America (Bering Connection) or Europe and North America (DeGeer Passage) helped to contribute to a shared community structure across the Holarctic at this time (McKenna 1980).

At the end of the Cretaceous and extending into the Early Tertiary, global temperatures declined, but they increased again rapidly in the Late Paleocene and Early Eocene (Martin 1994). This is theoretically important because modern disease groups all derive from species living under tropical conditions at this time. We can postulate that tropical infections are older than other modern pathogenic systems. Our primate ancestors occupied a rainforest canopy that was nearly worldwide. Like modern canopy vertebrates they were characterized by very small body size (under 100 g in some cases). Like all primates they were specialized herbivores feeding on the reproductive parts of plants and occasionally taking an insect. They probably traveled in small family groups that gathered in individual trees when flowers or fruits became available. There may have been little close bodily contact between unrelated individuals. Their droppings fell to the forest floor with little chance of contact with other individuals. They drank water from individual dewdrops or transitory accumulations of water in accidental pockets where there was little opportunity for contamination and the water did not remain long enough for pathogens to multiply significantly or complete their life cycle. With primates, the young would accompany the mother, and there would be little extended occupation of burrows and nests, another potential source of repeated infection. Many of the vectors that we associate with modern diseases were not really operative in this environment.

Chronic disease that relied on insect vectors or sexual contact would seem to be most effective in the Paleogene, and it has been noted (Lockhart *et al.* 1996) that sexually transmitted disease seems more prevalent in mammals, a group that was largely arboreal in the Early Tertiary. This can be understood in terms of widespread prevalence at the base of the mammalian radiation where they could become established in the basal groups of the later radiations. There are also deep tropical roots to the pathogens carried by mosquitoes and those arthropods that parasitise our skin, but these conditions did not continue everywhere. The history of the Cenozoic is characterized by global cooling resulting in a gradual contraction of tropical environments towards the equator (Martin 1994). New communities formed at high latitudes. Global cooling resulted in drying environments and a worldwide reduction in tree cover. During the Oligocene, we see the beginning of savannas as trees became scattered, with intervening grassy intervals. Primates began to come down from the trees and cross these open areas. Body size increased and we have primates that we might readily call 'monkeys'. For the first time we might expect them to encounter their own feces or drink from standing water, providing new sources of infection. Aggressive interactions may have become more common and trauma more serious. Burrowing also becomes an important aspect of the mammalian community structure at this time (Martin 1994), and the subsequent development of a stable nest location lent new opportunities to parasites and other pathogens.

Apes evolved a little later (Miocene) and represent an adaptation to the margin between forests and more open areas. This is the natural evolutionary station of the large cat-like predators and cats, who pose a special threat to apes and their kin. In this increasingly hostile environment, large body size evolved. Humans are specially adapted apes that walked away from the trees and survived in open country. In fact, the fundamental adaptation in becoming human seems to be the evolution of the human gait with the enlarged brain developing much more slowly. While moving into this open area, humans became the first primates to develop an intimate association with the large social ungulates. This opened new opportunities for exposure to pathogens that evolved in herd animals where rapid dispersal could result from shared air, drinking water, or the natural spread of feces on the grass that they grazed. In the latter case, it is easy to see how diarrhea might be a suitable dispersal mechanism or how a virus might find the nasal membranes a good target in animals that travel close against each other. It was not much like the original primate conditions, but for emerging disease it was in a direction that became familiar as humans increased their own social density.

Social ungulates and humans both started as small groupings of herbivores that occupied the interface between forests and more open vegetation. In both cases, they began by utilizing an under-story vegetation that was becoming more abundant as forests converted to savannah. In more forested environments, the under story is patchy in distribution, and groups of herbivores move in isolation from one patch to another. This was true of ungulates and some primates. Small groups are reservoirs for chronic infection, but lack the size to maintain a virulent strain. As grasslands developed further, herds became larger and began to show migratory patterns. This probably had its roots in the Late Miocene and reached a zenith in the Holocene when vast, nearly monotypic, stands of plants became common (steppes and open savannah). The development of great single species herds of ungulates in the Holocene followed a considerable decrease in the diversity of large mammals (End Pleistocene Extinction). At this time migrations began to concentrate huge numbers of individuals of ungulates from all over the species range. Virulence could increase under these conditions, and direct contact with discharges (urine, feces, air) from other individuals ceased to be rare. While these changes were taking place in the large mammal populations, early humans increased their meat intake, initially through scavenging and later through active hunting. At this time, they probably picked up parasites and infections already cultivated for millions of years by other large predators.

Evidence for this has been presented for parasites. Despres *et al.* (1992) give molecular evidence that schistosomes shifted to hominids from ruminant and rodent definitive hosts, as humans became more carnivorous. A recent article (Hoberg *et al.* 2000) also using molecular techniques, gives evidence that the tapeworm genus *Taenia* began its association with humans in the Plio-Pleistocene when humans began to hunt and scavenge. In this system, wild canids, felids, and hyaenids served as primary hosts and bovids as intermediate hosts. As they put it (Hoberg *et al.* 2000), 'Species of *Taenia* are historical ecological indicators of the foraging behavior and food habits of hominids during the diversification of *Homo* spp.'

The interaction between humans and rodents is especially interesting. Prior to the Late Eocene, most small mammals lived in trees or in the leaf litter on the forest floor. Few of these animals burrowed. As the trees disappeared and the grasses began to predominate, small animals had to look for new refuges to escape their enemies and a secure place to raise their young. With the advent of greater seasonal extremes, they also needed safe places to estivate or hibernate. In a sense, they created trees in inverse. By digging burrows, they provided security and a certain amount of climate control. The cost included the creation of a stable location where disease spores and parasites would have an opportunity for repeated infection. The energetics of burrow excavation limited body size and most of these animals were small and easy to kill if you were clever enough. They must have been among the earliest of human prey and are still a staple in hunter-gatherer diets. Human occupation of permanent shelters was rapidly exploited by rodents. From the viewpoint of disease dispersal, human dwellings are very similar to nests and burrows, and there would have been some sharing of disease as associations with rodents developed. This relationship was most marked in murids (rats and mice), a group of tropical rodents with members that have learned to share the artificially tropical environments of human dwellings. The high densities of these rodents in cities created a new incubation chamber for increased virulence in diseases shared with their human hosts.

2.4 Civilization as a symbiotic organism

Human interactions with other animals took an entirely new turn during the Holocene as humans progressed from predators on ungulate herds to members of these herds through domestication. In a very real way, we can think of humans and their domestic animals as a symbiotic organism. In the end, the wild species that provided the original stock for domestication could not compete with this new symbiosis and became extinct. Humanity and their domestic animals had become superorganisms that we term civilizations. In the end, the population of domesticates far exceed any number or density found in their wild progenitors. The history of ecosystems since the rise of civilizations shows a gradual elimination of anything that competes with a given civilization. This special relationship between humans and domesticated animals resulted in new feedback loops for the origination and dispersal of disease. The inclusion of a major predator, the wolf, into this symbiotic bond created an unusual social mix of predator and prey. The predators brought with them nearly 35 million years of interactive evolution with their pathogens and introduced humans to cycles of parasitism and other disease that were unlike any in their more herbivorous past. Dog domestication goes back to somewhere in the Late Paleolithic to Early Neolithic, some 15 000–10 000 years ago. Cats were domesticated later probably within the last five or six thousand years.

When we talk about domestication, we should not exclude the development of human societies. Humans are as much domesticated animals as are their dogs and cattle. Human societies routinely sacrifice the good of the individual for the good of the group. In the end, the forces of natural selection on the individual are mitigated by the common support of the group. Some aspects of our treatment of this process are confusing. For instance, why is an asphalt highway less of a 'natural' object than is a beaver dam, and why do we make a distinction between 'natural' and 'artificial' selection? Much of Darwin's evidence for the change possible in evolution came from the results of 'artificial' selection on domesticated organisms. The one real difference that I can see between natural and artificial selection is that any change in a naturally occurring population must occur within a framework where the whole animal benefits from increased fitness. In other words, the behavioral, physiological, and anatomical support system for the new adaptation must already be in place, requiring a complex and improbable series of events. Improbability in earth history is measured in time. Given enough time, the simply improbable becomes certain. Each step of the transformation is adaptive to the individual as a whole. Changes that outpace their support system are eliminated. In human terms, most natural selection is too slow for observation. We can look at changes in populations, but these must reflect ordinary fluctuations in a world of changing environments. The one evolutionary change that we do observe regularly is extinction.

In artificial selection, the society engaged in artificial selection fills in the missing support structure. Coat color need no longer be protective, as protection is provided. More energy can be devoted to reproduction, as the group will support and protect all of the offspring. Variability may increase, as the special requirements of the variables will be met by the group. Natural resistance to disease becomes less critical when the group can provide artificial resistance. Individual genetic problems are tolerated by the group and become more widespread in the population. Sexual selection and the maintenance of social bonds become more important than the survival skills of physical protection and food procurement. Our domesticated animals have become less an integral part of our struggle for survival and more surrogates in our need for social fulfillment. As such, they become even more intimately part of our extended families. They enjoy our medical system and become potential vectors of pathogens at levels previously known only in members of our own species. It is clear that we cannot limit our understanding of disease in this framework to any individual species. We must look at the symbiotic superorganisms that compose human societies in terms of complex interspecies interactions.

2.5 The tropical roots of disease

Almost everything discussed up to this point began in the African continent. The distribution of primates was limited to near tropical conditions because of food requirements. They required plants in some form of reproduction the year around. The retraction of the tropics in the late Miocene/Pliocene resulted in barriers that prevented further exchange of tropically adapted animals across high latitudes. Primates evolved in fixed locations with occasional extinctions, eliminating some of them from their former range. Meat changed the equation for humans. Meat could stand in for fruits and buds. A hunting primate had the potential to cross the coldest wastes in the world. This was a crucial innovation and probably represents the first time that technology changed human evolution. Meat is a fibrous, elastic material that is very difficult to pull apart. It has to be cut. It seems likely that one of the earliest uses of stone tools was simply to act as a stone carnassial, cutting off each bite of meat as it was held by the teeth. Cooking breaks up the fibers and makes it possible to remove bites without tools. It confers another interesting advantage in that most consumed calories in mammals go into maintaining body temperature. Cooler environments require greater food intake, but this can be supplemented by eating hot food. Clothes also help heat retention but reduce the amount of skin exposed to the sun. High latitudes carried a risk of rickets. It seems obvious that the earliest humans were deeply pigmented, and that this pigmentation was rapidly lost as higher latitudes were colonized. The Neanderthals of Europe were probably blond and blue eyed. This loss of pigmentation would have resulted in an increase in the importance of skin cancers. Sleeping in small closed spaces to conserve heat would have facilitated spread of respiratory affliction. Shared clothing or bedding became a possible vector for parasites and disease. Changes in social behavior as cold climates were colonized would have changed the nature of human disease.

Association with social ungulates and the development of a more carnivorous habit brought humans into two additional pathogenic communities, but it seems unlikely that diseases could have maintained much virulence in scattered human populations. We might expect chronic infections surviving a long time without the need to spread. This was true when isolated family groups became founders in broad new territories. Much of the original spread of *Homo sapiens* may have taken place by boat and/or through small isolated bands crossing regions of inhospitable, terrestrial habitat. Either route introduces a long quarantine that would have weeded out all but the most persistent infections. In this way, many diseases of humanity's long African tenure could be lost by the vagaries of founder effect during human colonization of the rest of the world. Today we still think of the tropics as 'disease ridden'. Compared with the more lately colonized portions of the world, this is true. This disease load is probably more detrimental to modern densely packed populations than it was to our more scattered ancestors. Modern travel opportunities with same day arrival from any part of the world allows us to reassemble the isolated

remnants of human disease until we confront a disease load equivalent to or greater than any faced by our ancestors. Our super-dense populations allow virulence to soar above anything that they would have faced. Ewald (2000) has written a book 'Plague Time' about modern vulnerability to infection, and he is probably right that the present is a unique time for the origin and spread of especially dangerous disease.

The development of long-distance commerce changed some of the rules of disease transmission. The creation of ships with relatively large crews coupled a small population that could act as a disease reservoir with reasonably swift dispersal. Before ships, dispersal of disease over long distances was slow and required a larger host population to remain intact until it reached a new population. The original human dispersal involved archaic humans. Spreading out of Africa, it crossed southern Eurasia, proceeding south to Java. It involved humans with low technological achievements and was largely restricted to primate-friendly climates (warm and moist). Lowered sea levels exposed vast areas of continental shelf during glacial periods and connected many parts of the Far East that are now separated by enormous water gaps. Disease strains were part of the baggage carried on this dispersal, but the resulting populations were small and isolated. Pathogens had to maintain themselves over long periods without much hope of further dispersal.

2.6 Ice Age geography and the emergence of disease

By the last glacial maximum some 18 000 years ago, the archaic humans were extinct or assimilated. *Homo sapiens* occupied essentially its modern range, but consisted of isolated populations separated by physiographic features produced by the ice expansion. The ice itself was a formidable barrier, separating North America south of Canada from Siberia. Most of Canada and the Northeastern United States was covered by ice. Cold air drained off of the southern margin of the ice front, creating a narrow belt of tundra with tundra muskox and collared lemmings. South of the tundra was a broad belt of spruce forest with the moose-like woodland muskox, *Bootherium* and the extinct stagmoose, *Cervalces*. Below the spruce forests, the eastern United States contained a deciduous forest refugium and below that the subtropical deciduous forests of Florida and the gulf coast. The western United States was a vast region of complex coniferous parklands, characterized by the American camel, *Camelops*. Martin and Neuner (1978) proposed a system of faunal provinces to characterize these ice age co-occurrence patterns. Martin and Hoffman (1987) showed that these faunal provinces were the Pleistocene equivalent of biomes, and finally, Rogers *et al.* (1990) demonstrated that native American linguistic groupings fit the provincial pattern, suggesting that humans were in North America at least early enough Rogers *et al.* (1994) to have been affected by the ice age biomes. This suggests that these human ethnic associations had ecological boundaries and that these boundaries might give insight into the emergence and spread of disease. At the very least, they should help us to model the ecological parameters associated with fossil evidence for pathogens. We cannot emphasize enough that Late Pleistocene full glacial conditions were the most isolating faced by modern humans at any time in their history. Whatever validity there is to racial types, their morphological basis must have been established at this time (Rogers 1986). Not only were the ice sheets themselves isolating, mountain ranges were more impressive barriers because the depression of snowlines closed all but the lowest and most southerly mountain passes. Vast inland lakes formed, including one that may have filled the entire Amazon basin restricting the Amazonian biota to the surrounding highlands. We need to put the origins of human diseases within the context of these special times.

For humans the most significant of these inland freshwater seas was Lake Mansi in Central Asia (Rogers 1986). It formed as the spreading Siberian ice sheet dammed the north-flowing rivers and encompassed most of Central Asia to the foothills of the Himalayan mountains. Europe and Scandinavia lay on the west side of Lake Mansi. Rogers has suggested that this region was occupied by the progenitors of the modern Uralic Caucasians (Finns, Lapps, and Hungarians). On the east side of the great lake was the homeland of the Altaic

speaking mongoloid peoples and the progenitors of the Native Americans. Part of that population extended through the cold, dry mammoth steppe up across the exposed Bering Shelf and into Alaska, where its expansion was stopped by the continental ice. Rogers *et al.* (1992) thought on linguistic grounds that the main colonization of North and South America south of the ice sheet took place by boat along a continental shelf expanded by sea level lowering. The discovery of Monte Verde, one of the earliest known Native American sites near the coast of Chile, is consistent with their model. Turner (1986) argued for three very distinct immigration events into North America while Rogers *et al.* (1991) showed how *in situ* evolution might have resulted in the same pattern. In either case, we have three isolated core populations that must have had their own epidemiology and the introduction of humans to two whole continents where pathogens had a long evolutionary history that did not include humans. The question then becomes one of distinguishing the contribution of the endemic biota from pathogens that accompanied the original human colonizers.

At the end of the Pleistocene, environmental reorganization may have permitted major shifts in the distribution of the original human populations. Rogers *et al.* (1990) using linguistic data proposed a series of shifts for Early Holocene North America. These suggested shifts provide potential models for understanding the prehistoric distribution of disease. Similar shifts in populations occurred in Eurasia at about the same time.

2.7 Centers of disease origin and spread

Eurasia maintained a remarkable degree of isolation from the Western Hemisphere during most of the Holocene, and the great epidemic diseases of the Old World (smallpox, measles, plague, etc.) were unknown in the Americas until European contact. These diseases are thought to have arisen from a combination of domestication, agriculture, and urbanization that apparently did not exist in the New World. At least no similar diseases lurked to decimate the colonizers and their homelands as the European plagues nearly exterminated the Native Americans. Had the pattern of discovery and colonization been reversed there is no reason to expect a great impact on European populations and the colonization would probably have failed in the face of the highly contagious diseases then current in Eurasia. I was interested to learn that this viewpoint is very well expressed in a book on the competition between human cultures by Jared Diamond (1997). This does not mean that the colonizers of the Americas escaped entirely. The venereal disease, syphilis, has a long history in the Western Hemisphere and little evidence for a pre-Columbian occurrence elsewhere (Rothschild *et al.* 2000). The antiquity of treponemal disease in North America was demonstrated in a Late Pleistocene short-faced bear from Illinois, and it is likely that yaws was present in North America by Late Pleistocene whereas syphilis is American in origin (Rothschild *et al.* 2000). Rheumatoid arthritis is another affliction that seems to have an American origin. According to Rothschild (2000), it shows a pattern of spread from Eastern North America. The absence of a known etiology makes the interpretation of the origin of this disease more problematic.

Recently, tuberculosis has been identified in a variety of Late Pleistocene bovids from North America (Rothschild and Martin 1993) but is absent in the native antilocaprids, horses and camels. This would suggest an introduction corresponding to the migration of Eurasian bovids (bison, muskoxen, sheep) across Beringia to North America. Its presence in the pre-Columbian human population (Ewald 1998) raises the possibility that the pre-Columbian population contracted tuberculosis from these bovids. Similar pathology has been identified in fossil bison remains from Europe. One infected specimen (a 17 000 year old bison metacarpal) was extensively sampled for evidence of DNA from the pathogen. The sample was split and studied independently by laboratories in London and Jerusalem. The two laboratories independently recovered DNA of *Mycobacterium tuberculosis* complex (Rothschild *et al.* 2001), which was sequenced. Spoligotyping was preformed on this DNA showing that it was a separate strain from *Mycobacterium bovis*, the species that might have been expected. Additional sampling revealed that *Mycobacterium* was restricted to the pathologic lesion and absent from other parts of the same bone

or from soil samples and stratigraphically contemporaneous samples of wolf (*Canis lupus*) and extinct horse from the same deposit. No exact match was found in modern samples of *Mycobacterium*, suggesting that the strain may have changed in the last 17 000 years.

Bison, like other cattle, travel in large herds. They spend considerable amounts of time closely packed against each other. Arial transmission of a pathogen causing a lung disease is possible in such conditions. It is also easy to see how herds of grass-eating organisms disturb the soil, raising it as dust that could be inhaled. Under those conditions, *Mycobacterium* strains that lived in the soil would be confronted with new opportunities as respiratory parasites. The development of grasslands in the Late Eocene–Oligocene may mark the beginning of many pathogenic lineages in groups that include soil bacteria. Pollution of the graze by urine or feces would be an effective mode of transmittal, especially if feces were composed of a watery discharge (diarrhea) rather then discrete and easily avoided lumps. In this case, the host's need for a rapid removal of bacteria and their associated toxins might also serve the pathogen's need for dispersal. In the latter case, we might predict that there should have been an increase in the occurrence of gastrointestinal disease as grasslands expanded during the late Cenozoic. Such a trend has been demonstrated for spondyloarthropathy (Rothschild *et al.* 1993a), an affliction often associated with gastroenteritis. The origin and spread of tuberculosis should also fit into this overall pattern.

Most models for the origin and spread of tuberculosis in humans relate it to the beginning of cattle domestication, but earlier contact with wild cattle or bison now seems equally likely. The close similarity between the fossil strain and *Mycobacterium africanum* (Rothschild *et al.* 2001), along with the broad taxonomic occurrence (bison, muskox, bighorn sheep), suggest that the strain may have developed early in the bovid radiation. This hypothesis could be tested by examining Miocene and Pliocene collections of African antelope. Conversely, tuberculosis could have developed in the Holarctic. In the latter case, we should seek relevant fossil examples in China and Russia. *M. tuberculosis* itself could still have evolved within the context of cattle domestication (Zias 1998). Transmission of infection in the mother's milk is a uniquely mammalian vector and one that would only have become important in the Cenozoic (the age of mammals). Only humans would be in danger of cross-species infection from milk-transmitted disease. The important point is that we now seem to have the tools to address these questions directly. The presence of good osteological markers (Rothschild and Martin 1993) coupled with a wealth of information on the modern genome and demonstrated recoverability of DNA makes this one of the most promising targets for paleopathological studies. Tuberculosis is also one of the most important and dangerous of the diseases that are presently undergoing re-emergence. The study of this material combining many disciplines and laboratories is also an apt example of the value of the multidisciplinary approach. Multiple evaluation of the same material from widely different viewpoints is not additive in its value, but is rather exponential. It is the substance from which all good science is made.

The diseases that seem to have a Western Hemisphere component are often serious, but tend to be chronic over long periods of time. They are suited for dispersal from one small isolated population to another, but do not generally kill large numbers quickly. Although some of the population centers in Central America were among the largest in the world during pre-Columbian times, they were isolated from other similar centers. A highly infectious disease could burn itself out without reaching other similar centers of population. The absence of maritime commerce in the Americas may be an especially significant factor in the absence of evidence for plague-forming disease. Large ships can carry a sufficient number of infected people to sustain a disease over a long distance. There are no urban myths of plague ships in Native American cultures, although they are common in European cultures.

2.8 Conclusions

Humanity has only had a real comprehension of what makes us sick for a little over a hundred years. The discovery of microbes and their role in disease must stand as one of the greatest achievements of all time. Humans are the first organisms to recognize the existence of a vast array of enemies too small to be seen, but with more potential impact than all the armies in history. For the last century, we have devoted time and treasure towards learning how to kill these microorganisms and in many cases enlisting their own enemies in the microscopic world for materials toxic to them. Until recently, much of this research was little more than 'trial and error'. While these crude approaches have yielded some major success, eventually, we must understand why they are there. We must know the conditions that convert relatively harmless microbes to mass murderers. We need to understand the role of variability in the virulence of pathogen populations and discover how disease emerges and spreads. These questions are too big to fit into a laboratory. We must go into the field to address them, and we cannot wait for plagues to occur in order to make observations. Instead, we study the historical record of disease and finally the record preserved in the prehistoric bones themselves.

The value of paleopathology rests ultimately on the strength of the hypotheses that we test. Ewald's attempt to model the population biology of disease in terms of host populations, virulence, and vectors provides a basis for a reinterpretation of human history for the past few centuries. The development of agriculture and population centers changed the parameters for disease and the pathogens changed in response. One theme is recurrent in Ewald's scheme: mobility of the host. Modern mobility may change the scope of disease as much as did earlier cultural changes. This is a new pathogenic landscape characterized by rapidity of change, but with roots extending into the dim origins of humanity. The basic premises of population biology can provide a roadmap to even this little traveled realm. Our examination of the last 65 000 000 years permits a series of probable scenarios from which we can extract hypotheses.

Destruction of the major ecosystems and extinction of the dominant vertebrates and invertebrates at the end of the Cretaceous should be reflected in a similar extinction and re-organization of pathogenic populations. This constitutes a baseline for the maximal age of most modern pathogens.

The Early Tertiary was characterized by global tropical conditions. Tropical disease is potentially older and more basic than the other pathogenic systems that constitute the modern disease load.

Early Tertiary primates existed in small arboreal bands giving little opportunity for direct transmission of pathogens from one individual to another. Invertebrate vectors and sexual infection may dominate in this system.

In the Oligocene, cooling and drying in mid to high latitudes reduced the continuity of the tree cover and promoted the evolution of larger primates that spent more time on the ground going from tree to tree where they were more likely to come into contact with standing water or feces. The importance of water and feces for disease transmission should have increased at this point. There would also have been opportunities for foraging on the ground where predators were larger and more common. As terrestrial foraging changed from an occasional activity to the normal lifestyle for some primates, larger body size resulted in part to resist predators. Humans carried ground foraging to an extreme and began to systematically exploit small animals and carcasses. A small amount of these activities has always been a primate activity, but in humans it became a way of life leading to hunting of even the largest prey. At this point, the disease parameters of humans began to take a number of important turns away from the general primate model.

• The consumption of meat introduced early humans to diseases, including parasites, that had evolved within the ecology of carnivorous mammals. This would have included some forms of bacteria and toxins characteristic of decayed flesh.

• As hunting progressed to the highly social bovids, humans came into contact with disease that had evolved in animals who formed dense, but highly mobile, herds. These conditions resemble the dense highly social and mobile societies eventually achieved by humans. Humans aggravated the pathogenic aspects of this relationship by incorporating many kinds of social animals into the general structure of humanity through domestication.

• Domestication also included carnivores associated with humans and brought their diseases and parasites into the pathogenic mix.

• The spread of modern humans during the Late Pleistocene diluted and isolated the disease load. The founder populations were in some ways healthier than the parent population because of loss of part of their disease load during dispersal.

• The development of symbiotic social structures with domesticated animals in the Holocene formed new feedback loops for the emergence and spread of disease.

• The development of sedentary centers of population with high density and regular communications with similar centers elsewhere permitted disease virulence that was probably unprecedented.

• Modern rapid travel connecting all areas where humans reside has assembled the diseases lost by individual founder populations and created a potential disease load greater than that faced by any previous organism.

Although unknown factors may confound this sequence, it is at least a reasonable time line organizing the appearance of disease systems and providing a theoretical basis for their origination. We can test its authenticity through direct assessment of the fossil record and the relative divergent times extracted from the molecular systematics of pathogen strains.

The origin and spread of any disease has a history. They have not always been with us nor do they appear from an ecological vacuum. They result from opportunities that were exploited by the pathogen. We need to understand how these opportunities come about. The most effective way to treat a disease is to suppress its origin or prevent its spread. If we understand the ecology of emergent diseases, we are better armed to fight them at this basic level. Humanity is no longer composed simply of individuals, but must be thought of in terms of cultural groups composed of a complex association of organisms united in a symbiotic bond. These cultural groups compete with each other for a share of dwindling resources. They combine pathogenic relationships that evolved independently in social herbivores, carnivores, and primates over millions of years, resulting in a new ecological mix for pathogens. We should not be surprised if this has resulted in novel diseases. These cultural units were previously somewhat isolated by time and space, but modern transportation has removed such barriers. The result may be catastrophic.

Acknowledgements

I have benefited from the insights and encouragement of C. Greenblatt, and especially B. Rothschild, who has been my inspiration and guide to the fossil history of disease. The support of the Center for the Study of Emerging Diseases, Jerusalem, is gratefully acknowledged.

CHAPTER 3

Bacterial symbionts of protozoa in aqueous environments—potential pathogens?

Hans-Dieter Görtz and Rolf Michel

Finally, microbes and vectors swim in the evolutionary stream, and swim much faster than we do. Bacteria reproduce every 30 minutes. For them, a millennium is compressed into a fortnight. They are fleet afoot, and the pace of our research must keep up with them or they will overtake us. Microbes were here on earth 2 billion years before humans arrived, learning every trick for survival, and it is likely that they will be here 2 billion years after we depart. (Richard M. Krause, Emerging infections, 1998)

3.1 Introduction—Legionnaires' disease as a warning

The public was shocked in 1976 when a number of American veterans suffered from an enigmatic pneumonia and many of the patients died. The so far unknown type of pneumonia was then termed Legionnaires' disease. It was caused by an intracellular bacterium that was designated *Legionella pneumophila*. It was found that *Legionella* can exist in protozoa such as *Tetrahymena* and *Acanthamoeba*, and the bacteria were isolated from small amoebae from freshwater and soil (Rowbotham 1980, 1986; Barbaree *et al.* 1986; for a review see Jeon 1991).

The first occurrence of Legionnaires' disease was a warning that newly established technical facilities in human environments may be of epidemiological significance. New facilities may provide suitable habitats for microorganisms or their reservoir hosts. In the specific case of this disease it would be thin water films found in air conditioners resulting from condensing water. Also, the significance of protozoa and invertebrates as reservoir hosts or even vectors of intracellular pathogenic microorganisms had not been realized before. Although

various types of infection were known in protozoa, the connection with human disease had not been established. The significance of protozoa and invertebrates as reservoir hosts or even vectors of intracellular pathogenic microorganisms was not realized. Instead, bacterial infections in protozoa and invertebrates were regarded as exotic rather than of general interest. This chapter will concentrate on bacterial infections in protozoa as hosts of microorganisms, although infections in multicellular invertebrate animals are certainly important too.

Until this decade little was known about the relationships of intracellular bacteria in protists with pathogenic or environmental bacteria. After the emergence of Legionnaires' disease, these relationships became more interesting and new molecular methods became available to study them. It appears obvious that if bacteria in protozoa were related to pathogens, they could become infectious for humans under certain conditions. The question arose as to whether protozoa can become reservoir hosts for human pathogens under suitable conditions. Meanwhile it is evident and we must be aware that even small changes in local environments may cause new or ancient pathogens to emerge. To evaluate the potential risk for humans it is neccessary to study the variety of bacterial infections in protozoa. Since protozoa are older than metazoan animals, they should have been colonized and exploited by microorganisms long before metazoa evolved and it appears not unlikely that certain pathogens of animals have derived from intracellular bacteria in protozoa. How do protozoa get along with their intracellular bacteria? Are all of these infections old associations or are new infections found in protozoa as well? Are there new risks of bacterial infections in protozoa for humans, and if so, under what conditions? We shall try here to give some answers to these questions.

3.2 Bacterial symbionts of protozoa—exotic adaptations

Like those invertebrate animals that are filter feeders (e.g. sponges) protozoa use microorganisms as their prey. This is probably one of the reasons why bacterial infections of various types are frequent in protozoa. Bacteria that are ingested by a protozoan cell are immediately killed by oxygen radicals that are released into newly formed phagosomes (food vacuoles). The phagosome is then acidified and lysosomal enzymes are released into the phagosome in order to digest the prey bacteria. This is what normally occurs. Some bacteria, however, may resist all attacks in the phagosome and persist in the host cell (Fig. 3.1).

Intracellular pathogens show a diversity of life strategies, but at first all intracellular bacteria have to invade their host cells, have to defend themselves against host cell attacks, and have to develop means to move in host tissues. The mechanisms they have evolved to survive in a hostile environment may, however, differ markedly from case to case, resulting in a great diversity of bacterial symbionts of protozoa. Some of the fascinating adaptations and strategies of bacterial symbionts of protozoa will be presented here.

Traditionally, intracellular bacteria in protozoa have been called symbionts without referring to the particular nature of the association, such as mutualism or parasitism. In fact, it is often difficult to judge what is the significance of intracellular bacteria for the host. Permanent mutual symbionts as well as highly pathogenic bacteria are found in protozoa, but in most cases we still do not know the nature of the associations (for reviews see

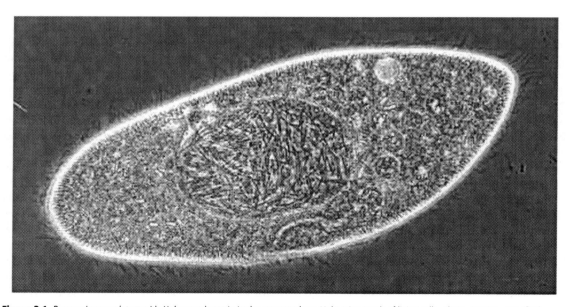

Figure 3.1 *Paramecium caudatum* with *Holospora* bacteria in the macronucleus. Light micrograph of living cells, phase contrast, microflash.

Heckmann and Görtz 1991; Jeon 1991). Whereas some infections appear to be accidental, sometimes fatal to the host, certain bacterial symbionts may provide their hosts with vitamins and a few are even essential for the host cells.

The best studied example of an essential bacterial symbiosis in a ciliate is that of *Polynucleobacter necessarius* in *Euplotes aediculatus*. Like this ciliate, freshwater species of a major group of the genus *Euplotes* also depend on symbionts. *P. necessarius* and related bacteria cannot be grown outside their hosts. Obviously they need a complex substrate with metabolites that they normally obtain from their hosts. This is similar to other symbionts of ciliates and obligately intracellular agents like rickettsias. Like those, *P. necessarius* has a small genome size of only 0.5×10^9 Da (reviewed in Heckmann 1983). *Polynucleobacter* and its relatives are intimately adapted to their hosts and may function like an organelle. However, it is not known what metabolites are provided by the symbionts. When the symbionts are removed by antibiotics, the host cell can no longer divide and dies. Reinfection with *P. necessarius* may save the ciliates, and in certain cases, microinjection of *P. necessarius*-like symbionts from a different host species has also been successful (Fujishima and Heckmann 1984). This is the best argument for an identical function of the bacteria in different *Euplotes* species and for a common origin of the symbionts in various species of the host genus. The initial infections may go back to a time long before the evolution of metazoa and the symbiosis is reminiscent of the symbiotic origin of mitochondria and chloroplasts (Heckmann 1983).

Essential bacterial symbionts are also found in other protozoa, e.g. in *Crithidia* and *Blastocrithidia* (Jeon 1991); these kinetoplastid flagellates dwell in the gut of certain insects and are related to *Leishmania* and *Trypanosoma*. There are always two symbionts found in each individual flagellate and the symbionts have therefore been called diplosomes. The bacteria provide their host cells with essential metabolites. e.g. lysin and haemin. Whereas the diplosomes, too, appear to be ancient symbionts, this is not the case for the intracellular symbionts found in *Amoeba proteus* by Jeon (1983). He observed bacterial infections in his laboratory cultures. The bacteria were harmful to the amoebae for a period of more than a hundred fissions of the amoebae. Thereafter, the amoebae became dependent on the bacteria. Thus, within a relatively short period a pathogenic microorganism became an essential symbiont. This phenomenon turned out to be reproducible when other *A. proteus* cells were infected by the bacteria. The bacteria containing a plasmid were found to provide the host cell with a 29 kDa protein. However, the mechanisms by which the host became dependent have not been discovered. The intracellular bacteria in *A. proteus* are obviously a recent infection probably originating from free-living contaminants in the culture medium, as the infection has not been observed before. This indicates that new bacterial infections in protozoa may result in stable intracellular associations under certain conditions. The new symbiosis in *A. proteus* also shows that even today intracellular bacteria may develop into organelle-like symbionts.

A considerable number of bacteria in ciliates confer killer traits upon their hosts. In the presence of cells bearing such bacteria other cells that are not infected are killed—a phenomenon analysed in detail by Sonneborn and Preer (for a review see Preer *et al.* 1974). A number of different symbionts are known that provide their host cells with a killer trait. In co-culture with symbiont-free paramecia the latter are killed by a toxin released from the killer-paramecium. Thus, infected paramecia have a competitive advantage over symbiont-free cells in laboratory cultures (Landis 1988). The bacteria may therefore be regarded as mutualistic. When the bacteria are destroyed by antibiotics or other means the former host cells become sensitive themselves to the killer toxin. They may be reinfected in various ways and again become killers that are resistant themselves. The ecological and evolutionary significance of symbionts that confer killer traits upon their hosts appears obvious and at least in some cases Landis (1988) observed a strong selective advantage for killers (infected cells) and a disadvantage for sensitives in competition studies in conditions found in nature.

Toxicity as well as pathogenic potency may directly increase the fitness of intracellular bacteria. It has been suggested that genes relevant to pathogenicity may be clustered in genomic islands termed pathogenicity islands. Genomic islands are blocks of DNA that function as mobile elements and may be transferred horizontally (for a review of genomic islands and pathogenicity see Hacker and Carniel 2001). Just as pathogenicity islands have most likely been positively selected, there may be toxicity islands in intracellular bacteria as well.

Taking a closer look, the killer trait in some cases is found to be more complex. Intracellular bacteria must be regarded as a costly burden for host cells. In *Paramecium* thousands of these bacteria colonize the cytoplasm or even the nuclei, and infected cells have a slower growth rate than uninfected cells. For the killer symbiont *Caedibacter caryophilus* living in the macronucleus of *Paramecium caudatum* it has been shown that the bacteria continue growing under unfavourable conditions e.g. during mild starvation. The bacteria even overgrow the host nucleus, finally resulting in the death of the host cell (Schmidt *et al.* 1988). There is no question that individual cells suffer severely from the infection, and the bacteria may be regarded as parasitic. So, it is the resistance to the killer toxin of their own symbionts rather than an advantage over competitors (which are killed) that makes the ciliates dependent on their bacteria. Consistently, once an infection by these bacteria is found the whole population is infected. During fast growth of infected paramecia, single cells may lose their symbionts, but those that do are eventually killed.

Nevertheless, it is found that after rapid growth, a certain percentage of cells in laboratory cultures may outgrow their bacteria and become free of the symbionts. Why are these cells, which are sensitive to the toxin, not killed immediately? It has been found that the bacteria may only provide the toxic principle when they contain unique protein bodies termed R-bodies (for a review see Preer *et al.* 1974). The proteins of these R-bodies are encoded by plasmids or phage genomes (Pond *et al.* 1989). Once R-bodies are present the bacteria can no longer divide. During rapid growth of host cells and their bacteria the plasmid or phage genes are poorly expressed and therefore no R-bodies are found. In rapidly growing populations, cells that have become free of bacteria may therefore survive;

they are, however, killed when conditions become worse and the R-body proteins are expressed again. The killer trait in paramecia is a good example that plasmids and phages of intracellular bacteria may be of great significance for host cells and host cell populations. Intracellular bacteria may show new physiological and behavioural traits once they are infected with plasmids or phages.

While killer symbionts such as *Caedibacter* species provide their host with toxins against competitors, Rosati and coworkers investigated a unique symbiosis of true advantage for the host ciliate. The symbionts are epibiotic bacteria on the marine ciliate *Euplotidium* (Petroni *et al.* 2000). Though this chapter focuses on intracellular bacteria, the epibionts of *Euplotidium*—called epixenosomes by the authors—will be considered as they show peculiar adaptations. Epixenosomes appear firmly attached on the surface of the ciliates. The complex life history of the bacteria establishes two main stages. Stage I has a typical prokaryotic structure and multiplies by binary fission. Stage II does not divide; it contains a complex extrusive apparatus with a proteinaceous matrix that rapidly elongates at ejection. The extrusive apparatus is encircled by a basket of bundles of tubules. The tubules, with a diameter of about 22 nm, are sensitive to inhibitors of tubulin polymerization and react positively with different antitubulin antibodies. There is clear evidence that stage II epixenosomes have a defensive function for the host cell.

All the bacteria described above are not, or are only mildy, infectious. Other bacteria are highly infectious, such as the species of the genus *Holospora* infecting the nuclei of *Paramecium* (Fig. 3.1) and some other ciliates (for a review see Görtz 1986). These bacteria will be described below in some detail to show that even intracellular bacteria of protozoa may have developed sophisticated mechanisms for infection and maintenance in their host cell.

Holospora bacteria are both host specific and nucleus specific (Fig. 3.2). They undergo a developmental cycle with a specialized infectious form and a morphologically and physiologically different reproductive form. After each division of the host cell, the infectious forms are released into the medium, while the reproductive forms remain in

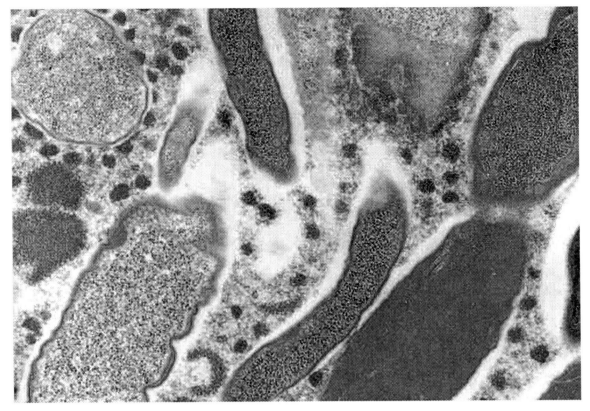

Figure 3.2 Different stages of *Holospora* bacteria in the macronucleus of *Paramecium*. Not much chromatin appears to be left due to many bacteria colonizing the nucleus. Electron micrograph.

the nuclei until the next division of the host cell, when some of them develop into the infectious form again. A new infection may begin with the ingestion of an infectious form by an uninfected paramecium. The bacterium leaves the phagosome, is transported through the cytoplasm and introduced into the host nucleus by a fusion of the two membranes of the transport vesicle with those of the nuclear envelope (Görtz and Wiemann 1989).

The periplasm of the infectious form contains a number of stage-specific proteins that are lost from the protein pattern in the early steps of infection (Fujishima *et al.* 1990; Görtz *et al.* 1990). This is what would be expected if such proteins were used for communication with membranes and cytoskeleton of the host cell. Released periplasmic proteins could also protect the bacteria against lysosomal enzymes or could inactivate such enzymes.

Holospora bacteria do not normally kill their hosts. Symbionts and host appear to be well adapted, although the host cells may die under unfavourable conditions. After infection of the generative micronucleus, successful sexual reproduction is no longer possible. This may be compared with the parasitic castration of metazoa caused by some trematodes or crustacea. However, natural populations of *Paramecium* are not normally infected to more than 10% by the micronucleus-specific bacteria. The low infection rate seems surprising in view of the high infectivity observed. One reason may be that uninfected paramecia propagate more quickly than infected cells under suboptimal conditions. Moreover, micronuclear infections may occur only in populations of low population density, where the propability of new infections will be lower. Also, paramecia seem to have mechanisms to cure themselves of endonuclear symbionts. It has been discovered that *Holospora* may be lysed in the host nuclei of certain strains of *Paramecium* after an infection (Fokin and Skovorodkin 1997). After lysis of the bacteria the host cells remain viable. The lysis may be due to a special defence mechanism evolved in ciliates against endonuclear infections. It would be of medical interest to understand the mechanisms protozoa may use to cure themselves of intracellular infections.

Compared with the *Holospora* bacteria there are no pathogens of humans or animals with similar behaviour. No bacterial pathogens are known that specifically infect the nuclei of animal cells and no bacteria show a similar developmental cycle with an infectious form being much larger than the reproductive form and having a voluminous periplasm filled with proteins. Chlamydias, which also show a developmental cycle, have infectious forms (elementary bodies) that are considerably smaller than the reproductive forms (reticular bodies) (Fields and Barnes 1991).

3.3 The famous relatives: phylogeny of symbionts of protozoa

As a rule, intracellular microorganisms show low abundance and are difficult to detect. In most cases they may hardly grow *in vitro* but only in their specific host cells. In the past most of the methods to detect intracellular microorganisms were based upon microscopical investigations or indirect methods, such as polymerase chain reaction (PCR) from tissue samples, not revealing the exact site of the pathogens. Medical investigations are usually based on certain clinical features, whereas most detections of microorganisms in protozoa and lower animals have been incidental. Appropriate dyes for nuclei acids like orcein or diamidinophenylindole (DAPI) have been proven to be useful, but staining with such dyes does not give specific information on the nature of the microorganisms.

In recent years PCR amplification of microbe-specific sequences has been used in medical diagnostics where it has become a powerful tool for detection of pathogens. The detection of unknown microorganisms in nature by PCR amplification bears a high risk of confusion with contaminants. It is frequently observed that primers used for PCR-amplification give a better match for gene sequences of certain contaminants than for unknown microorganisms. The PCR detection of new intracellular microorganisms therefore is only reliable with subsequent verification of the result. This verification may be achieved by fluorescence *in situ* hybridization (FISH) using fluorochromated oligonucleotide probes designed according to the PCR-amplified sequence (Amann *et al.* 1991). Genes like the small subunit rDNA have highly conserved regions, but also variable sequences, which may show species-specific differences. Probes may be species specific if they are designed to complement a sequence of appropriate variability. Probes designed according to more conserved regions of a gene may rather be used to identify various species of a genus or even taxa of higher order.

The introduction of *in situ* hybridization using fluorochromated specific oligonucleotide probes to detect intracellular bacteria was certainly a milestone in this field of research. In protozoa this method was first used by Amann *et al.* (1991) for *Holospora obtusa* in *Paramecium*. Amplified rDNA of *H. obtusa* was sequenced to analyse its phylogenetic position. As most intracellular bacteria do not grow *in vitro*, bacteria must be isolated from their host cells before amplification or, as an alternative, bacterial rDNA may be amplified *in situ* in infected cells using appropriate primers. Protozoa feed on free-living bacteria, e.g. *Enterobacter aerogenes*, and contamination by other environmental bacteria are frequently found in the cultures as well. It is therefore neccessary to prove that the amplified rDNA belongs to the intracellular bacteria and not to prey organisms or any contaminants that were amplified. A probe to match the sequence obtained must be designed, fluorochromated, and used for *in situ* hybridization to verify that the amplified rDNA belongs to the intracellular bacteria (Figs 3.3 and 3.4). Prey bacteria in the phagosomes of the host cells must not be labelled. The latter bacteria may, however, be labelled in a control after *in situ* hybridization using a probe specific for eubacteria.

Using this method the phylogenetic positions of a number of intracellular bacteria have been revealed in the past few years. Quite unexpectedly intracellular bacteria in protozoa belong to different groups of bacteria (Table 3.1; Fig. 3.5). *Holospora* bacteria were found to be related to rickettsias and

FISHing for intracellular bacteria

Intracellular bacteria may be identified and classified by fluorescence *in situ* hybridization (FISH).

Each bacterial cell contains thousands of ribosomal RNA (rRNA) molecules. The sequence of ribosomal genes is unique for each taxon (e.g. species, order, phylum). Complementary to certain regions of their rRNA, oligonucleotide probes are designed and fluorochrome-labeled.

These probes hybridize with (bind to) complementary sequences of the specific rRNA. Due to the fluorochrome bound to the probes, rRNA of this type may be detected by fluorescence microscopy.

By labelling different probes with different dyes, two or more bacterial species may be identified simultaneously.

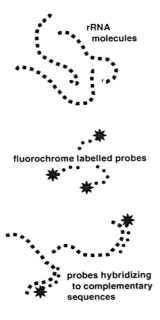

rRNA molecules

fluorochrome labelled probes

probes hybridizing to complementary sequences

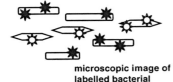

microscopic image of labelled bacterial

Figure 3.3 The method for fluorescence *in situ* hybridization (FISH).

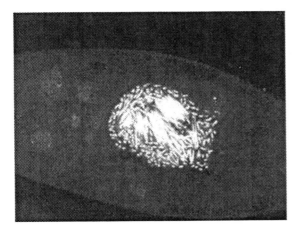

Figure 3.4 *In situ* hybridization of a *Holospora* species in the macronucleus of *Paramecium caudatum* using a probe specific for *Holospora* designed by Amann *et al.* (1991). Only *Holospora* bacteria in the nucleus are labelled. The host cell and some phagosomes with prey bacteria are faintly visible.

closely related to the killer symbiont *C. caryophilus*, being biologically and morphologically completely different from the holosporas (Springer *et al.* 1993). Like *Holospora* and *C. caryophilus*, the rickettsias are

obligate intracellular bacteria and are occasionally also found within the host nucleus. *Rickettsia* species are causative agents of typhus and various other diseases in humans and animals (for a review see Weiss 1992).

A more systematic search for further *Holospora* species resulted in the identification of nine species that are host specific to various paramecia (Fokin *et al.* 1996). This example of many *Holospora* species in only one genus of ciliates indicates that the variety and abundance of intracellular bacteria in protozoa may be unexpectedly great.

Typically, rickettsioses are louse-, flea-, and tick-borne diseases, but although transmission by protozoa has not been observed, the biology of rickettsias resembles that of protozoan symbionts. Rickettsias such as *Rickettsia typhi* or *Rickettsia prowazekii* enter host cells by phagocytosis. Just like holosporas, they escape from the phagosome into the cytoplasm. It has also been shown that *Ehrlickia risticii* may be capable of inhibiting phagosome–lysosome fusion. We do not know the nutritional metabolites that holospora bacteria utilize from their host nucleus,

Table 3.1 Examples of bacterial symbionts in Protozoa and some related pathogens of phyla shown in Fig. 3.3

Symbiont phylum	Bacterial symbiont	Protozoan host	Related pathogen[a]
α-Proteobacteria	Holospora	Paramecium (cil.)	Rickettsia
	Caedibacter	Paramecium (cil.)	Ehrlichia
	Paracaedibacter	Acanthamoeba	
	Ehrlichia-like symbiont	Saccamoeba	
	Odyssella[b]	Acanthamoeba	
β-Proteobacteria	Polynucleobacter	Euplotes (cil.)	
	Crithidia symbiont	Crithidia (kin.)	
γ-Proteobacteria	Sarcobium lyticum	Small amoeba	Legionella
			Vibrio
CFB group	Amoebophilus[b]	Acanthamoeba	Cytophaga sp.[c]
Chlamydiacea	Neochlamydia	Hartmanella (am.)	Chlamydia
	Parachlamydia	Acanthamoeba (am.)	
Verrucomicrobia	Epixenosomes	Euplotidium (cil.)	

[a]Pathogens found in the same phylum or sub-phylum.
[b]Designations of these symbionts was provisionally proposed as Candidatus [Candidatus odyssella (Birtles et al. 2000); Candidatus amoebophilus (Horn et al. 2001)].
[c]A Cytophaga sp. that had been spread by an air cooling system, implicated as causative agent of a lung disease; am., small amoeba; cil., ciliate; kin., kinetoplastid flagellate.

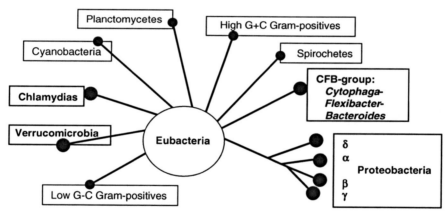

Figure 3.5 Bacterial phyla (bold) with intracellular bacteria of protozoa and some other eubacterial phyla in an arrangement showing their approximate phylogenetic positions.

but nucleoside triphosphates seem good candidates. Interestingly, rickettsias, having an ATP–ADP translocase, make use of ATP supplied by their host cell. From their biology the holosporas not only resemble rickettsias but, similarly to Listeria and some other pathogens (Falkow et al. 1992), Holospora appears to make use of the cytoskeleton of the host cell to move from the phagosome to the nucleus, to move in the nucleus during nuclear division, and to leave the host cell without destroying it.

Not only ciliates are infected by bacteria. In recent years a number of intracellular bacteria has been detected in small free-living amoebae (Fig. 3.6). Springer et al. (1992) identified the intracellular bacterium Sarcobium lyticum, symbiotic in a Saccamoeba species, as belonging to the γ-group of proteobacteria, related to Legionella pneumophila. Bearing in mind that L. pneumophila may grow in protozoa (Rowbotham 1980, 1986; Barbaree et al. 1986; Pasculle 1992), this is a further indication

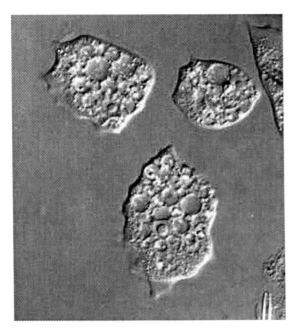

Figure 3.6 Three small free-living amoebae newly isolated from a local pond in Stuttgart. Light micrograph of living cells, differential interference contrast, microflash.

that pathogenic legionellae may originally have been protozoan symbionts. Various symbionts from marine invertebrates, too, were found to belong to the γ-group of proteobacteria, e.g. the symbiont of *Riftia pachyptila* (hydrothermal vent worm) and methano- and thio-autotrophic bacteria in a hydrothermal vent mussel (Felbeck and Distel 1992; Distel *et al.* 1995).

The classification of an intracellular bacterium in an *Acanthamoeba* as *Parachlamydia acanthamoebae* (Amman *et al.* 1997) is another example that symbionts of protozoa may be related to pathogens (Fig. 3.7). Chlamydias are causative agents of such severe diseases as psittacosis, pneumonia, trachoma (the major cause of blindness on earth), and venereal urethritis and cervicitis. As true intracellular microbes, these pathogenic chlamydias may also be capable of infecting protozoa, given appropriate conditions.

Of course, not all intracellular bacteria in protozoa are closely related to pathogens. *P. necessarius* in *Euplotes* belongs to the β-group of proteobacteria. They do not appear close to pathogens, their closest

relatives being the genus *Alcaligenes* (Springer *et al.* 1996). Symbionts of Euplotes rather provide a good example that bacteria other than the ancestor of mitochondria (α-group) also had the capability to develop into essential symbionts. Even methanogens were found as intracellular symbionts in anaerobic protozoa (Finlay and Fenchel 1992) and may not be assumed to be related to pathogens or become harmful.

Epixenosomes of *Euplotidium* were found to belong to the recently discovered *Verrucomicrobia* (Petroni *et al.* 2000). Representatives of the *Verrucomicrobia* were isolated from fresh water and wet soil; epixenosomes of *Euplotidium* are the first symbiotic bacteria of this division. This is another fascinating example that symbiotic bacteria may originate from a variety of prokaryotic taxa; the phylogenetic diversity of protozoan symbionts seems much greater than that of human pathogens and it is quite likely that some of them could have the potential to colonize human cells, too.

3.4 A home for the unwanted? Protozoa are reservoirs of pathogens

Most parasites are host specific. This holds true also for many bacterial pathogens. Only a few animals are appropriate victims of human pathogens. Because of the diversity of protozoa and, even more, because of their higher plasticity (one may also say lower complexity) compared with animals, certain protozoa may serve as reservoir hosts for various human pathogens. Free-living protozoa such as ciliates and amoebae are important members of microbial communities acting as predators of bacteria. Many protozoa are not very selective but also ingest opportunistic pathogens. By this behaviour protozoa may overcome the host specificities of certain pathogens. Some species and strains of free-living amoebae (FLA) are known to cause acute or chronic forms of meningoencephalitis and severe keratits (Visvesvara 1995). Isolates of amoebal strains from environmental and clinical sources, from drinking water systems as well as from corneal scrapings of keratitis patients were found to be infected by intracellular bacteria (Fritsche *et al.* 1999; Horn *et al.* 2001).

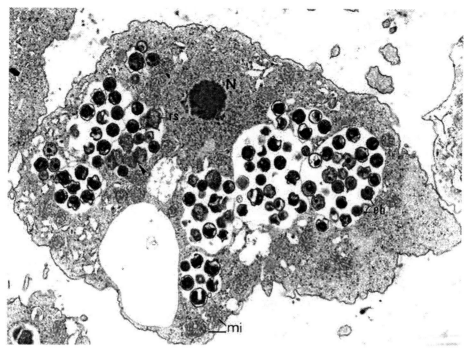

Figure 3.7 *Acanthamoeba castellanii*, strain C3, harbouring chlamydia-like parasitic endosymbionts, strain Berg17, of the genus *Parachlamydia* (Parachlamydiaceae). Elementary stages (eb) and reticulate stages (rs) are found within vacuoles of the host amoeba. Binary fission of the latter can be observed (arrowhead). This infected amoeba was isolated from human nasal mucosa. N, nucleus of the host amoeba; mi, mitochondria.

Intracellular bacteria themselves could cause infections additional to those by their amoebal hosts and on the other hand also influence the pathogenicity of the protozoa. How to prevent the risks?

Small FLA such as acanthamoebae, naegleriae, and hartmanellae are known to harbour pathogenic bacteria. Consequently, FLA contribute to the dispersal of bacteria such as *L. pneumophila* and *Chlamydia*-like bacteria (Rowbotham 1980, 1986; Horn *et al.* 2000; Schmid *et al.* 2001). Specifically, acanthamoebae are reservoirs as transitory hosts for a great variety of clinically relevant bacteria (Harb *et al.* 2000). *Legionella* can exist in protozoa such as *Tetrahymena* or *Acanthamoeba* which may harbour these bacteria and serve as reservoir sources for human infections (Barbaree *et al.* 1986). A number of pathogens do not multiply but are able to survive in protozoa such as ciliates and amoebae. Fritsche *et al.* (1999) found intracellular bacteria in about one of four acanthamoebae from clinical isolates and from natural environments. This is consistent with the observations by Horn *et al.* (2000, 2001), and the spectrum of pathogens able to infect and multiply to various extents within amoebae includes rickettsias, *Legionella* spp., *Ralstonia* (*Burkhoderia*) *pickettii*, *Listeria* spp., *Vibrio cholerae*, *Francisella tularensis*, *Mycobacterium avium*, and *Chlamydia pneumoniae* (literature cited by Fritsche *et al.* 1999).

Although only a few of the many intracellular bacteria in protozoa and invertebrates are classified as yet, the variety appears to be great, and we have to realize that protozoa and lower animals offer appropriate conditions for all types of microorganism. With the new methods of *in situ* amplification and *in situ* hybridization, we now have excellent tools to detect, identifiy and classify intracellular bacteria in protozoa as well as in tissues of metazoa. The great number of intracellular symbionts that are highly host specific appears unlikely to be of potential risk to humans, even in close contact. However, others may be infectious for humans, and unexpected reservoir hosts of pathogens may be found as new vehicles. Among the many protozoa that are found eventually to harbour intracellular bacteria, some may well be suitable hosts even of mycobacteria, enteropathogenic *Escherichia coli*, or other pathogens. Various marine animals are infected by *Vibrio* species (Farmer and Hickman-Brenner 1992). Many of them are parasitic, some are bioluminescent symbionts in light organs. We must be aware that animals or protozoa may even serve as reservoirs of pathogenic *Listeria monocytogenes* and *V. cholerae* strains (Ly and Müller 1990; Thom *et al.* 1992). Recent findings reported by Broza and Halpern (2001) indicate that chironomid egg masses may serve as an intermediate host reservoir for *V. cholerae*, facilitating its survival and mutliplication in freshwater bodies.

3.5 New bacteria—how to limit the risks of emerging pathogens

High population densities of a host may favour propagation of infectious bacteria, but bacterial infections in protozoa and invertebrate animals have not been observed as a specific danger to humans in the past even in cities or countries with dense populations. Rather, the diseases originating from protozoa harbouring pathogenic bacteria result from environmental changes. Civilization and new technologies may be creating favourable conditions for various parasites and pathogens. The best example so far may be *Legionella*. This bacterium may be spread by aerosols produced by insufficiently serviced air conditioners.

L. pneumophila, like other intracellular bacteria, cannot be grown easily in laboratory cultures. Media must be extremely complex according to their natural habitat, which is represented by a living cell. Legionellae may live in air conditioners because host amoebae or even *Tetrahymena* find appropriate conditions in water films originating from condensing water due to high humidity. Similarly, cooling towers, swimming pools, showers, or humidifiers may provide suitable habitats for protozoa possibly harbouring intracellular bacteria. A large outbreak of Legionnaires' disease affected many visitors to the West Frisian Flower Show near the town of Bovenkarspel in the Netherlands (Den Boer *et al.* 2002), where 188 visitors became ill. Two whirlpool spas of the exhibition and a sprinkler were culture positive for *L. pneumophila*. Both whirlpool spas and sprinklers may be regarded as appropriate habitats for bacterially infected FLA and perfectly suited for the spreading of these organisms with aerosols. It is a pity that the localities of the flower show were obviously checked only for the presence of *Legionella* but not for FLA as potential hosts.

Protozoa in sewage plants may frequently harbour intracellular bacteria (H.-D. Görtz, unpublished). Pathogens, howsoever having found their way into sewage plants, may hide in protozoa, many of which are present there. Certainly, we must not let pathogens escape from hospitals with waste water or garbage, in order not to expose them to protozoa in sewage plants or dumps. It is recommended that the efficiency of sewage plants at eliminating pathogens should be controlled. With the new techniques of PCR and *in situ* hybridization this may be done accurately. The same holds true for cooling towers, room humidifiers, and so on.

Can we predict future risks? Protozoa may generally be adequate hosts of various microorganisms that may be potential pathogens of humans. New conditions may introduce protozoa to new types of free-living microorganisms that could adapt mechanisms to prevent digestion and colonize the protozoa. Preconditions for bacteria to become intracellular organisms in protozoa are much the same as those for infection of human cells. Among these preconditions are the resistance to low pH, resistance to oxidative stress and lytic enzymes, and tolerance to moderate starvation. Many bacteria have these features. To become infectious, bacteria in addition must have the ability to pass through membranes. Accidental uptake of plasmids or bacteriophages containing so called pathogenicity islands including genes for protein secretion (Vogel *et al.* 1998; Thanassi and Hultgren 2000; Hacker and Carniel 2001) in their genomes may switch free-living bacteria into pathogens. Earlier evidence for the significance of plasmids comes from the bacteria that became essential for certain strains of *A. proteus* (Jeon 1983).

Polluted environments may selectively promote growth of infected protozoa. One example has been described by Hagnere and Harf (1993). While small amoebae are found in all types of environments, habitats polluted by heavy metals are unfavourable for amoebae as for many other organisms. The authors found that amoebae bearing intracellular bacteria showed higher tolerance to heavy metals than uninfected ones and therefore could grow in polluted habitats. Certain technologies may introduce protozoa into existing environments that have not previously been inhabited by those species. It has been reported that ballast water may contain invertebrates and protozoa that become neozoa or neoprotozoa, respectively, where the water is released (Galil and Hülsmann 1997). It is known that intracellular bacteria are even maintained in resting cysts of protozoa. Special risks must be expected if new technologies not only provide favourable habitats, but also promote contact between infected protozoa and humans. The conclusion must be to constantly monitor changing environments of human populations for bacterially infected protozoa. We should then be able to detect and eventually identify emerging pathogens early enough to prevent new diseases.

Care in preventing escape of pathogens into free-living protozoa is reasonable not only because of the immediate risks of infection of humans. It has been pointed out that free-living protozoa from soil and water that are professionally ingesting and destroying bacteria of a wide phylogenetic spectrum serve as a natural testing ground for evolutionary experiments in intracellular survival as has been considered repeatedly (Barker and Brown 1994; Steinert et al. 1998; Fritsche et al. 1999). Upon living in a host cell bacteria may even be gaining new adaptations resulting in an even higher danger of the germs for humans.

3.6 Conclusions

With billions of people on earth, population densities increasing, overcrowded regions extending and the danger of epidemics and pandemics growing, old diseases such as tuberculosis are returning and new diseases are emerging. One such new disease is legionellosis. Where do new pathogens come from? Are there potential pathogens in other organisms? If so, why do they threaten now and not a thousand years ago? We can now answer some of these questions, estimate the risks to some extent, and give advice.

3.6.1 New bacteria—new risks?

Now that we have appropriate methods for detection, characterization, and phylogenetic classification, the number of bacteria we are finding in ciliates and other protozoa is increasing. The more we look the more new species we find. There are a considerable number of new species that have been described in small free-living amoebae, and amoebae may be even more relevant as reservoirs of pathogens than ciliates.

3.6.2 Exotic adaptations of symbionts

Intracellular pathogens of higher organisms show a diversity of life strategies. Though there are necessarily similarities, as intracellular bacteria have to invade their host cells, have to defend themselves against host cell attacks and have to develop means to move in host tissues, the mechanisms evolved may differ considerably. The diversity of bacterial symbionts of protozoa, however, is an even greater.

3.6.3 Phylogeny of bacterial symbionts of protozoa

It has been revealed that many bacterial symbionts of protozoa are related to pathogens. Interestingly, the phylogenetic diversity of protozoan symbionts seems even greater than that of human pathogens, and we cannot at all exclude an occurrence of completely new pathogens.

3.6.4 Protozoa—reservoirs of intracellular bacteria

Most parasites are host specific. This holds true to some extent also for bacterial pathogens. Only a few animals are appropriate victims of human pathogens; because of the diversity of protozoa and, even more, because of their higher plasticity (you may also say lower complexity) compared with animals, certain protozoa may serve as reservoir hosts for various human pathogens.

3.6.5 How can we prevent the risks?

Knowing the risk of occurrence of potentially pathogenic symbionts in protozoa it seems necessary to constantly monitor changing environments of human populations for bacterially infected protozoa. Also, care must be taken avoiding intense contact between humans and protozoa that are potential hosts of pathogenic microorganisms.

CHAPTER 4

The microbiology of amber: a story of persistence

Raúl J. Cano

A hen is only egg's way of making another egg. (Samuel Butler, Life and habit, 1877)

4.1 Introduction

Until 1984 (Higuchi and Wilson 1984), the isolation and characterization of fossil DNA was considered unattainable, as the methodologies for extracting minute quantities of partially degraded DNA and their subsequent enzymic amplification were not available. With the advent of the polymerase chain reaction (PCR) (Mullis *et al.* 1986), a new analytical tool became available for the molecular study of fossils. It is now possible to conduct molecular studies of extinct organisms or fossils, utilizing their DNA to unravel biological and evolutionary questions.

The value of fossil evidence is that it may demonstrate the condition of taxa before evolutionary divergence obscured phylogenetic relationships. Because they are older, ancient fossil DNA sequences should be less divergent than extant sequences and should therefore have value for relating more derived extant taxa. When compared with extant DNA, ancient DNA sequences may also provide an insight into the pattern of molecular evolutionary change through time. In this light, ancient DNA could provide invaluable information on the evolution and epidemiology of infectious diseases. Understanding the evolution of virulence and the acquisition of virulence factors by microbial populations could help in identifying microorganisms that are likely to become 'emerging' pathogens, as well as in creating methods to arrest the development of pathogenic potential for those organisms.

A body of scientific evidence already supports the use of DNA from extinct animals and plants for phylogenetic studies. Higuchi and coworkers (Higuchi *et al.* 1984) demonstrated that the remains of a mammoth and the extinct quagga species contained fragments of the original DNA. Pääbo (1989) reported the extraction of clonable DNA from the 2400-year-old mummy of a child. Subsequent DNA analysis revealed fragments measuring ~3.4 kb. Thomas *et al.* (1989) isolated DNA from hair found in a century-old untanned hide and a piece of dried muscle collected from an extinct marsupial wolf. This DNA was later enzymatically amplified by PCR and phylogenetic studies were made. More recently, Handt *et al.* (1994), Rollo *et al.* (1994), and Cano *et al.* (2000) isolated and characterized ancient DNA from the 5000-year-old Tyrolean Ice Man. Similarly, Poinar *et al.* (1998) extracted DNA from sloth coprolites and assessed the type of vegetable matter that comprised the sloth's diet. Ubaldi *et al.* (1998) isolated and characterized the DNA from a pre-Colombian Andean mummy.

There have also been reports of isolation and characterization of ancient DNA from much older fossils, although these reports are still controversial. Golenberg *et al.* (1990) reported the isolation and characterization of *Magnolia* chloroplast DNA from a Miocene *Clarkia* deposit dated as 17–20 million years old. Cano *et al.* (1992) reportedly isolated and characterized DNA from the extinct bee *Proplebeia dominicana* in 25–40 million-year-old Dominican amber. DeSalle *et al.* (1992) employed DNA extracted from fossil termites to resolve phylogenetic relationships between the termites, cockroaches, and mantids. Cano *et al.* (1993) reported the extraction of DNA from a 120–135 million-year-old nemonychid weevil in Lebanese amber and showed, by nucleotide sequence alignments and phylogenetic inference analyses, that the fossil weevil was most closely related to the extant nemonychid weevil *Lecontellus pinicola*. Poinar *et al.* (1993) used DNA sequences from the extinct legume *Hymenaea protera* in Dominican amber in a bio-geographical study in which they showed that the extinct *H. protera* was most closely related to the extant African species *Hymenaea verrucosa*, as morphological studies had suggested. Finally, Cano *et al.* (1994) used DNA sequences from 25–40-million-year-old *Bacillus* spp. in Dominican amber inclusions to study a symbiotic relationship between *Bacillus* and the now extinct stingless bee *P. dominicana*. The isolation of ancient DNA from amber, however, is irreproducible (Austin *et al.* 1997) and therefore does not appear to be a reliable source for biological and evolutionary studies.

4.1.1 Strategies of analysis

In terms of the analysis of ancient DNA, the single most important technology is the PCR. This exponential amplification produces enough copies of the target strand of DNA for it to be manipulated and analyzed using standard molecular techniques, such as cloning and enzymically directed sequencing.

As important as PCR is, however, the success or failure of such an assay depends upon the quality and care of the sample preparation. New and refined techniques for extraction of biomolecules have been developed, but these have been optimized largely for the extraction of modern DNA. Such methods must be optimized and the solutions modified for the extraction of small fragments of DNA that may have undergone some diagenesis and for the removal of PCR inhibitors.

New techniques of DNA detection, such as the fluorogenic 5′ exonuclease chemistry (Holland *et al.* 1991) and oligonucleotide probe array strategy (Lipshutz *et al.* 1997), could revolutionize the field of paleomicrobiology. Such strategies have the potential to detect a large number of genes at once and potentially arrive at a nucleic acid sequence of the extracted DNA without extended amplification protocols. Kaplan *et al.* (2001) recently introduced a PCR-based method to assess microbial communities. This method has been used to characterize the microbial flora in rat feces (Kaplan *et al.* 2001) and the microbial diversity of Middle Eastern amber (Greenblatt *et al.* 1999).

Analytical software is available—e.g. CLUSTAL (Higgins *et al.* 1992) and FASTA (Miller *et al.* 1991)—which allows the sequences obtained by the above methods to be matched against homologous sequences from other species that have been deposited in a data bank. Statistical analyses can then be performed and estimates of relatedness and genetic distance can be obtained. Phylogenetic trees based on sequence data can be constructed, using software packages such as MEGA (Kumar *et al.* 1993), PAUP (Swofford 1990), and PHYLIP (Felsenstein 1989). This allows for the objective placement of an organism within the framework of known taxa (Fig. 4.1). It also allows any modern

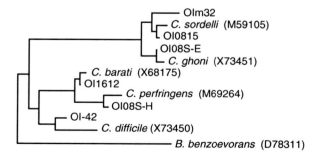

0.1

Figure 4.1 Maximal likelihood tree of amplicons from the Ice Man's colon and their corresponding modern taxa, based on 16S rRNA gene sequences.

DNA that may be contaminating ancient tissues to be characterized and possibly recognized.

4.1.2 Preservation of potential of biomolecules

It is a commonly held belief—based on experimental evidence as well as from extrapolated predictions based on studies of DNA in aqueous solution—that nucleic acids do not survive in fossil remains on a geological time-scale. The taphonomy of the ancient DNA sample, however, is more important than the geological age of the sample itself. For example, amino acid racemization studies indicate than certain tissues of insects entombed in amber curtail amino acid racemization, while 5000-year-old bone amino acids are almost fully racemized (Poinar *et al.* 1996).

The double stranded, helical structure of DNA is more resistant to damage than single stranded RNA, but its structure and chemistry make it susceptible to certain types of diagenetic damage over time. Conversion of bases through hydrolytic deamination (guanine changes to xanthine, cytosine to uracil or its derivatives) and depurination (removal of the bases guanine and adenine from the sugar–phosphate backbone) affect the informational content of the molecule.

Exposure to oxygen free radicals or ultraviolet (UV) radiation also damages DNA strands (Eglinton and Logan 1991). Mechanisms have evolved in living organisms to repair such DNA damage as it occurs, maintaining genetic information and preventing accumulation or error (Grossman 1991). With the death of the organism, this self-repair process stops, while enzymic attack and exposure to water, oxygen, and ultraviolet radiation continue with advancing decay.

Exposure to water is probably the single most destructive force acting on the DNA molecule. Water has been shown to initiate strand breaks by attacking the base–sugar bonds. Where the base is lost, the chain is weakened and eventually cleaved (Eglinton and Logan 1991; Lindahl 1993).

Oxidation is another source of DNA damage, and removing DNA from water—as in amber or bone—does not protect the molecule from oxidative attack. Oxygen in its molecular state does not attack DNA; it is the formation of oxygen free radicals that attack

the nitrogenous bases. Oxidative attack would be rapid at first, but then would level off (Pääbo 1989). It is proposed that chelation of copper or other metal ions (Eglinton and Logan 1991) enhances the preservation of this molecule by contributing to a reducing environment and compensating for the production of oxygen free radicals

Exposure to UV light also causes extensive damage and degradation of DNA. Rapid burial or encapsulation of organisms in protective matrices is important to minimize the consequences of UV damage to DNA.

4.1.3 Problems of working with 'fossil' biomolecules

The extreme sensitivity of PCR, which opens the door to the direct analysis of DNA obtained from ancient materials, also poses the most complications. The fact that PCR technology can amplify as little as one molecule of DNA means that minute amounts of contaminating DNA from modern sources, such as bacteria, soil fungi, or human skin cells, can also be amplified. Indeed, any such modern contaminant would probably be amplified preferentially over ancient target molecules, owing to the probable state of degradation of the latter.

Ultimately, however, the proof of the authenticity of any DNA presumably obtained from ancient materials comes from careful analysis of sequence data. If phylogenetic analysis of the sequences does not agree with predicted relationships based on morphological evidence, then the DNA data must be carefully re-evaluated. In addition, at least two different genes or gene regions should be analysed, and the results of both should show similar or identical phylogeny, before any claims can be made regarding the sources of the DNA.

4.2 Microorganisms in amber: studies on the evolution of host–parasite interactions

Advances in molecular and cell biology have contributed greatly toward our understanding of host–parasite interactions. There are reports in the literature describing the role of one or more genes in promoting or maintaining the pathogenic or parasitic stage of microbes in their arthropod hosts. *Yersinia pestis* has been reported to possess a unique gene (*pla*) that encodes for coagulase and fibrinolysin, two proteins involved in the transmission of plague by *Xenopsylla cheopis* (McDonough *et al.* 1993). This gene, encoded in the 'pesticin' plasmid pKYP1, has been associated with the deleterious effects of *Y. pestis* on fleas that have been associated with flea blockage and plague transmission. In similar studies, Schwan *et al.* (1995) demonstrated that ticks influence the induction of an outer membrane protein on *Borrelia burgdorferi*.

Leishmania major has also been shown to express a metalloprotease only in the promastigote form in the insect vector but not in the amastigote form in the mammalian host (Schneider *et al.* 1992). This suggests that the protease is involved in the development of the parasite into an infectious form within the insect vector, rather than an integral part of the infectious process in the mammalian host.

Microorganisms have also been shown to induce gene expression in their arthropod hosts. Albuquerque and Ham (1995) reported the activation of the prophenoloxidase cascade in the hemolymph of *Aedes aegypti* in response to *Brugia pahangi* microfilariae. Similarly, a number of bacteria-induced proteins in *Manduca sexta* have been shown to be differentially synthesized (Spence *et al.* 1992). One of these bacteria-induced proteins is scolexin, one of several immune proteins produced by insects or their larvae, in response to colonization by *Bacillus thuringiensis*.

As can be seen from these studies, the arthropod–microbe interaction is an essential one for the survival and development of the pathogen within its host. In most instances, one or more microbial genes are expressed only as a result of the interaction between the microbe and the arthropod. From these reports it can be extrapolated that most if not all human-pathogenic microbes with a life cycle that includes an arthropod host possess essential genes allowing them to survive or interact within their arthropod host. If detected, these genes could represent ideal targets for the development of antimicrobial agents through rational drug design or through natural product mechanism-based screens. By using ancient DNA or microorganisms entombed in amber inclusions, it could be possible not only to detect these targets, but also to assess the amount of molecular change through time and to identify conserved regions within the gene that might be suitable targets for vaccines or chemotherapeutic agents.

Evolutionary studies of the host arthropod and its symbiont(s) can also be helpful in assessing the nature and scope of host–parasite interactions. Moran *et al.* (1993) elegantly demonstrated that endosymbiotic bacteria of aphids co-evolve with their hosts, and calibrated a molecular clock for the 16S rRNA of these bacteria using their insect hosts as 'clock' calibrators. This work provided evidence that it is possible to determine the rate of evolution of ribosomal genes of bacterial symbionts, utilizing their host to calibrate the evolutionary time.

4.2.1 Recovery of ancient microorganisms from amber

Cano and Borucki (1995) isolated a strain of *Bacillus sphaericus* from the abdominal contents of an extinct stingless bee (*P. dominicana*) in 25–40 million-year-old amber from the Dominican Republic. They showed that this bacterium was of ancient origin based upon methodological, phenetic (biochemical and enzymic) profiles, sequence comparison, and molecular clock studies (Moran *et al.* 1993; Cano and Borucki 1995; Lambert *et al.* 1998). These studies indicated that the ancient isolate, although morphologically similar to modern *B. sphaericus* isolates, exhibited unique enzymic activities and had rDNA sequences that were ancestral to their modern counterparts as shown by phylogenetic analyses. DNA hybridization studies showed that it had 80% homology with group III *B. sphaericus* (Yousten and Rippere 1996). Since the original report by Cano and Borucki (1995), this finding has been independently verified by DiTullio *et al.* (personal communication) and Lambert *et al.* (1998). Greenblatt *et al.* (1999) reported the culture of bacteria from Dominican and previously untested 120 million-year-old Israeli amber. In this study, 27 isolates from the amber matrix were characterized by fatty-acid methyl ester (FAME) profiles and/or 16S rRNA sequencing. They also performed a terminal restriction fragment (TRF) pattern analysis of the original amber before prolonged culture by consensus primer amplification of the 16S rRNA, followed by restriction enzyme digestion of the amplicons. Sample TRFs were consistent with a sparse bacterial assemblage and included at least five of the isolated organisms.

Lambert *et al.* (1998) reported the isolation and characterization of a novel *Staphylococcus*-like bacterium from amber. The bacterial isolate, designated AMG-D1, was isolated from 25–35 million-year-old Dominican amber. Biochemically, AMG-D1 most closely resembled *Staphylococcus xylosus* physiologically, but differed significantly from the staphylococci in cell wall and fatty acid composition. Phylogenetic analysis of 16S rRNA sequences indicated that AMG-D1 was most closely related to *Staphylococcus xylosus*, *Staphylococcus equorum*, and *Staphylococcus saprophyticus*. DNA–DNA hybridization under stringent conditions supported the phylogenetic analysis and revealed homologies of 38% with *S. equorum*, 23% with *S. xylosus*, and 6% with *S. saprophyticus*. These results suggested that isolate AMG-D1 was a novel organism, tentatively named *Staphylococcus succinus*. sp. nov. Lambert *et al.* (1998) hypothesized that the physiological and phylogenetic characteristics indicate that AMG-D1 might be a branchpoint organism between the corynebacteria and the staphylococci (Fig. 4.2).

FAME analysis has been used for the identification and classification of bacteria (Dees and Moss 1975; Kaneda 1977; DeBoer and Sasser 1986; Garcia *et al.* 1987). Using standardized growth and analytical conditions, reproducible fatty acid profiles suitable for multivariate statistical analysis can be obtained, and thus identification systems such as Microbial ID, Inc. (MIDI, Newark, DE) have been established. Putative ancient isolates of *B. thuringiensis*, ranging in age from 2 to 26 million years old, were characterized by FAME analysis and matched using similarity indices to MIDI's library of 44 *Bacillus* species. Isolate AG4 and AG567 were identified as good matches to *B. thuringiensis*, while AG187 and AG262b were identified as atypical strains of *B. thuringiensis*.

The FAME data obtained were used in principal component analysis to assess species diversity among the various isolates of *B. thuringiensis* using Mintab 10.Xtra to analyze the data. The results suggested that although all isolates were identified by FAME analysis as *B. thuringiensis*, all modern isolates formed a well-defined cluster, while the four ancient isolates were clustered in separate groups outside the centroid for the modern isolates, but not forming a cluster among themselves. These results support the hypothesis that AG4, AG187, AG262b, and AG567 were isolates obtained from inclusion in amber and not modern laboratory contaminants.

Johnsonbaugh and Cano (unpublished results) isolated a new strain of *B. sphaericus* from 40 million-year-old Dominican amber and coded it BCA17. The sequence analysis revealed that this isolate was most closely related to BCA16 (Cano and Borucki 1995) and other extant strains of *B. sphaericus* isolated from both soils and insects. The phylogenetic tree based on 16S rRNA sequences and constructed using the maximal likelihood algorithm indicates that both BCA16 and BCA17 are ancestral to extant *B. sphaericus* when rooted by *Sporosarcina ureae*.

More recently, Greenblatt *et al.* (1999) were able to determine the microbial composition of Middle Eastern amber by performing TRF pattern analysis (see above) and were able to correlate the presence of the bacteria in TRF patterns with those isolated from the amber.

Ancient microbial isolates have been recovered from Tertiary ambers (20–40 million years old) and Colombian copal (≤2 million years old). All microorganisms recovered were Gram-positive

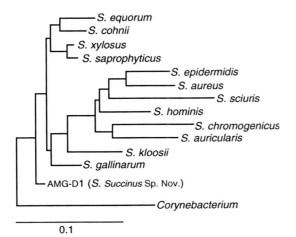

Figure 4.2 Phylogenetic position of AMG-D1 in relation to other staphylococci.

spore-formers, most of which were in the genus *Bacillus*, with the remaining microorganisms being other Gram-positive bacteria (e.g. *Arthrobacter*, *Sporosarcina*, diphtheroids, and actinobacteria).

4.2.2 Natural history of *Bacillus*–bee symbiosis and population genetics

It is well established that there is a symbiotic relationship between *Bacillus* species and many species of bees (Gilliam and Taber 1991; Gilliam *et al.* 1990a,b). *Bacillus* spp. (*Bacillus subtilis*, *Bacillus megaterium*, *Bacillus pumilus*, *B. sphaericus*, and *Bacillus circulans*) have been consistently isolated from the abdominal cavity, glandular secretions, pollen, and larval provisions of stingless bees (e.g. *Trigona*, *Melipona*, *Plebeia*, *Centris*, and *Anthophora*) and honeybees (e.g. *Apis*). It is hypothesized that the larval provisions are inoculated with *Bacillus* endospores maintained in the crop of adult worker bees. The inoculated endospores germinate and have a fundamental role in the metabolic conversion, fermentation, and preservation of the food of perennial colonial insects that rely on stored food (Machado 1971). Microscopic examination of abdominal contents from modern and amber-entombed bees reveals the presence of a large number of endospores, along with few cells (Shimanuki 1978; Gilliam *et al.* 1990a,b; Gilliam and Taber 1991; Cano *et al.* 1994), supporting the hypothesis that symbiotic *Bacillus* spp. are stored in the crop of workers primarily in the form of endospores, which could then be used to seed brood provisions (see below).

It appears that certain strains of *Bacillus* spp. are selected by the bee colony based on their metabolic activity, growth in the presence of acid byproducts of fermentation, and production of enzymes that can digest the collected provisions. Gilliam *et al.* (1990b) found that the *Bacillus* spp. isolated from stingless bees produced a variety of enzymes, including esterases, lipases, proteases, aminopeptidases, phosphatases, and glycosidases, that could convert food into more digestible products for storage. Gilliam *et al.* (1990a,b) also reported that the lipases most often associated with tropical bees are caprylate esterase lipase (CEL) and butyrate esterase, and that CEL activity was detected in all bee isolates evaluated.

A preliminary study conducted in our laboratory compared the production of CEL lipase by *B. megaterium* isolates from stingless bees (*Plebeia frontalis* and *P. dominicana*) and *B. megaterium* from soils (both modern and in amber inclusions). The results, summarized in Table 4.1, indicate that CEL was produced by all the *B. megaterium* symbionts tested but only in 2/13 (15%) of the soil isolates, and that the levels of CEL produced (measured in nanomoles) was greater in the symbionts (3.8 ± 0.79) than in the free-living strains (0.3 ± 0.68).

These results support those published by Gilliam *et al.* (1990a,b). Thus the bee-derived *B. megaterium* can be discriminated from those of soil origin by the level of expression of the CEL gene. It is not clear from the results, however, whether the increased level of activity is due to an increased copy number of the gene in bee symbionts, or as a result of mutational events affecting the regulation or primary structure of the CEL gene and leading to a more efficiently expressed gene.

Both solitary and social bee species have physical, physiological, behavioral, and chemical adaptations that control spoilage of food stores rich in protein, lipid, and carbohydrate (Roubik 1989). These adaptations, in addition to potential mutualistic relationships with microbes and other organisms, may be particularly important in perennial bee species that rely on stored food for survival.

The bee–*Bacillus* association is well described (Gilliam and Taber 1991; Gilliam *et al.* 1990a,b), and the bacterial spores appear to be ubiquitous in the abdominal cavity of worker bees. Additionally, fossil bees are commonly found as inclusions in Dominican amber, thus affording a readily available and abundant source of specimens for study.

Table 4.1 Production of CEL by *B. megaterium*

Source of isolates	Number of isolates		nmoles
	CEL+	CEL–	
Modern sources			
Bee isolates	14	0	3.8 ± 0.79
Soil isolates	2	11	0.3 ± 0.68
Amber sources			
Bee isolates	4	0	3.5 ± 0.96
Soil isolates	1	2	-0.25 ± 0.5

There is evidence that the symbiotic relationship between *Bacillus* species and many species of bee dates back millions of years, since *Bacillus* DNA has been amplified from the abdominal tissue of 25–40 million-year-old bees that were preserved in amber (Cano *et al.* 1994). Furthermore, the tissue of these amber-entombed bees may harbor viable *Bacillus* endospores preserved in a desiccated state for millions of years. The successful germination and culture of these ancient endospores would allow the study of the physiology and may provide insights into ancient symbiotic relationships.

Bacillus spp. have been consistently isolated from the abdominal cavity, glandular secretions, pollen, and larval provisions of stingless bees (e.g. *Trigona, Melipona, Plebeia, Centris,* and *Anthophora*) and honeybees (e.g. *Apis*) and appear to play a vital role in the production, metabolic conversion, and/or preservation of larval provisions. In particular, *B. subtilis, B. megaterium, B. pumilus, B. sphaericus,* and *B. circulans* are frequently associated with bees. Unlike honeybees, stingless bees use mass provisioning, and each larva is provided with the total amount of food required for adult development before the cell is sealed. As noted above, it is hypothesized that the *Bacillus* endospores maintained in the crop of adult worker bees germinate and play a fundamental role in metabolism.

Studies by Gilliam *et al.* (Gilliam *et al.* 1990a,b; Gilliam and Taber 1991) reported a striking similarity between the microbial contents of different types of food from two social bees. They hypothesized that this similarity may reflect similar metabolic roles of *Bacillus* species that have evolved in the nutrition of bees. The results of a study by Machado (1971) supported this hypothesis. His study uncovered an association between the pollen of *Melipona quadrifasciata*, a stingless bee, and a *Bacillus* species resembling *B. pumilus*. Machado reported that the elimination of this *Bacillus* species from pollen stores caused the eventual death of the colony. As noted above, the abdominal contents from modern and amber-entombed stringless bees are rich in endospores, and this is also the case for honeybees (Shimanuki 1978). Therefore, *Bacillus* acting as endospores in the crops of worker bees, probably function to seed brood provisions.

4.3 The bacterial endospore: a time capsule for surviving extended dormancy

Bacteria of the genus *Bacillus* can initiate the process of sporulation, generally when one or more nutrients becomes limiting for growth. Spores are much more resistant than their growing cell counterparts to a variety of environmental stresses, including heat, radiation, and chemicals, and can survive in a metabolically dormant state for extremely long periods of time (Setlow 1994, 1995).

An early event in sporulation is an unequal division of the cytoplasm, giving rise to small and large progeny, each with a complete genome. The small compartment, termed the forespore, is destined to become the mature spore; the large compartment, termed the mother cell, engulfs the forespore, resulting in a cell (forespore) within a cell (the mother cell). Eventually, after a series of further morphological and biochemical changes, the mother cell lyses, releasing the mature spore into the environment (Setlow 1995). From the outside in, the layers of the spore include the exosporium, which has not been well characterized, the proteinaceous spore coat, the cortex—composed of peptidoglycan with a structure similar but not identical to that of cell wall peptidoglycan—and finally the central core, the site of most spore enzymes, ribosomes, and of course the DNA (Setlow 1994). The spore coat and cortex are relatively impermeable, thus restricting access of potentially toxic molecules into the spore core (Setlow 1994).

A major difference between the environments within a growing cell and a dormant spore is the amount of water present. Growing cells are 75–80% water, or 3–4 g water per g dry weight. Values for the water content of the spore coat and cortex are similar to those for intact growing cells. However, the spore core has much less water per gram dry weight: from 0.5 to 1 g water per g dry weight, depending on the species examined. Thus, the spore core has much less free water than a cell (Setlow 1995), and consequently mature spores have no detectable metabolism (they are dormant). The reduced amount of water in the cell is also largely responsible for enzyme inhibition, which is required for a spore's enzymic dormancy and vital to the long-term survival of the spore (Setlow 1995). In their state of dormancy, spores neither contain nor make ATP.

Ultimately, spore DNA must be protected so that germination can occur without error. One of the early steps in germination is the initiation of hydrolysis of the spore cortex; the breakdown of the spore cortex allows a tremendous influx of water into the spore, increasing the volume of the spore core more than twofold, thus rehydrating the spore and restoring its metabolic activity (Setlow, 1995).

The major factor preventing damage to spore DNA is the saturation of this DNA with a novel group of small, acid-soluble proteins (SASPs) of the a/b type, whose binding greatly alters DNA's chemical and enzymic reactivity as well as its UV photochemistry. Binding of these proteins is thus a key factor in spore DNA resistance to UV radiation, depurination, and oxidizing agents (Potts 1994; Setlow 1994).

The other two mechanisms protecting spore DNA from oxidative damage are decreased spore permeability to oxidizing agents and decreased spore water content. Hydrogen peroxide can cleave the DNA backbone but, due to the decreased permeability of the spore cortex to hydrogen peroxide, less hydrogen peroxide can get into the spore and cause oxidative damage. With the decreased water content of a spore, there are fewer hydroxyls to cause oxidative damage in the spore (Lindahl 1993). Finally, SASP DNA binding proteins protect spore DNA from hydroxyl radical cleavage. These proteins saturate the spore chromosome and protect the DNA backbone against hydroxyl radical cleavage (Setlow 1995). Protection of DNA from free radical damage by a/b-type SASPs could be a major component of spore longevity (Setlow 1995).

Therefore, while one might infer that a dormant cell, with no means of DNA repair, would incur significant DNA damage, there are several unique mechanisms that spores have for protecting their DNA. The bacterial endospore is a biological strategy that overcomes the intrinsic instability of DNA (Lindahl 1993). Additionally, amber-entombed endospores also benefit from the unique preservative qualities of the amber.

It has thus been noted that spores and amber provide an environment of partial dehydration (Stankiewicz *et al.* 1998). In addition, spores are largely impermeable to oxygen (oxidizing agents), while amber achieves the same end by completely sealing off the inclusions from oxygen (Lindahl 1993). Amber inclusions are not exposed to microbial contamination (Lindahl 1993), have high osmolality (Cano 1994), which reduces depurination 10-fold (Lindahl 1993), and a neutral pH environment, all of which enhance the resistance of biomolecules to denaturation and hydrolysis. Furthermore, Dominican amber mines located 50–100 feet underground, provide an environment for fossilized tissues of moderate stable temperatures and protect them from DNA-damaging UV irradiation. These five factors—dehydration, high osmolality, oxygen-free environment, protection from UV radiation, and stable ambient temperatures—make endospores preserved in amber a source of microorganisms for evolutionary studies of symbiosis.

In a recent review on the long-term survival of microorganisms in fossils, Morita (1999) suggests that H_2 is the energy source that microorganisms use to carry on the necessary repairs during cryptobiosis. In his review, Morita explains the discrepancy (time factor) between the finding of viable bacteria in ancient material and the racemization of amino acids and depurination of DNA that would have contributed to their death. He argues that the omnipresence of H_2 in the biosphere since life began, its ability to penetrate the microbial cell, its low energy of activation, its ability to form protons and electrons in the presence of Fe^{2+}, and its role in many biochemical reactions make H_2 the best candidate as the energy of survival for microbial cells. Although the concentration of H_2 in most environments is below the threshold level for microbial growth, the surviving cells have a long period of time to carry out the necessary metabolism to offset the racemization and depurination processes.

4.4 Can insect symbionts become pathogens?

Virulence, at least in bacteria, requires that the potential pathogen survive both within and outside the host and that it possess traits not found in nonpathogenic strains, whose expression causes adverse physiological or anatomical alterations in the host. These traits could be adhesiveness, toxin production, antimicrobial resistance, or invasiveness; and they can be acquired through mutation, transformation, transduction, or conjugation.

Pathogenicity islands (Mecsas and Strauss 1996) are virulence 'cassettes' found in the chromosome of many pathogens. These islands can be of various lengths, but are generally large clusters of genes contributing to a particular virulence phenotype. For example, the pathogenicity island found in enteropathogenic *Escherichia coli*, named LEE, is a 35 kb region, which mediates attaching and effacing lesions on intestinal epithelial cells (McDaniel *et al.* 1996). This region of chromosomal DNA is absent from avirulent *E. coli* strains and codes for an outer membrane protein that promotes contact with host cells and a type III secretion system that exports proteins that modify the host cell (Mecsas and Strauss 1996). As can be construed from this, acquisition of a pathogenicity island by a microorganism could potentially convert a commensal symbiont into a parasite.

Among the Gram-positive bacteria, many of which can be isolated from bees (Gilliam *et al.* 1990a,b; Gilliam and Taber 1991), there are reports of virulence factors involved in pathogenicity. A new chromosomal locus has been identified in *Listeria* (Rouquette *et al.* 1996). This island, involved in the virulence of this intracellular pathogen, displays the same genetic organization as the *clpC/mecB* locus of *B. subtilis*. It contains a thermoregulated operon of four genes, one of which, the ClpC ATPase, is a general stress protein involved in intracellular growth and *in vivo* survival of this pathogen in host tissues.

Fully virulent *Bacillus anthracis* are encapsulated and toxigenic. These bacteria contain two transferable plasmids, pXO1 and pXO2, carrying genes for toxins (*pag*, *lef*, and *cya*) and for capsular biosynthetic enzymes. Isolates lacking both pXO1 and pXO2 are indistinguishable from *B. cereus*. The former species is a common cause of highly fulminant post-traumatic and metastatic endophthalmitis. Virulence factors for this organism are multifactorial, but a hemolysin (HBL) is thought to be involved.

The basic taxonomic unit in bacterial systematics has long been recognized as the species, and yet there remains controversy as to the most appropriate way to define this concept in bacteria. The traditional view is that a species comprises a group of strains that show a high degree of overall phenotypic similarity and that differ considerably from strains in related groups (Logan 1994). Gordon extended this definition by suggesting that the description of a species should include both freshly inoculated strains and laboratory cultures, since strain phenotypes can be modified on repeated subculture on laboratory media (Gordon *et al.* 1973). Based on genomic differences, a bacterial species generally includes strains with 70% or greater DNA–DNA relatedness (Wayne *et al.* 1987); by this criterion, all of the members of the *B. cereus* group (*B. cereus*, *B. anthracis*, *B. thuringiensis*, and *Bacillus mycoides*) belong to one species (Sneath 1986). However, *B. cereus* differs from both *B. anthracis* and *B. thuringiensis* by pathogenicities for various hosts, while *B. cereus* and *B. mycoides* differ by morphological characteristics, and thus the four taxa are usually considered to be distinct species. But are they? Is it possible that the transfer of genetic information between and among these four taxa could be the determining factor for their phenotypic differences? If this is the case, it is entirely possible that a *B. cereus* bee symbiont can turn into the insect pathogen *B. thuringiensis* or the human pathogen *B. anthracis*.

The four species of the *B. cereus* group are often indistinguishable. *B. anthracis* is the causative agent of anthrax, and is distinguished from *B. cereus* based on its pathogenicity for man and animals; however, according to Gordon *et al.* (1973), a strain of *B. anthracis* that had lost its virulence would be identified as *B. cereus*. Similarly, *B. thuringiensis* is distinguished from *B. cereus* based on its pathogenicity for insect larvae; however, non-pathogenic strains are indistinguishable from *B. cereus*. In addition, *B. cereus* grown in mixed culture with *B. thuringiensis* can take on the characteristic insect pathogenicity. *B. mycoides* is distinguished from *B. cereus* based on its rhizoid growth morphology, and this characteristic is also unstable, so that non-rhizoid strains of *B. mycoides* are indistinguishable from *B. cereus* (Sneath 1986). In addition, Priest (1993) recognizes that the rDNA sequences of representatives of the four (sub)species are virtually identical within the variation expected of a single species. As a result, Gordon *et al.* (1973) suggest that *B. cereus* is the parent species or stable species and the other three species are varieties.

Specifically, it is the relationship between *B. cereus* and *B. thuringiensis* that is important to this thesis. *B. thuringiensis* differs from *B. cereus* in producing a crystal protein toxin that is pathogenic to various insect larvae, notably Lepidoptera. The parasporal crystal is formed outside the exosporium and separates readily from the liberated spore (Sneath 1986). The parasporal body appears microscopically as lozenge shaped, square, round, or of indefinite shape (Sneath 1986). In all other respects *B. thuringiensis* and *B. cereus* are virtually identical, and there are no differential features as there are for distinguishing *B. anthracis* from *B. cereus*. Pyrolysis gas chromatography, in which whole cells are pyrolized in an inert atmosphere, can distinguish the two species, but the distinguishing feature is based on pyrolysis of the toxin components (Priest 1993); again, it is the crystal protein that distinguishes the two species.

The stability of crystal production is controversial; only 9/15 of Gordon's strains produced the protein crystal (Gordon *et al.* 1973). In addition, the capacity to form crystals may be lost by laboratory cultures (Sneath 1986), because in most cases the toxin genes of *B. thuringiensis* are located on large conjugative plasmids. By definition plasmids are not unconditionally required for the survival and replication of individual bacteria (they would be chromosomes if they were): in some situations they are essential, while in others they enhance the ability of bacterial populations to colonize and compete in natural communities (Levin 1993). In an environment where selection does not favor maintenance of the plasmid, the plasmid may be lost, and thus this defining characteristic of *B. thuringiensis* may be lost (Levin 1993). In addition to the inconsistent presence of this plasmid in *B. thuringiensis*, through conjugation the plasmid can be transferred to *B. cereus*, which may then synthesize the crystal proteins, leaving the dividing line between *B. thuringiensis* and *B. cereus* even more dubious (Priest 1993). In conclusion, according to Gordon *et al.* (1973), 'crystal formation as a taxonomic characteristic is unreliable'.

More recently, in 1994 a study on the genetic diversity of 24 strains of *B. cereus* and 12 strains of *B. thuringiensis* showed that a high level of variability among these two species was observed, but no consistent differences were found (Carlson *et al.* 1994). Strains were analyzed by restriction enzyme digestion profiles (using pulsed field gel electrophoresis) and were examined for variation in 15 chromosomal genes encoding enzymes (by multilocus enzyme electrophoresis). These researchers also concluded that *B. cereus* and *B. thuringiensis* should be considered a single species (Carlson *et al.* 1994).

In view of these observations, it is possible to study those factors that contribute to the conversion of a commensal (or even a mutual) into a parasite. If indeed microorganisms can be reliably and reproducibly isolated from insect amber inclusions, it would provide a powerful tool to study the evolution of virulence in microbial populations.

CHAPTER 5

Evolution of arthropod disease vectors

William C. Black IV

In the beginning, God created the earth, and he looked upon it in His cosmic loneliness. And God said, 'Let Us make living creatures out of mud, so the mud can see what We have done.' And God created every living creature that now moveth, and one was man. Mud as man alone could speak. God leaned close as mud as man sat up, looked around, and spoke. Man blinked. 'What is the purpose of all this?' he asked politely. 'Everything must have a purpose?' asked God. 'Certainly,' said man. 'Then I leave it to you to think of one for all this,' said God. And He went away. (Kurt Vonnegut, Cat's Cradle)

5.1 Introduction

Fossils resembling arthropods first appeared during the late Proterozoic about 600 to 540 million years ago. Today the phylum Arthropoda contains ~80% of all extant metazoan animal species, and arthropods have come to occupy virtually every marine, freshwater, terrestrial, and aerial habitat on earth. Arthropods are also essential components of most of the major food chains. In many ecosystems arthropods consume and recycle detritus and the associated bacteria, algae, and fungi. Many species consume either the living or dead tissues of terrestrial and aquatic plants and animals. Many are voracious predators or parasites. Given this huge taxonomic and ecological diversity, it is not surprising that a small proportion of arthropods have evolved the ability to utilize a rich and abundant source of nutrients, the blood of vertebrates.

Hematophagy, the habit of blood-feeding, has evolved at least 21 times in disparate arthropod taxa. Within the single insect order Diptera, hematophagy has probably evolved independently in nine families.

We will use the term 'parasite' in a generic way in this chapter to encompass all viral, bacterial, and eukaryotic organisms transmitted by arthropod vectors. Hematophagy was probably quickly exploited by parasites both as a means of increased

mobility and, more importantly, as a means to find and occupy novel vertebrate hosts. Great benefits accrued to parasite species or populations that possessed the correct combination of morphological, physiological, and biochemical characters necessary to survive in hematophagous hosts. These parasites abruptly acquired many adaptations beneficial to their own survival and dispersal. These included physiological and behavioral mechanisms evolved for host location and precise morphological and neurosensory mechanisms for proximal location of blood within the vertebrate vascular system. In addition, vector-borne parasites became transmitted via saliva that contained immunomodulatory components that increased the chances of their transmission and survival in a new vertebrate host.

Gaining access to an arthropod vector was only half of the adaptive process. The parasite was next faced with the problem of invading the arthropod gut, disseminating to various organs, and eventually ending up in an organ of transmission without compromising host survival. Most parasites targeted the arthropod salivary gland for oral transmission. Alternatively, some escaped through arthropod waste or secreted products onto the surface of a vertebrate host. Many arboviruses and rickettsiae exploited the female reproductive system to become transmitted either transovarially or via transovum. Similarly some arboviruses exploited male reproduction and became venereally transmitted. Simultaneously, parasite adaptation in a hematophagous host required either being undetected, or minimizing the impact on host survivorship and fecundity or, if the first two strategies were not possible, evading destruction by the arthropod's innate immune system. Thus, a useful way of understanding the evolution of any parasite–vector system is through a consideration of how parasites progressively adapted to and exploited specific morphological, physiological, and biochemical aspects of a hematophagous host without simultaneously compromising the fitness of that host.

This chapter reviews the evolution of disease vectors as a two-step process. First we explore the many ways that different arthropod lineages have independently overcome a common set of developmental, morphological, behavioral, physiological, and biochemical barriers to become hematophagous. Second, we explore the rather sparse, but recently growing literature that describes the ways in which parasites have come to exploit hematophagous arthropods for their own development, propagation, dispersal, and transmission.

We will purposely avoid the terms 'co-evolution' or 'co-adaptation', because these terms imply that a parasite–arthropod association is either mutually beneficial or antagonistic and has evolved via stepwise, compensatory pathways. The concept of 'host specificity' is also implicit in the conceptual framework of co-evolution or co-adaptation; suggesting that a parasite has narrowly adapted to a specific host species. We believe that the evolution of host specificity would have been maladaptive, restricting a parasite to one or a narrow set of hosts to maintain itself in natural enzootic cycles. Brooks and McLennan (1991, 1993) in an extensive review of literature on host–parasite evolution, also provide abundant empirical evidence to dismiss the concept of host specificity. They conclude that the perception of host specificity usually arises in associations with limited historical opportunities for a parasite to transfer to alternative hosts. Klompen et al. (1996) argued that host specificity in ectoparasites is often an artifact of oversampling of common hosts and inadequate sampling of rare or uncommon hosts.

We therefore prefer the term 'co-speciation' (Brooks and McLennan 1993), which implies that parasite phylogenies parallel host phylogenies simply because the parasites have not had the opportunity for host transfer. In this chapter we will approach the problem of parasite–arthropod evolution as a process that arose through the evolution of hematophagy followed by adaptation of the parasite to its invertebrate host.

5.2 The evolution of hematophagy

We will now trace the adaptations associated with vertebrate host location, proximal localization of shallow capillary beds, blood acquisition through either pool feeding (telmophagy) or probing (solenophagy), the role of saliva in disrupting hemostasis, reducing pain and inflammation at the feeding site, and immunomodulation. We will then examine the adaptations for processing the blood-meal both through digestion and excretion and the use of the digested blood in maturing eggs for subsequent generations.

All of these adaptations for hematophagy have arisen on at least 21 independent occasions in arthropods. In every case, the pre-hematophagous ancestral lineages faced a common set of problems. Six basic arthropod mouthparts have been modified to enable the various forms of feeding. Mouthpart evolution has followed very different paths to derive a common set of phlebotomist tools; initially to permit telmophagy and eventually to enable solenophagy. Recent research has shed light on the biochemical adaptations that have occurred in the saliva of hematophagous arthropods to overcome the common problems of hemostasis, vasoconstriction, pain sensation, and inflammation. As with mouthparts, virtually every arthropod examined to date has enlisted a different biochemical solution for each of these problems.

The overriding message in considering all of these adaptations is that no general, consistent morphological, physiological, or biochemical adaptations for hematophagy have been detected among all of the hematophagous arthropod lineages. However, as a general theme, the arthropods, when faced with a common set of problems associated with gaining access to vertebrate blood, have taken up many independent but ultimately convergent paths.

5.2.1 Host location

Three different strategies to locate suitable vertebrate hosts have evolved in hematophagous arthropods. Some arthropods evolved a highly sensitive neurosensory apparatus for long distance, followed by proximal, host location. Other arthropods bypassed the need for long distance host localization by adopting a nidiculous ('nest-dwelling') lifestyle in which they lived on or very near their hosts. There are many free-living arthropod vectors that evolved an intermediate host-location strategy. These species maximized the opportunity for host contact by positioning themselves in the host's habitat without actually becoming ectoparasites or adopting a nidiculous habit.

Flies (Order Diptera) are undoubtedly the masters of long-distance host location. Adult mosquitoes (Culicidae), blackflies (Simulidae), sandflies (Psychodidae), biting midges (Ceratopogonidae), horse/deer flies (Tabanidae), and the hematophagous cyclorrhaphous flies only associate with their vertebrate prey when in need of a bloodmeal. Once physiological signals indicate the need for a bloodmeal, these flies actively locate and orient on suitable hosts through a number of cues.

Visual cues, odors, heat, and other stimuli act along with an imposed diurnal rhythm to give a unique feeding pattern for each vector. For instance, *Aedes aegypti* is chiefly a crepuscular mosquito that also bites, albeit with less intensity, in the middle of the day. It quests for prey between 0.5 and 1 m from the ground. It senses CO_2, water vapor, and lactic acid, which are all heavier than air and thus close to the ground. The mosquito does not seek objects with temperatures higher than 37°C, but is attracted to moving objects. In contrast, *Culex* mosquitoes will quest high in a room for ascending warm air currents given off by a warm body and then dive in on those currents.

5.2.2 Proximal blood location

Once an arthropod has landed or crawled on a host's skin, a different set of proximal stimuli are used to identify optimal locations for bloodmeals. Chemoreceptors located either at the tip of the mouthparts or in the antenna, inspect the skin surface for appropriate 'flavors'. Mechanoreceptors are present in the tips of mouthparts and signal appropriate positioning of the feeding apparatus for penetration.

5.2.3 Blood acquisition

Snodgrass (1943) performed an intensive comparative analysis of hematophagous mouthparts and concluded that arthropod lineages have modified the basic mouthpart blueprint in a wide variety of ways to enable hematophagy. He identified two general hematophagous strategies. The first, and probably the most primitive, he referred to as telmophagy or 'pool-feeding' in which the arthropod mouthparts slice through superficial capillary beds in the skin and then lap or suck up the blood that emerges at the surface. Telmophagy is probably an evolutionarily intermediate step towards a more derived form, called solenophagy, in which the mouthparts penetrate the vertebrate skin and either severe and rest within a capillary bed or actually cannulate individual blood vessels.

It easy to envision telmophagy arising from an ancestral species that occasionally fed on blood exposed through a superficial wound. The next step probably entailed using mouthparts to abrade and irritate the surface of wounds, mucous membranes, or skin to expose superficial blood vessels. This behavior is actually seen among cyclorrhaphous flies in the families Muscidae (*Musca*, *Hydrotaea*, and *Morellia* genera) and Chloropidae (*Hippelates* and *Siphulunculina*). Eventually rasping mouthparts would give rise to cutting mouthparts and telmophagy. However, telmophagy is very irritating to the vertebrate host, and was limited to hosts without much defensive behavior or restricted to arthropods sturdy enough to withstand being beaten by an ungulate's tail or even being rolled upon. Alternatively, many telmophagous arthropods fed at a location on the host that could be groomed, for example, along the midline of the belly, around the anus, or in the ears. Restricted host range, damage due to mauling and preening all probably generated selective pressure for gradual modification and elongation of mouthparts towards solenophagy.

Entomologists subdivide insect mouthparts into six operational structures (Fig. 5.1). These parts show an amazing diversity among the arthropod vectors, and adaptations appear independently. Barbed mandibles anchor in the skin, hooked maxillae curve through the skin, and maxillary bundles cannulate blood vessels. Some vectors were pre-adapted for solenophagy because their ancestors fed through sucking-piercing stylets on plants or captured and sucked nutrients from other invertebrate fauna.

The most primitive hematophagous insects are true bugs in the order Hemiptera. The subfamily Triatominae (Family Reduviidae: vectors of *Trypanosoma cruzi*, causative agent of Chagas' disease) and *Cimex* ('bedbugs': Family Cimicidae) are hematophagous hemipterans. Both families arose from ancestors pre-adapted for solenophagy because most hemipterans feed through sucking-piercing stylets on plants or capture and suck nutrients from other invertebrate fauna. However, the two families arose independently in Hemiptera and therefore arrived at hematophagy via different routes. There are fundamental differences in the manner in which the fascicle operates. In Triatominae the barbed mandibles anchor into the superficial skin tissue, and the maxillary bundle cannulates a blood vessel. The tips of the maxillae differ: one is hooked, and the other spiny so that they curve through the skin when sliding over one another. In *Cimex* the basic details of mandibular and maxillary construction are quite similar, but it is the mandibles that penetrate deeply into peripheral blood vessels. In triatomines the labium is swung forward but does not bend

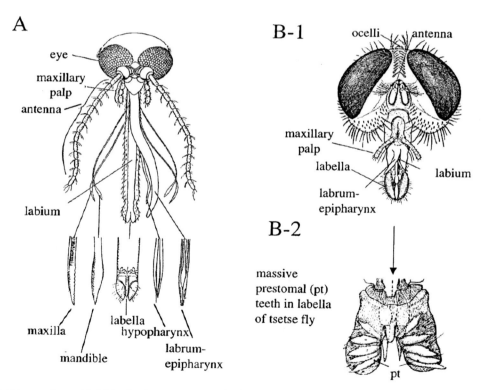

Figure 5.1 Biting mouthparts. (A) Culicine female mosquito. Generalized pattern of insect mouthparts including all the basic elements. (B-1) Sucking mouthparts of Muscomorph fly based on the development of the labium. (B-2) Further development of prestomal teeth for penetration of the skin in a biting fly. Modified from W. B. Herms (1950). *Medical entomology*, p. 70. The Macmillan Company, New York.

during feeding. In *Cimex*, the labium folds back at the basal segments.

Hematophagy next arose in the sucking lice (Order Anoplura). Entomologists generally agree that the sucking lice arose from the chewing lice (Order Mallophaga). The Mallophaga suborder Amblycera contains many genera (e.g. *Menancanthus*) that gnaw through skin and feed on blood and lymph and may thus represent the ancestor to Anoplura. Mouthparts of all extant Anoplura are distinctly adapted for solenophagy, but the stylets lie in a sac concealed within the head. The opening to this sac is at the extreme anterior portion of a proboscis. The proboscis is eversible and is thought to consist of the labrum, the tip of which contains small recurved hooks. These are pushed forward into the host skin by muscular action until firm attachment is achieved. The piercing fascicle consists of three stylets that lie within a long internal sac and are comprised dorsally of united maxillae and the hypopharynx. The labium is attached posteriorly to the walls of the sac. The mandibles are vestigial. The apposed maxillae form the food duct, and the hypopharynx forms the salivary channel. Salivary secretion is poured into the wound, and the cibarial and pharyngeal pumps draw blood.

Hematophagy arose many times independently during the evolution of the flies (Order Diptera). On the other hand, all evidence suggests that arachnid ancestors (subphylum Chelicerata) never possessed mouthparts for mastication and probably subsisted on liquids. All arachnids have a highly developed sucking apparatus and paired chelicerae that are used for grasping, holding, tearing, crushing, or piercing food items. The chelicerae also serve as cutting structures that gain entrance into the vertebrate host. All hard and soft ticks have a highly sclerotized structure bearing many recurved denticles called the hypostome that anchors the mouthparts in place. Extensive laceration of blood vessels is characteristic of tick feeding, leading ultimately to localization of the chelicerae in a superficial capillary bed. Ticks do not cannulate individual vessels but by the same token do not bring blood to the surface of the wound. Soft ticks (Argasidae) generally feed rapidly; hard ticks (Ixodidae) characteristically remain attached for a period of 3–7 days and fix their mouthparts in

place using a cement substance secreted from the salivary glands.

The mouthparts of mites resemble those of ticks but without the hypostome. Mites tend to feed on lymph and tissues other than blood. Among trombidiform mites ('chiggers') the mouth parts became progressively adapted for piercing by a transformation of the movable digits into hooks or stylets.

5.2.4 The evolution of hematophagous saliva

Once an arthropod was physically capable of withdrawing blood, a host of additional problems presented themselves. Vertebrate hemostasis is a highly redundant phenomenon in which platelet aggregation, blood coagulation, and vasoconstriction prevent blood loss from injured tissue. Consequently, there must have been strong selection pressure for the evolution of salivary antihemostatic components to locate blood during probing or for maintaining flow during feeding (Ribeiro 1987b). All hematophagous arthropods salivate during intradermal probing of vertebrate prey before ingesting blood. Anti-platelet, anticoagulant, and vasodilation agents have been found in the saliva of all bloodsucking arthropods that have been examined to date. In addition to adaptation to hemostasis, there must also have been intense selection to reduce pain and inflammation at the feeding site.

The saliva of hematophagous arthropods also contains potent immunomodulators and selection pressure must have been intense for the evolution of these compounds. Before the evolution of immunomodulators, vertebrates probably produced antibodies against salivary components. Thus, only immunologically naïve ('unbitten') vertebrates would have been suitable blood sources. Proteins that prevented inflammatory/immune responses from a host to the various salivary proteins would have had an immediate selective advantage both for individual insects and for a species. In a sense these substances maintained the population density of susceptible vertebrate hosts.

Physiological and biochemical adaptations in response to hemostasis, pain, and inflammatory/immune responses from a host have resulted in the evolution of a huge diversity of pharmacologically active substances in hematophagous arthropod saliva. Most of what we know of these substances comes from the pioneering work of Drs Jose Ribiero, Richard Titus, Steve Wikel, Tony James, and Jesus Valenzuela.

As arthropods penetrate or lacerate their host's skin, they salivate and taste whatever is available. Purine nucleotides provide a positive stimulus for most bloodsucking arthropods. Depending upon the arthropod group, adenosine, AMP, ADP, ATP, or 2,3-diphosphoglycerate have all been shown to stimulate the uptake of a warm saline meal through a membrane. The chemoreceptors for purine nucleotides are usually located high in the food channel, near the cibarial pump. Because most purine nucleotides are inside red blood cells, it is presumed that shearing forces acting at the pump and the shredding action of the cibarial and pharyngeal armatures and other papillae and spines in the foregut release nucleotides from the erythrocytes.

Anti-hemostatic agents

Anti-platelet and anticoagulant agents have been found in the salivary glands of all hematophagous arthropods. Apyrases are enzymes that hydrolyze nucleotide di- and triphosphates to orthophosphate and mononucleotides. Arthropod salivary apyrases thus prevent ADP-induced platelet aggregation at the site of the bite. At least two families of enzymes, belonging to the 5'-nucleotidase and to the actin/heat shock protein 70/sugar kinase super-family, have been enlisted by arthropods to perform the apyrase function. Interestingly, the apyrase of the bedbug (*Cimex lectularius*) had no sequence similarity to any other known arthropod apyrases, but similar sequences were found in the genomes of the nematode *Caenorhabditis elegans*, mice, and humans (Valenzuela *et al.* 1998). Apyrase was subsequently purified from female salivary glands (Champagne *et al.* 1995b) and found to be similar to a ubiquitous family of 5'-nucleotidases. The authors made the important point that an enzyme involved in a common metabolic event at the cellular level had been enlisted by the mosquito for use in anti-hemostasis. A different anti-hemostatic mechanism evolved in *Anopheles* mosquitoes. An

anti-thrombin peptide (anophelin) was isolated from the salivary glands of the mosquito *Anopheles albimanus* (Valenzuela *et al.* 1999). The purified peptide inhibited thrombin-induced platelet aggregation, thrombin esterolytic activity on a synthetic substrate, and thrombin cleavage of fibrinogen. A number of other clotting activities have been identified in vectors (Stark and James 1995, 1996, 1998) and analysis showed that the AFXa gene product has similarities to the serpin superfamily of serine protease inhibitors and may represent a novel, highly diverged member of this family.

Mosquitoes in the genus *Culex* have a strong tendency to ornithophagy and appear to have only recently adapted to mammals. Thus, they may not yet have evolved efficient mechanisms to counteract mammalian platelet responses, while birds only have relatively inefficient thrombocytes. Accordingly, Ribeiro (2000) compared the probing behavior of *Culex quinquefasciatus* with two other mosquito species from different backgrounds. *Culex* took much more time to find blood on a mammalian host (human or mouse) when compared with the two other mosquito species, but did not differ in probing behavior when feeding on a chicken.

Vasodilators

The first report of a vasodilatory peptide came from the salivary glands of the sandfly *Lutzomyia longipalpis* (Ribeiro *et al.* 1989). Chromatographic analysis, antibody reactivity, and bioassays of a potent salivary erythema-inducing factor indicated striking similarity with human calcitonin gene-related peptide. Lerner *et al.* (1991) isolated the vasodilator and reported that it had 500 times the vasodilatory activity of calcitonin gene-related peptide, previously the most potent vasodilator peptide known. These authors named this novel peptide maxadi-lan (MAX).

A number of other active vasodilators have been identified, among them, as follows.

1 Large amounts of adenosine and 5'-AMP in sandfly saliva. These purines, 75–80% of which were secreted from the glands following a blood meal, have vasodilatory and anti-platelet activities (Ribeiro *et al.* 1999).

2 Nitrovasodilator activity in R. prolixus saliva (Ribeiro et al. 1993). This vasodilator is a nitrosylheme protein with an Fe(III) heme that binds nitric oxide (NO) reversibly.

3 Nitrophorins (NO-carrying proteins) from salivary glands of the kissing bug (*Rhodnius prolixus*) (Champagne et al. 1995).

4 A peptide was of the tachykinin family from salivary glands of *Ae. aegypti* (Ribeiro 1992).

5 Salivary NO synthase activity in R. *prolixus* and *Cimex* with similarity to the vertebrate constitutive NO synthase (Ribeiro and Nussenzveig 1993; Valenzuela et al. 1995).

Immunomodulators

Gillespie et al. (2000) is an excellent synopsis of recent progress in our understanding of the immunomodulatory activities of sandfly and tick saliva. Readers are encouraged to consult this article for information that we will only briefly summarize below. Gillespie et al. (2000) argued that inflammation is a common part of any immune response and that cytokines are one of the principal mediators of this response. Thus, they argued that inflammation inhibitors and cytokine modulators could also prevent vertebrates from becoming sensitive to the saliva. Their review focused largely on MAX, as one of the best examples of a vasodilator that also acts as an immunomodulator. The authors made the important point that vasodilators and anticoagulants from many other arthropods may also act as immunomodulators. Insufficient research has been completed on this fascinating topic.

Discovery of the immunomodulatory effects of MAX began with the observation that millions of parasites are usually required to induce infection with *Leishmania major* in mice when the parasites were injected by syringe (Warburg and Schlein 1986) but that only 1–100 parasites were required when mice were infected with L. *major* in the presence of L. *longipalpis* saliva (Titus and Ribeiro 1988). Furthermore, lesion size increased as did parasite burden within the lesions in the presence of L. *longipalpis* saliva. In fact, Samuelson et al. (1991) and Lima and Titus (1996) noted that lesions persisted over the lifetime of mice treated with saliva. Titus and Ribeiro (1988) noted that the parasite burden in saliva-treated lesions was 5580-fold

greater than the parasite burden found in saliva-free inoculum. Since the initial work with L. *major*, it has been shown that sandfly saliva increases infection with all species of *Leishmania* tested (Samuelson et al. 1991; Theodos et al. 1991; Warburg et al. 1994; Lima and Titus 1996).

Attention next focused on identifying the salivary factors responsible for this effect. The recombinant MAX when injected into the skin induced the same long-lasting erythema as found with injection of whole saliva or associated with the bite of a sandfly (Ribeiro et al. 1989; Lerner et al. 1991; Lerner and Shoemaker 1992). Most importantly, when co-injected with L. *major* into the footpads of mice, MAX enlarged lesion size and increased parasite burden within the lesions to the same degree as co-injection with salivary gland (Lerner et al. 1991). Thus, MAX appeared to be the principal peptide that enhanced *Leishmania* infection. It has also been discovered that MAX inhibits the functions of T cells and macrophages but that the effect on T cells is mediated indirectly through macrophages (Theodos and Titus 1993). MAX inhibits the production of NO, H_2O_2 and TNFα by macrophages, all of which are associated with *Leishmania* killing by the macrophage (Titus et al. 1994).

The extended feeding time in hard ticks allows inflammatory cells to invade the site of feeding. In fact, Wheeler et al. (1989) observed a concentration of neutrophils at the tick feeding site. The presence of T and B lymphocytes at the feeding site suggested the possibility of a specific immune response (Brossard et al. 1982; Brown 1982; Wikel 1984). Indeed, specific anti-tick immune responses are now well documented (Wheeler et al. 1989; Whelen and Wikel 1993; Borsky et al. 1994; Mbow et al. 1994). This suggests that there has been and continues to be strong selection for the evolution of anti-inflammatory immunosuppressive elements in hard tick saliva. The first evidence for this came from the work of Steve Wikel (Wikel et al. 1978; Wikel and Osburn 1982; Wikel 1982b) who demonstrated that ticks feeding on a host had a systemic immunosuppressive effect. This effect was subsequently demonstrated *in vitro* with the saliva of many different hard tick species (Wikel et al. 1978; Wikel 1982a; Ribeiro et al. 1985a; Ramachandra and Wikel 1992; Urioste et al. 1994;

Fuchsberger *et al.* 1995; Ramachandra and Wikel 1995; Ferreira and Silva 1998).

This immune response is directed by salivary gland components (Wikel *et al.* 1978; Wikel and Osburn 1982; Wheeler *et al.* 1989; Ganapamo *et al.* 1997). Ribeiro (1987a) and Ribeiro *et al.* (1986, 1990) reported that tick saliva inhibits neutrophil function and interferes with the complement system. There also exists *in vitro* evidence of natural killer (NK) cell (Kubes *et al.* 1994; Kopecky and Kuthejlova 1998) and macrophage activity modulators (Ramachandra and Wikel 1992, 1995; Urioste *et al.* 1994; Ferreira and Silva 1998). The high concentrations of prostaglandins (PGE2, PGF2α, and PGI2) in tick saliva (Bowman *et al.* 1996) are likely to have some direct effects on the immune system of the host. It is possible that tick prostaglandins serve as vasodilators at the site of attachment (Champagne and Ribeiro 1994; Bowman *et al.* 1996).

Gillespie *et al.* (2000) pointed out that anticoagulants might also inhibit the development of an immune response. For example, simulidin from the blackfly *Simulium vittatum* is an anticoagulant that inhibits α-thrombin, and α-thrombin is chemotactic for immune cells such as macrophages (Abebe *et al.* 1995). The same would be true for the thrombin inhibitors anophelin described earlier in *An. albimanus* (Francischetti *et al.* 1999) and americanin isolated from the hard tick *Amblyomma americanum* (Zhu *et al.* 1997).

Thus, to date three different immunomodulatory factors have been identified that target different components of the immune system: macrophages (MAX), IL-4 (*Phlebotomus papatasi* saliva) and IL-2 (*Ixodes scapularis* saliva). We echo the view of Gillespie *et al.* (2000) that much more work needs to be done with regard to the other hematophagous arthropods discussed in this review. Evidence of immunomodulatory saliva is by no means limited to sandflies and hard ticks. For instance, the saliva of *Aedes triseriatus* enhances the infectivity of Cache Valley virus Bunyaviridae (Edwards *et al.* 1998).

We highlight the recent results of Lanzaro *et al.* (1999) who found extensive amino acid sequence differentiation among MAX alleles from different populations. The vasodilatory activity of MAX was equivalent among recombinant MAX variants. However, profiles based on the more antigenic index differed among peptides. The authors argued that this extensive variation could arise through diversifying selection to prevent development of inflammatory/immune responses by individual vertebrate hosts and within host populations of the species. Diversifying selection is a type of density-dependent selection that usually selects against the most prevalent form of a gene. Diversifying selection accelerates the rate of nucleotide, and presumably, amino acid substitutions in a gene. There is no reason to believe that the effects of diversifying selection should be limited to MAX. However, at this point, MAX is the only salivary gland gene that has been examined for evidence of diversifying selection.

Pain suppression

Charlab *et al.* (2000) characterized salivary gland products of a Latin American sandfly and found adenosine deaminase (ADA) activity. They suggested that these invertebrate proteins may exert their actions through adenosine depletion. Subsequently, Ribeiro *et al.* (2001) identified high ADA activities in certain mosquitoes and proposed that ADA activity may help blood-feeding by removing adenosine, a molecule associated with both the initiation of pain perception and the induction of mast cell degranulation in vertebrates, and by producing inosine, a molecule that inhibits production of inflammatory cytokines.

Other functions

Among other functions the following have been noted.

1 Hyaluronidase activity (Ribeiro *et al.* 2000a) in sandflies and blackflies. They speculate that hematophagous pool-feeding insects may secrete this enzyme to help the spread of salivary anti-hemostatic agents in the vicinity of the feeding lesion;

2 salivary anti-complement may contribute to successful feeding (Ribeiro 1987a);

3 IgG binding proteins (IGBPs) in both tick hemolymph and saliva may be present to eliminate host IgG from the host (Wang and Nuttall 1994);

4 esterase and phosphatase activity in salivary gland lysates of adult *Ae. aegypti* (Argentine and James 1995).

5.2.5 Water balance

Once arthropods have taken an enormous amount of blood, locomotion or flying becomes difficult. The volume of blood ingestion is regulated by stretch receptors located in the gut or crop. Normally, as the blood meal distends the abdomen, signals are sent to the brain, which arrests the feeding reflex. When the abdominal nerves to stretch receptors are severed in mosquitoes and *Rhodnius*, the insects feed until the abdomen literally bursts. Once stretch receptors are stimulated, the insect ceases feeding, withdraws it mouthparts, and flees from the host. The engorged insect seeks a suitable resting place to initiate digestion and water loss.

Hemipterans, lice, and cyclorrhaphous flies feed exclusively on blood, whereas mosquitoes, blackflies, and sandflies drink water and nectar and are more flexible in their ability to maintain water balance. Ticks have developed a unique method to capture water vapor from ambient air. They secrete a hygroscopic saliva that is spread over their palps. This saliva captures moisture from the atmosphere and is then imbibed. Soft ticks can remain alive without desiccating for more than 10 years.

Because most of the lipids and carbohydrates in bloodsucking insects must be made from amino acids, excess nitrogen has to be excreted. In all arthropods uric acid is usually secreted by the excretory organs (Malpighian tubules) into the insect rectum. Diuretic hormones have been isolated from Malpighian tubules of *Rhodnius* (Maddrell and O'Donnell 1992). These hormones are present in the bug's hemolymph a few seconds after ingestion of the meal so that the insect begins urinating on the host while still feeding. Hormonal regulation of water transport from the midgut to the hemolymph and from the hemolymph to the Malpighian tubules is under the control of a number of neuropeptides and the role of serotonin in this process has been recently characterized (Spring 1990; Wheeler and Coast 1990).

Ticks eliminate excess water in a different way. Soft ticks, which feed for no longer than an hour, have evolved a unique filtration apparatus called the coxal gland. A dilute saline solution and some protein are eliminated from the coxal glands while the tick is feeding. Alternatively, the gut of hard ticks absorbs water from the bloodmeal into the hemolymph. Excess water is pumped back into the host in saliva.

5.2.6 Digesting the blood meal

As a blood meal is ingested in mosquitoes, sandflies, blackflies, and fleas it travels directly to the gut. Tsetse flies instead store the blood in a diverticulum before gradually shunting the meal into the midgut. The anterior midgut of *Rhodnius* serves the same function. A chitin–protein web, known as the peritrophic matrix, is secreted around the blood meal within a few hours of ingestion. This structure is thought to serve a protective function in arthropods feeding on meals containing abrasive, or coarse substances. It is not unique to hematophagous arthropods.

The presence of a food bolus in the midgut causes digestive enzymes to be secreted into the lumen of the gut. Within a few hours or days after taking blood, protease levels can rise 20-fold. In *Aedes* spp., a small amount of one type of early trypsin is produced following ingestion of free amino acids, but the main trypsin is not produced if the meal does not contain protein. Early trypsin is apparently a part of a unique signal transduction system (Noriega *et al.* 1994, 1996, 1997, 1999). A large pool of transcribed message resides in the midgut of newly eclosed adults. Translation of early trypsin is not induced by nectar feeding but is induced by blood-feeding. Its function may be to 'taste' the incoming meal to determine whether there is sufficient protein to support a gonadotrophic cycle. If so, the signal transduction pathway activates late trypsin transcription to digest the blood meal.

Most hematophagous arthropods digest blood proteins using trypsins. Further digestion is accomplished by amino- and carboxypeptidases. These enzymes typically have optimal activity at neutral-to-alkaline pH. In contrast, triatomine midguts have sulfhydryl proteolytic enzymes with pH optima of around 5.0 similar to the lysosomal cathepsins (Gooding 1972; Billingsley 1990; Terra 1990). Ticks secrete a hemolysin, and then slowly pinocytose the gut contents by intestinal cells, resulting in the intracellular digestion of the meal by lysosomal enzymes.

Blood is rich in proteins and essential amino acids but deficient in carbohydrate, fat, and adequate amounts of many B vitamins. Accordingly, all insects that feed exclusively on blood, such as the triatomines, lice, bed bugs, and cyclorrhaphous Diptera, have bacterial symbionts that either supplement the blood with B vitamins or balance the diet by converting proteins to carbohydrates (glyconeogenesis). Without these endosymbionts the insects die or are infertile. These endosymbionts live in the insect gut, either free (as in triatomines), or in mycetomes, special structures of the anterior midgut (as in tsetse or lice).

5.2.7 Growth and egg development

The products of blood digestion accumulate in the arthropod's hemolymph. In females, products are rapidly sequestered into the tissues or cells known as the fat body. After a period of time, a substance known as juvenile hormone (JH) stimulates the female fat body to start vitellogenin synthesis and prepares ovaries to receive vitellogenins. Vitellogenins are large glycolipophosphoproteins. After vitellogenin passes through spaces between follicular cells in the oocyst, it binds with receptor proteins on the oocyte surface membrane and is taken in by a process of receptor-mediated endocytosis. As vitellogenesis proceeds, the oocyst grows and activates stretch receptors that signal the arthropod brain to biosynthesize and release allatostatin. One of the activities of allatostatin is to block JH production, which then decreases vitellogenesis.

In mosquitoes, adult eclosion causes JH synthesis from a portion of the arthropod brain. JH then circulates to the ovary where it induces growth, signals synthesis of vitellogenins, and initiates secretion of another hormone known as ecdysone. JH presence also triggers feeding and mating behaviors. *Culex* mosquitoes will not bite in the absence of JH and have two egg production cycles. The first cycle is induced by release of JH following adult emergence, and leads to primary growth of oocysts but discontinues until blood-feeding. After a meal, vitellogenesis occurs in the oldest, primary oocysts, and another hormone (20-hydroxyecydysone) stimulates the separation of secondary

follicles from the germarium in preparation for the second gonadotrophic cycle. After mating, the female *Culex* brain is stimulated to produce and release egg development neuro-hormone (EDNH), which circulates and activates the ovaries to begin biosynthesis and release of ecdysteroids, which in turn trigger vitellogenin synthesis by the fat body, which is quiescent from the earlier JH peak and is receptive to EDNH. The post-vitellogenic ovary then resignals the brain to stop EDNH production and end the stimulation of the ovary. In Triatominae, an anti-gonadotrophic hormone is produced in the abdomen when mature eggs are present in the ovaries. This blocks the effect of JH on the follicle cells and prevents vitellogenesis.

5.3 Parasite adaptation to hematophagous hosts

We have described the many adaptations in arthropods associated with the evolution of hematophagy. We will now discuss the ways in which parasites have adapted to hematophagous arthropods as a means of increasing their own reproduction, dispersal, and survival. We will only discuss interactions that provide evidence of actual adaptation by the parasite. We will not, for example, discuss the benefits accrued to the parasite by the immunomodulatory effects of saliva. We have discussed the abundant evidence that parasites benefit from immunomodulatory saliva, but believe that these properties evolved through diversifying selection on the arthropod host and do not represent adaptation in the parasite.

There are two fundamental ways that parasites have adapted to life in their hematophagous arthropod host. Arboviruses, most *Rickettsia*, some bacterial species, *Leishmania*, and *Trypanosoma* spp. evolved to invade, replicate within, and become released from their invertebrate host without being detected or by having such a small impact on host survival that the vector did not react to their presence. Alternatively, in some protozoa and nematodes, the arthropod frequently attempts to kill or reject the parasite. This is not surprising, because it is well documented that infections with these organism reduce vector fitness and reproduction (Berry *et al.* 1986, 1987a,b, 1988). In these cases the parasite has had to evolve adaptations to avoid detection or destruction.

It is also important to emphasize that there are some, probably recently evolved, vector–parasite interactions that are not benign for either the vector or the parasite. Consider the interactions between *Rickettsia prowaseki*, the etiological agent of louse-borne typhus and its primary vector, *Pediculus humanus*, the human body louse. The midgut epithelial cells of the louse are destroyed by the rickettsial infection, and the louse dies. *Rickettsia* are only transmitted through the feces of the louse, or via ingestion of all or part of the louse body. Inhalation of dust or dried body parts is also a common route of infection.

5.3.1 Vector competence

Vector competence refers to the intrinsic permissiveness of an arthropod vector for infection, replication, and transmission of a vertebrate parasite. From the parasite's perspective, a competent vector has what the parasite needs to replicate, survive, and be transmitted and doesn't mount a defense against the parasite. Alternatively, an incompetent vector either does not have what the parasite needs and, even if it does, mounts such an effective defense that the parasite must fight to use the vector.

Vector competence may be manifest as differences in competence of two closely related vector species for a particular parasite. For example, *Anopheles stephensi* is highly susceptible to *Plasmodium berghei* whereas *Anopheles gambiae* is nearly incompetent. Similarly, extracted salivary glands from a competent vector, *Anopheles dirus*, are infected by *Plasmodium knowlesi* when they are transplanted into an incompetent vector, *Anopheles freeborni*, but are not infected when transplanted from *An. freeborni* into *An. dirus* (Rosenberg 1985). The human head louse *Pediculus capitis* does not become infected with *Rickettsia prowaseki*, the etiological agent of louse-borne typhus, whereas its sister subspecies *P. humanus* becomes readily infected and, as discussed above, dies. In other cases, different populations of a single vector species may differ in their ability to transmit a particular parasite. This is the case among different populations and subspecies of *Ae. aegypti*. When populations of West African *Ae. aegypti formosus* are orally infected with yellow fever virus in the laboratory, only ~30% become capable of oral transmission. In contrast, no fewer than ~60% of *Ae. aegypti aegypti* collected from most locations throughout the world transmit flaviviruses when orally infected.

Plasticity in vector competence is consistent with a general genetic model in which multiple structural or biochemical factors in the arthropod vector must be present for successful completion of a parasite's life cycle. The absence of any one of these factors renders a vector species incompetent. Alternatively, if a parasite is deleterious to vector survival or reproduction, the vector may well have evolved an active resistance for rejection or destruction of a parasite. Some vector species may be fixed for the presence of resistance factors, while others may lack one or more of these factors. Variation in vector competence among populations of a single species could arise due to genetic factors being fixed in some populations and at variable frequencies or altogether absent in others.

5.3.2 Arbovirus adaptation

There are six potential barriers to transmission that an arbovirus faces in being transmitted by an arthropod. The virus must first attach and penetrate the mosquito's midgut epithelial cells and then replicate to a high titer within those cells. Factors that block either of these two events constitute a midgut infection barrier (MIB). Next the virus must pass through the basal lamina surrounding the midgut, and infect and replicate in surrounding tissues. Anything that prevents these two events acts as a midgut escape barrier (MEB). Finally, the arbovirus must infect and possibly replicate within the salivary gland before it can be shed into the lumen of the gland for final transmission in the subsequent bite. Factors that suppress these phases serve as a transmission barrier (TB).

We can further subdivide any of these barriers at a finer, biochemical level. A vector protein or substance produced by a protein is required for each step of viral penetration and replication in midgut tissues. It may be that every vector species has most of these factors but only competent species have all. Alternatively, it may be that susceptible populations have all of the required factors, while those with intermediate susceptibility are polymorphic for a required factor and refractory populations lack the factor altogether. For example, a MIB for a positive sense RNA virus may exist in the lumen of the midgut and involve reduced ability to cleave proteins on the viral coat before attachment or penetration can occur. In addition, there may be receptors on the surface of the midgut cell that are necessary for attachment of the virus. Once inside the cell, host factors may be necessary for uncoating and then translation of non-structural viral proteins (e.g. RNA-dependent polymerases). Host factors are probably involved in proper transcription and translation of viral structural proteins. In the final stages, host factors are undoubtedly involved in proper packaging and assembly of virus.

For some arboviruses (e.g. Bunyaviruses and Orbiviruses), there is now evidence that proteolytic processing of virion surface proteins is necessary for efficient vector midgut cell interaction (Ludwig *et al.* 1989, 1991; Mertens *et al.* 1996; Xu *et al.* 1997). Interestingly this proteolytic cleavage is not a pre-requisite for successful infection of cells in secondary target organs, only for the midgut epithelial cells that are exposed to the proteolytic milieu of the midgut lumen. Laminan has been proposed to be a mosquito cell receptor for alphaviruses (Ludwig *et al.* 1996). Recent research in our department indicates that the midgut trypsins in *Ae. aegypti* are necessary for infection of midgut epithelial cells (Bennett *et al.*, unpublished).

5.3.3 Bacterial and rickettsial adaptation

In fleas infected with the *Yersinia pestis* bacillus, the bacilli accumulate in the flea's proventriculus. The massive accumulation of bacilli forms a plug that disrupts the feeding process of the flea vector. When an infected flea attempts to feed on a new vertebrate host, this blockage causes the flea to regurgitate the bacteria into the next vertebrate host. Because the flea cannot feed, it becomes permanently hungry and contacts many more hosts than it would otherwise and increased disease transmission results. Hinnebusch *et al.* (1996) showed that blockage was dependent on the hemin storage (*hms*) locus in *Y. pestis. hms* mutants established long-term infection of the flea's midgut but failed to colonize the

proventriculus. Thus, the *hms* locus markedly alters the course of *Y. pestis* infection in its insect vector, leading to a change in blood-feeding behavior and to efficient transmission of plague.

These patterns may, however, be specific for the flea species. Engelthaler *et al.* (2000) used a quantitative competitive PCR assay to quantify *Y. pestis* loads in fleas and bacteremia levels in mice used as sources of infectious blood meals. *Xenopsylla cheopis* achieved higher infection rates, developed greater bacterial loads, and became infectious more rapidly than *Oropsylla montana*. Their results suggest that at the time of flea feeding, host blood must contain $>10^6$ bacteria/ml to result in detectable *Y. pestis* infections in these fleas, and $>10^7$ bacteria/ml to cause infection levels sufficient for either species to eventually become capable of transmitting *Y. pestis* to uninfected mice. *Y. pestis* colonies developed primarily in the midguts of *O. montana*, whereas infections in *X. cheopis* often developed simultaneously in the proventriculus and the midgut.

Relapsing fever *Borrelia* spirochetes have exploited the water elimination anatomy (the coxal glands) and physiology of soft ticks as a means of transmission. Spirochetes are found in the excreted coxal fluid and are infectious either through the wound at the tick feeding site or through other skin lacerations or mucous membranes. Schwan and Piesman (2000) and Schwan *et al.* (1995) showed that *Borrelia burgdorferi* undergoes changes in expression of important outer surface proteins in the midgut of *I. scapularis*. The authors speculated that these changes may be important in the development of virulence of *B. burgdorferi* for the vertebrate host.

5.3.4 Plasmodium adaptation

After an anopheline mosquito takes a gametocytemic bloodmeal from a vertebrate host, the gametocytes must exflagellate and then fuse to form an ookinete. The ookinete must then successfully penetrate the mosquito's midgut epithelial cells before forming an oocyst between the midgut cells and the basal lamina. Vinetz *et al.* (2000) showed that the *Plasmodium* ookinete produces chitinolytic activity that allows the parasite to penetrate the chitin-containing peritrophic matrix.

Factors that block any phase of ookinete development and cell penetration constitute an MIB. A number of factors have been identified that block *Plasmodium* invasion and development. Invertebrates are capable of both cellular and humoral reactions to combat infections. More recently it has been shown that mosquitoes also express several elements of vertebrate-specific immune responses (Barillas-Mury *et al.* 1996, 1999). It has been shown that ookinetes penetrating midgut follicular epithelial cells induce various *Anopheles* species to produce defensin, a Gram-negative bacteria binding protein and NO synthetase, and to initiate other enzymic pathways that may ultimately lead to apoptosis (Han *et al.* 2000). In general, this MIB is so potent that only a small percentage of mature ookinetes reach the basal lamina to form oocysts.

The oocyst must next undergo sporogony and burst through the basal lamina. Anything that prevents these events acts as an MEB. An MEB has been observed in strains of *An. gambiae* that have been artificially selected to encapsulate, melanize, and thereby destroy oocysts (Collins *et al.* 1986). Finally, the sporozoites must infect the salivary gland and make their way into the lumen of the glands for final transmission in the subsequent bite. Prevention of salivary gland infection would constitute a TB. Susceptible vector species may have failed to develop one or all of these resistance factors whereas refractory species may have potent or constitutively expressed resistance factors or, as with viruses, lack factors that are essential for *Plasmodium* survival.

5.3.5 Filarial adaptation

Microfilariae (mf), each in a sac-like sheath, are ingested from the peripheral blood of the vertebrate host during the act of feeding. Diurnal patterns of infection are suggestive of adaptation. The presence of the mf in the vertebrate host may be more prevalent during defined periods of a day. In *Wuchereria bancrofti*, for instance, microfilariae are present in significant numbers in the human peripheral circulatory system between 10:00 p.m. and 2:00 a.m., which overlaps with the preferred biting activity of the vector *Culex pipiens*. Thus, periodicity may have evolved in the filarid species as a response to the

biting activity of vectors. Biochemical or physiological adaptation to explain periodicity are obscure and do not involve the release of a new generation mf each day. Host-mediated cues including reduced arterial oxygen tension and lowered temperature may be involved. In endemic areas of the South Pacific, certain *Wuchereria* strains exhibit a diurnal presence in the peripheral blood and are referred to as subperiodic. Daytime feeding mosquitoes are the major vectors in these areas.

McGreevy *et al.* (1978) showed that mf of *W. bancrofti* and *Brugia pahangi* are killed by the shredding action mouthparts and other papillae and spines in the foregut of mosquitoes. *Anopheles* species (the primary vectors in nature) have well developed cibarial armatures and killed 36–96% of the ingested mf. *C. pipiens* has a poorly developed cibarial armature and killed only 6% of the mf. *Ae. aegypti* and *Ae. togoi* lack cibarial armatures but have the remaining foregut structures and killed only 2–22% of the mf. Having passed through this MIB, the mf shed their sheaths, and migrate through the walls of the midgut. Migration through the midgut wall can occur in minutes, although live and possibly unsuccessful mf may be found in the mosquito gut up to 4 days after ingestion. Practically all mf passing through the gut migrate into the thorax within 12 hours.

Following midgut penetration, development of mf takes place within the large thoracic flight muscles where the larvae become slightly shorter and much thicker. As with *Plasmodium*, refractory mosquitoes sense and treat the parasite as a foreign body and can encapsulate and melanize mf larvae in the hemolymph or in the Malphigian tubules (Chen and Laurence 1985). A number of internal changes and two molts occur during development in vector tissues, ultimately resulting in infective third-stage larvae. These changes are not accompanied by an increase in numbers. The infective larvae migrate with little difficulty to the proboscis, and when the mosquito feeds on a vertebrate these larvae emerge through the wall of the mosquito's labellum.

5.3.6 Trypanosomal adaptation

Once trypomastigotes of *Trypanosoma brucei* are ingested they undergo developmental changes into the procyclic form and then the mesocyclic form in the tsetse fly gut. The mesocyclic form then penetrates the midgut wall and migrates through the anterior end of the gut into the salivary glands where it undergoes further differentiation to the epimastigote, pre-metacyclic, and metacyclic trypomastigotes before oral transmission by the fly. Little is understood about vector competence in the tsetse host. However, many experiments have documented that developmental changes in the fly are essential for transmission. Only the metacyclic trypomastigotes are infectious and this appears to be associated with establishment of the basic antigen surface in the salivary glands.

Small numbers of trypomastigotes of *Trypanosoma cruzi* circulate in the peripheral circulatory system of humans. Once taken up in the triatomine bloodmeal, the organism replicates as a epimastigote in the mid- and hind gut of the bug. One to two weeks after infection, metacyclic trypomastigotes appear in the hindgut. *Trypanosoma cruzi* exploits the rapid flow of urine that washes out the triatome rectum during feeding. In this way literally thousands of metacyclic trypomastigotes are delivered to the vertebrate host's skin. Trypanosoma inhabit the proteolytic gut of their host and have dense surface glycolipidic coats that cannot be digested by their vector's enzymes. Ribeiro (1996) noted that the adaptations of these parasites to survive in a proteolytic environment probably pre-adapted them to the macrophage phagolysosome.

5.3.7 Leishmanial adaptation

Little is understood about vector competence in sandflies. The requirement for a sugar meal by the infected sandfly was demonstrated more than 50 years ago as being essential for *Leishmania* to become infective to vertebrates, but the underlying reason remains obscure. *Leishmania* also undergo developmental changes within their sandfly vector. The fly ingests amastigotes in its bloodmeal. As with the trypanosomes, *Leishmania* inhabit the proteolytic gut of their host and have dense, indigestible surface glycolipidic coats that probably protect them from the vector's enzymes. In the gut they go through a series of transformations to become infective. Schlein *et al.* (1991) showed lysis

of the chitin layer in the anterior region of the peritrophic membrane. This lysis permitted the forward migration of a concentrated mass of parasites. At a later stage the parasites concentrated in the proventriculus, which then lost its cuticular lining. They showed that chitinase and N-acetylglucosaminidase were secreted by cultured *L. major* promastigotes. Activity of both enzymes was also observed in a number of related forms. Subsequently, the chitinase also destroyed the valves associated with the feeding pump, presumably reversing the normal flow of pumping and injecting the parasites into the vertebrate host. The infected sandflies have difficulty feeding, which, as in the case of *Y. pestis* and *Plasmodium*, similarly increased transmission of the parasites. Charlab and Ribeiro (1993) showed that salivary gland homogenates of female *L. longipalpis* inhibit the *in vitro* multiplication of promastigotes of *Leishmania mexicana amazonensis*. The results suggest that vector saliva could influence the development of *Leishmania* parasites within the vector by inhibiting their growth and triggering them to a differentiation pathway.

5.4 General patterns in parasite adaptation to hematophagous arthropods

We have offered abundant evidence that the diversity of parasites have adapted in numerous independent ways to survive, and in many cases to replicate in and escape from their arthropod host. The examples discussed above display a continuum of arthropod–parasite interactions from the very minimal interactions seen between arboviruses and their hosts, to the defensive reactions elicited by *Plasmodium* and *Filaria*, to the lethal interactions seen between some *Rickettsia* and their lice vectors. As an explanation of this continuum we propose that the severity of arthropod–parasite interactions may reflect the number of generations over which these interactions have existed. This would suggest that arboviruses, most *Rickettsia*, and many trypanosome associations have existed for possibly millions and millions of generations, whereas *Plasmodium* and filarial associations with anopheline vectors may have been more recent, and *Rickettsia* and lice associations may be extremely recent.

The numbers of generations would be greatly accelerated in arboviruses, rickettsiae, and many bacteria that undergo many replication cycles within one arthropod body. In the case of RNA viruses, the process would be accelerated even further by the error-prone RNA-dependent DNA polymerase. Many parasitology textbooks discuss the variety of trypanosome kinetoplastids that are found in the guts of other, non blood-feeding arthropods. Arthropod–trypanosome adaptation may have been well advanced by the time Cimicids and Triatomines became hematophagous. Conversely, *Plasmodium falciparum* may have existed only as a species and have been associated with humans, and possibly *An. gambiae*, within the last 10 000 years (Tishkoff *et al.* 2001; Volkman *et al.* 2001).

These patterns suggest two general models for parasite adaptation to hematophagous arthropods. In the first model, we propose that hematophagous arthropods are constantly coming into contact with a wide diversity of potentially pathogenic organisms. In most circumstances the arthropod doesn't have what the parasite needs to replicate, survive, and/or be transmitted. Alternatively, the pathogen might elicit such a strong reaction from the arthropod that it is destroyed. In either case this terminates the interaction. Rarely, possibly accidentally, interactions give rise to a productive, non-hostile initial interaction between the parasite and arthropod and the processes of parasite adaptation to that arthropod host are allowed to proceed.

In the second model, hematophagous arthropod lineages evolve from non-blood-feeding ancestral lineages. We know that many protozoa, rickettsia, viruses, trypanosomes, and bacteria (e.g. the mycoplasmas) are frequently found in non-hematophagous arthropods. When these lineages subsequently gave rise to hematophagous arthropods, the parasite was already adapted to the arthropod host. In these cases, parasites became transmitted to the vertebrate host and may, in rare cases, have caused pathogenic effects in the vertebrate host. This may be the case with many new and emerging pathogens (see for example Raoult *et al.* 2001).

Acknowledgements

NIH Grants U01AI45430 and R01-AI49256 supported this research.

Part II

CHAPTER 6

The emergence and co-evolution of human pathogens

Jake Baum and Gila Kahila Bar-Gal

Nature (is) that lovely lady to whom we owe polio, leprosy, smallpox, syphilis, tuberculosis, cancer. (Attributed to Dr. Stanley N. Cohen)

6.1 Introduction

The great diversity of life that we see today has evolved through a complex history of selective interactions between organisms competing with each other for limited resources and striving to survive changes in the environment. Whilst our vision of this diversity is dominated by what we see with the naked eye, we should not forget the microscopic pathogens that plague everything from the single-celled prokaryote to multicellular plants and animals. As a group they are unsurpassed in the breadth of their life strategies and the biological niches that they occupy. Indeed even if we restrict our discussion to human pathogens, the consortium is still impressive, comprising a heterogeneous array of viruses, bacteria, fungi, protozoa, and metozoa, all living temporarily or permanently within our bodies, on our outer surfaces, in our body fluids, in our tissues, and cells and even in our genomes.

In this chapter we aim to introduce two key aspects relating to the diversity of human pathogens and disease. First, the origin of the breadth of human pathogens, focusing on the diversity that has arisen through cross-species transmission from other organisms, and second, the subsequent co-evolution of these pathogens with their human hosts. Some of these pathogens have probably been with us since before the last common ancestor of humans and chimpanzees, others are relatively new arrivals under continuing selective pressure to adapt to their new human host.

6.2 The emergence of the diversity of human pathogens

A number of pathogens are likely to have been with us throughout our recent evolutionary history from mammal to primate to lone bipedal ape. However, a pronounced change occurred in the makeup of human disease some 8000–10 000 years ago as humans made the transition from hunter-gatherers to settled small agricultural communities (Diamond 1997). The shift in mode of subsistence was the culmination of a sweep of radical behavioral changes. Humans had begun to domesticate animals as pets and livestock, to cultivate plants for food and to dramatically transform their environment. It may not have happened all at once, nor were the events necessarily dependent on each other (Diamond 1997); however, the changes to the local environment, the changes in the makeup of these early societies, and the altered diets each had dramatic effects on the type and diversity of human diseases. Staying still meant that human waste had to be dealt with, and where it accumulated the indirect contamination of water sources or the local environment with infectious organisms opened the door to transmission of a vast array of new pathogens, from the bacterial (e.g. *Vibrio cholerae*) to the parasitic worms (e.g. filarial and hydatid tapeworms). The close contact of humans with newly domesticated pets and livestock increased the chances for animal–human disease transmission (zoonosis). For example, flu from pigs, tuberculosis from cattle, and pertussis from dogs may all find their origins during this time. The increases in population size allowed diseases to spread rapidly, forming the so-called 'crowd diseases' (Diamond 1997), such as measles or smallpox, whose success depends directly on the size of the population. The disruption of the environment opened up breeding sites for vector-borne diseases such as malaria (Livingstone 1958), yellow fever, and leishmaniasis (Beran and Steele 1994).

This explosion in infectious disease diversity clearly marked a major transitional stage for the evolutionary forces acting on human populations. As we mastered our own local environment, infectious diseases became the main source of natural selection facing human populations (Haldane 1949), and even today they continue to be a major challenge to our survival and fitness. Moreover, as changes in our biology and behavior have led to the emergence of new diseases, these same diseases have played a significant role in shaping our societies, demography, and genetic makeup. And this is not a purely historical phenomenon.

As we enter an era where the optimism of the antibiotic age is slowly being replaced with a new-found respect for our long-term adversaries it is important to learn from the history of the human–pathogen relationship and to understand the evolutionary process that ties us together. The effect of this close relationship has been a mutual molding of genomes: the human host better adapted to protect itself from the onslaught of the ever-changing diversity of pathogens and the pathogen constantly adapting new methods of outwitting the vigilant host. This process of adaptation and counter-adaptation is termed co-evolution.

6.3 Co-evolution of humans and their pathogens

The word co-evolution is widely used to mean subtly different processes. The term was first coined to express the evolutionary relationship existing between plants and the insects that fed on them, often restricted to cases in which the interactions were beneficial to both species such as with pollinating insects and flowers. Then there are the broad ecological definitions of co-evolution, where

it reflects the fact that all organisms have trophic relationships with other organisms and therefore exert reciprocal evolutionary pressure on each other (the zebra's strips or the mollusk's hard shell against predators). Alternatively co-evolution may result from the very intimate interactions between specific pairs of species such as we find with host–pathogen associations (Kraaijeveld *et al.* 1998).

Here we will restrict our discussion to this last category of host–pathogen co-evolution, in particular to that of humans and their pathogens, although even here there are different levels at which co-evolution can be observed depending on the levels of dependency of the two species (Conway and Roper 2000). There are those pathogens that we rarely encounter, which once established do not transmit between individuals, for example, the rabies and hantaviruses that sporadically infect humans from animal reservoirs. Then there is the range of pathogens that can be transmitted within human populations. This includes a broad spectrum of pathogens, from those that are largely zoonotic in origin and which rarely infect human populations (e.g. the filoviruses, or trypanosomes), to those causing diseases that are exclusively endemic to human populations (such as *Mycobacterium tuberculosis*, *Plasmodium falciparum*, hepatitis B virus, and HIV-1 and 2 viruses). Finally there are those that have blurred the host–pathogen boundary altogether, living (usually) as partners within our bodies, in a more stable relationship, sometimes of mutual gain (e.g. *Escherichia coli* and the other commensals that inhabit the gastrointestinal tract) and sometimes less well-tolerated (e.g. the herpes simplex virus and *Helicobacter pylori*). The breadth of these associations and their complexity causes much of the confusion with defining host–pathogen co-evolution.

In its broad sense, co-evolution is inevitable since no organism exists in complete isolation and any changes in behavior, ecology, or biology of one organism will have direct and indirect consequences for others sharing the same environment. However, we will highlight four general categories where interactions may have co-evolutionary consequences.

Co-speciation between host and pathogen is perhaps the most fundamental interaction of these categories given that any host-dependent pathogen is likely to undergo speciation as a direct result of host speciation. This is particularly likely when populations become isolated on their way to speciation (allopatry) with pathogens mirroring host population separation. Pathogens are also likely to co-evolve with their host when the host solves a major adaptive problem enabling it to enter a new ecological niche. Any pathogen that can adapt to the new niche created by these changes will also benefit, resulting in apparent co-evolution. To the puritan, however, both of these two categories are governed by non-specific interactions where an opportunistic pathogen has either jumped host or hitched a ride with an adapting or speciating host. As such they are often excluded from the strict definition of co-evolution, although their consequences for evolution may be profound.

A third category of host–pathogen co-evolution is more specific, with changes in the host having a direct selective effect on the pathogen or vice versa. This type of relationship is exemplified by interactions typically seen in predator–prey conflicts. If the predator (here the pathogen) evolves a more successful adaptation for infection, the prey (here the host) must counter-adapt for its survival. This is akin to an arms race with each trying to outdo the other either in cycles equivalent to a game of scissors-paper-stone (where there can be no real winner) or in more antagonistic relationships escalating to all-out biological war. Some of these general interactions will involve multiple genetic loci in both organisms but there may also be cases of co-evolution where changes will arise in a single host/pathogen gene (Fig. 6.1). This type of interaction has been the hardest to demonstrate clearly in animal systems [the plant literature has a few good examples (Thompson and Burdon 1992; Mitchell-Olds and Bergelson 2000)]. Here, we will discuss each of these catergories separately.

6.4 Co-evolution by co-speciation: where the pathogen is isolated with the speciating host

Any disease that is intimately associated with its host, and shows strong host specificity, will be highly susceptible to co-speciation and co-evolution when host populations diverge on the road towards speciation. This host-dependent evolution is the basis for Fahrenholz's rule (Fahrenholz 1913), which predicts that the common ancestors of present-day pathogens are expected to have been the pathogens of the common ancestors of present-day hosts. Examples in the literature are few, especially with human pathogens. The small DNA viruses are one of the best examples and appear to agree with Fahrenholz's rule (Soeda *et al.* 1980; Shadan and Villarreal 1993). These viruses (from the Polyomaviridae, Papillomaviridae, and Parvoviridae) are generally of little consequence during infection and are carried by the host for life showing little evidence of disease. They are often highly species specific in their replication, making them particularly susceptible to host speciation. In the case of mammalian polyomaviruses a phylogenetic tree for the respective mouse, hamster, monkey, and human viruses show strong similarities in structure (congruence) with phylogenetic trees for the host species suggesting co-speciation of host and pathogen (Shadan and Villarreal 1993).

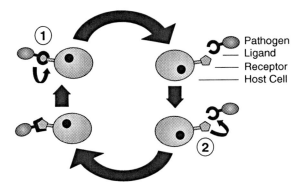

Figure 6.1 Host–pathogen gene for gene co-evolution. 1, Selection by parasite on host to develop strategies to counter parasite binding, and 2, Selection by host on parasite to overcome binding inhibition or change infection strategy.

Similarly the relationship among human and simian paroviruses matches the topology of the phylogeny for the host organisms based on mitochondrial DNA suggesting host-dependent evolution (Lukashov and Goudsmit 2001).

The primate lentiviruses (most pertinently the simian immunodeficiency viruses, or SIVs) also show evidence for co-evolution by co-speciation. The primate SIVs, although asymptomatic, are closely related to the human immunodeficiency viruses (HIV-1 and HIV-2) in both their genetic and structural makeup. As a group, all the viruses from any one (non-human) primate species are generally much more closely related to one another than to viruses from another species (Hahn *et al.* 2000). This close phylogenetic relationship implies that they have been infecting their respective primate hosts for a relatively long period of time, undergoing host-dependent evolution with the divergence of viral lineages reflecting, and most likely caused by, the divergence of host lineages (Fig. 6.2). For example, each of the four species of African green monkey, which reside in non-overlapping geographical locations, is infected with its own different SIV. Viruses from vervet monkeys, whether east or south African in origin, are more closely related to

each other than they are to viruses from grivet and tantalus monkeys from nearby sites in east Africa (Hahn *et al.* 2000), and each species specific virus is related to the other in a manner that matches the proposed radiation of the group (Fig. 6.2). This is a good indication that the host-specific relationship follows the evolution of the respective groups and not the species geographical distribution.

However, it is also clear that the transmission of SIVs across primate groups does not always follow speciation. Cross-species transmission can occur between groups and may explain why the Mandrill lentivirus (SIVmn) belongs in a sister clade to sun-tailed and L'Hoest monkey SIVs, which is at odds with what is known about the three species' evolutionary relationships (Fig. 6.2).

Data that are apparently in agreement with Farenholz's rule may result from the propensity of pathogens to jump into new species closely related to their old host, a process that is likely to require relatively minor adaptive changes. With humans this is most dramatically exemplified by the close relationships between human HIV viruses and other primate lentiviruses. The evidence suggests that HIV-1 originated from a zoonosis with a chimpanzee SIV well within the last 100 years [probably

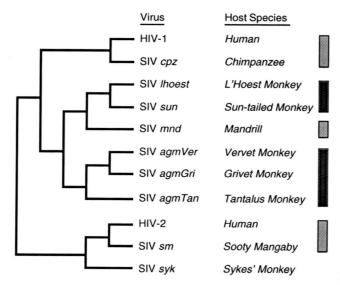

Virus	Host Species
HIV-1	Human
SIV *cpz*	Chimpanzee
SIV *lhoest*	L'Hoest Monkey
SIV *sun*	Sun-tailed Monkey
SIV *mnd*	Mandrill
SIV *agmVer*	Vervet Monkey
SIV *agmGri*	Grivet Monkey
SIV *agmTan*	Tantalus Monkey
HIV-2	Human
SIV *sm*	Sooty Mangaby
SIV *syk*	Sykes' Monkey

▨ Evidence for Origin by Cross-Species Transmission
■ Evidence for Host-Dependent Evolution

Figure 6.2 Phylogeny of primate lentiviruses (adapted from Hahn *et al.* 2000), showing both the host-dependent evolution (co-speciation) and cross-species transmission of viruses between groups. Branches are not drawn to scale.

in the 1920–30s (Korber *et al.* 2000)], not at the time of the chimpanzee–human split some 5–7 million years ago. Whilst it may be difficult to date the transmission event [since expansion could have predated or post-dated the zoonosis (Korber *et al.* 2000)] the data point to a recent origin and not co-speciation (Fig. 6.2). The close relationship of HIV-2 with SIV in the sooty mangabey provides another example of a cross-species transmission rather than co-evolution.

Whilst co-speciation may not be discernible for many human diseases, given the difficulties in tracing our associations with pathogens over long periods of time, local co-evolution of pathogens with geographically distinct human populations can be demonstrated and, on a micro-scale, is perhaps the first step in the long process of co-evolution by co-speciation. *H. pylori*, for example, shows some evidence for local co-evolution with host populations (Covacci *et al.* 1999). *H. pylori* is a Gram-negative bacteria that chronically infects nearly half of the world's population causing a spectrum of clinical outcomes ranging from asymptomatic inflammation of the stomach to gastritis, peptic ulcers, or gastric cancer. Given that its transmission is usually via the family unit, with children infected with the same strain as their parents, and the long permanence of each infection, *H. pylori* is particularly susceptible to host-dependent evolution. In the long-term it would be expected to show regional differentiation across different continents, and the genetic geography of *H. pylori* does indeed appear to coincide with that described in general for human populations. As such it is likely that *H. pylori* infection was already established in human stomachs before the beginnings of human migration out of Africa, and has been co-evolving with distinct human populations ever since (Covacci *et al.* 1999; Karlsson 2000), the first step on the road of co-evolution by co-speciation.

6.5 Co-evolution by adaptation to new niches: where the pathogen follows a host's adaptive changes

As mentioned before, the dramatic cultural and behavioral changes in human society during the last 10 millennia have been key determinants of human disease (Fig. 6.3). The shift from small nomadic populations to larger settled populations and the parallel changes in diet from a hunter-gatherer-based subsistence to the beginnings of agriculture affected both the plants and animals that were being domesticated and the human populations. For the domesticated crops and livestock, sedentism and its associated changes meant a radically altered environment, and selection for many new adaptations (Zohary *et al.* 1998). The new environment, the close contact with man, and the selective pressures caused major changes among plants and animals. The domestication of cattle, for example, led to dramatic changes in the longevity, herd sex ratio, reproductive biology (including changes in sexual maturation and fertility), and basic morphology and behavior of the animal (Hemmer 1990; Zohary *et al.* 1998), resulting in major differences between the domestic populations and the ancestral wild population.

These effects on crops and livestock are not surprising given the gamut of changes taking place in features such as population size, density and movement, degree of sedentism, settlement size, dietary staples and food preparation, trade systems, warfare, and population movements (Smith and Kolska-Horwitz 1998).

The effect of the neolithic revolution (this lifestyle transition) on human health status, was probably no less striking, particularly with respect to exposure to new diseases. The destruction of the local environment, with the clearing of land for agriculture and living space, profoundly changed the local ecology. Livingstone (1958) is credited with being the first major proponent of the idea that malaria in West Africa became hyperendemic following the deforestation of land for agriculture. According to this scenario the disruption of the environment originally by slash-and-burn farming and later the development of agriculture, would have led to an increase in the potential number of breeding places for the *Anopheles gambiae* mosquitoes (the most anthropophilic mosquito) and other vectors of the malaria parasites (and presumably other vector-borne diseases). The idea makes intuitive sense given that it is difficult to see how a disease like malaria could reach endemic proportions and successfully spread within the restricted population

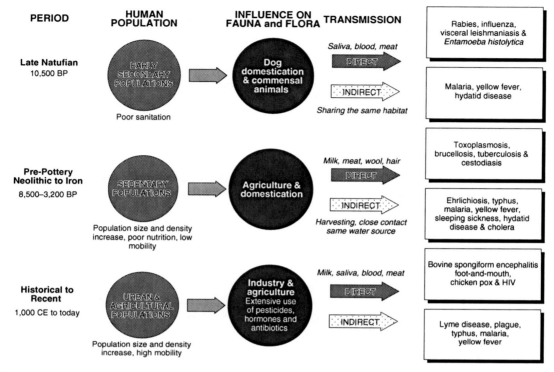

Figure 6.3 Scheme showing how cultural and behavioural changes in human society during the last ten millennia have been key factors influencing human disease.

sizes otherwise. Some support for this idea comes from the relationship between agricultural subsistence in East and West African communities and the prevalence of the sickle cell mutation in the respective populations (Wiesenfeld 1967). The sickle cell mutation, HbS (a valine to glutamic acid substitution in the β-globin chain of hemoglobin), is a deleterious mutation that is maintained at high frequencies in populations because of its ability to protect the heterozygote (the HbS carrier) from falciparum malaria (Allison 1954). Therefore, its presence should be indicative of the previous endemicity of malaria. As predicted there is good correlation between the dependence of the community on agriculture and an increased frequency of the HbS variant allele, demonstrating, somewhat indirectly, evidence for the influence of the local environmental factors on disease prevalence.

More recent work on the genetics of the three organisms involved (man, *Plasmodium falciparum*, and mosquito) further strengthens this proposed

link between agriculture, human and mosquito population expansion, and the subsequent increase in malaria prevalence. There seems to be good agreement in the time at which the most recent common genetic ancestors of the three converge. *Anopheles* mosquito populations [based on mtDNA, mircosatellites, and chromosomal variation (Coluzzi 1999)], *P. falciparum* populations [mtDNA, synonymous sites, and intron genetic variation (Volkman *et al.* 2001)] and genetic variation in the most common human enzymopathy (also thought to be maintained by malaria selection), glucose-6-phosphate-dehydrogenase, G6PD (Tishkoff *et al.* 2001), all indicate an origin of population expansion broadly placed within the last 10 000 years. Whilst the extremes of this range (some 3000–20 000 years) are difficult to reconcile with the archeological data or with what is understood about malaria transmission, respectively, the dates fit generally around the time frame of subsistence changes in human populations as proposed by Livingstone (1958).

More data will certainly help to clarify this point in the future.

Other diseases are also likely to have originated during this period. For example, given the very close genetic similarity between the agent of human tuberculosis (TB), *Mycobacterium tuberculosis*, and the closely related cattle pathogen *Mycobacterium bovis*, it is likely that the disease arose from the close association of humans with newly domesticated cattle. This idea is given some support from genetic data that place the common ancestor of the two to within the last 8000–10 000 years. TB currently infects one-third of the world's population and more than 3 million people die annually from it, making the bacterium the leading cause of death due to an infectious agent worldwide. Another possible demonstration of the potential evolutionary power that human innovations have had on our long-term burden of disease.

In evolutionary terms, TB and malaria along with other diseases thought to have arisen at this time [such as plague (Achtman *et al.* 1999) and possibly some mosquito-borne flaviviruses (Zanotto *et al.* 1996)] are relatively young pathogens. Other diseases are so recent that their emergence is within living memory, and yet are still the consequences of the changing impact of human behavior on the environment. HIV serves as a model example of how modern behavior has further added fuel to the speed at which diseases can overtake human populations. Whilst the exact causes and timing of the origin of the current HIV pandemic are somewhat contentious (Goudsmit and Lukashov 1999; Hooper 1999; Korber *et al.* 2000; Blancou *et al.* 2001; Poinar *et al.* 2001), the spread of the virus has certainly been aided by changes in behavior in modern human populations. HIV-1 and HIV-2 both represent zoonotic transmission from two different sources, namely chimpanzees (SIVcpz) (Gao *et al.* 1999) and sooty mangabys (SIVsm) (Gao *et al.* 1992), respectively. Transmission probably followed direct exposure of humans to animal blood or mucosal secretions, likely to arise during hunting or at butchering sites. In addition, other activities such as consumption of uncooked contaminated meat, together would appear to provide the simplest and most plausible explanation for the cross-species transmission of SIV into the new human host

(Hahn *et al.* 2000). The main alternative, and more sinister hypothesis for the origin for HIV-1, the result of the contamination of oral polio vaccines produced in the Belgian Congo (Hooper 1999), would appear to have been ruled out given the virus's recent origin (Korber *et al.* 2000) and a failure to detect chimpanzee DNA in the vaccine stocks (Blancou *et al.* 2001; Poinar *et al.* 2001). However, transmission alone is not enough to make a virus a global health threat. With HIV in particular the increase in regional travel, the increase in the average number of sexual partners, the use of intravenous drugs, the failures of the blood donation service to screen for the virus, and other consequences of modernity have all played a part in establishing HIV as a major emergent disease in the global human population.

Malaria, TB, and HIV highlight a repeating pattern of human behavioral change, environmental disruption, and the emergence of new, or re-emergence of old, diseases. Nature is opportunistic, so whenever a new niche is created, emergent and old diseases will inevitably seek to exploit the new opportunities. The fact that pathogens are able to maximize their potential very quickly following the opening of a new niche clearly serves as a warning of the power of humans as a selective force shaping disease diversity (Palumbi 2001).

6.6 Co-evolution and virulence

Although it is not a general rule, newly emergent pathogens like those that arise when a new niche is opened up are often very potent [as exemplified by the frightening virulence of some filoviruses and hantaviruses (Garrett 1994)], since their new host has had little time to adapt defense mechanisms to protect it from disease. Two classic examples will help to illustrate this.

The introduction of the myxoma virus to control European rabbit populations in Australia (May and Anderson 1990), whilst not a human pathogen, has become the gold standard in host-parasite co-evolution and deserves a mention here. When first introduced in the 1950s the virus caused 99% mortality in the rabbit population. However, within a relatively short period of time viruses isolated from the wild population were markedly less severe when infecting laboratory rabbits known to be highly susceptible to the virus. Clearly the virus had become less virulent. Furthermore, Australian rabbits infected with known virulent strains had acquired some form of resistance. Here the arms race between the rabbit and the virus reached a stable equilibrium with both evolving to tolerate the other, the virus maximizing its transmission potential by decreasing its virulence (virulent strains kill the host too quickly to be transmitted) and the rabbit tolerating the mild infection.

A second powerful demonstration of the potential potency of a new infection is seen with the introduction of established Old World pathogens (such as smallpox, measles, influenza, and typhus) to the New World by Columbus and other invaders following the arrival of Europeans in the Americas in 1492 (Desowitz 1997; Diamond 1997). The long-term association of the European populations with their particular pathogens had given ample time for host populations to develop immunity and evolve genetic means of protection from the various disease effects. However, when these same pathogens were introduced to the naive Native American populations, the consequences were devastating. Many more indigenous peoples died from the invading Eurasian diseases than died as casualties associated with battles against Spanish and Portuguese soldiers (Diamond 1997). This shows clearly how powerful the consequences of regional co-evolution can be when pathogens that have a long-term association with one host population are introduced to a new population that has never seen them before.

Tracing the long-term trends between humans and their pathogens is obviously more difficult than the example of the rabbit and the myxoma virus given our long generation times and the ethical difficulties of working experimentally with humans. As few such diseases have been followed over long enough time periods to fully understand the outcome of human–pathogen relationships, most of the work on them is predominantly theoretical or historical. Some devastating pathogens are certainly less virulent now [trends in the decreasing severity of symptoms associated with syphilis are a good demonstration of this (Diamond 1997)]. However, it would be wrong to conclude

that the end result of a long-term association between disease and host is necessarily the evolution of a more benign interaction (May and Anderson 1990; Ewald 1994; Levin 1996). On the contrary, pathogens will simply continue to evolve to maximize their transmission. If severity decreases the likelihood of transmission, for example, by reducing host mobility, then less virulent pathogens will certainly be favored. If virulence actually aids transmission, such as the extreme behavioral changes that are caused by rabies, the dramatic lesions of smallpox, or the extreme fluid release of certain diarrheal diseases, then the more virulent pathogens will clearly out-compete the others.

Clearly the development of new techniques with which to study pathogens from our distant past, most pertinently from the archeological record (discussed at length elsewhere in this book), will go some way towards filling in the gaps in our knowledge of the evolution of virulence.

6.7 Co-evolution of humans and their diseases: multigenic co-evolution between host and pathogen

So far we have dealt with the broad interactions between humans (in their behavior and biology) and the environment that have lead to the emergence and subsequent co-evolution of humans with their diseases. Now we will shift the focus to the more specific interactions between host and pathogen at the molecular level.

Upon exposure to a pathogen the principle human defense is the immune system, which through a complex process of genetic recombination and splicing generates a huge diversity of antibodies and T-cell receptors with which to target a potentially infinite number of pathogens. Like a good co-evolutionary system the pathogens also generate diversity, thus avoiding immune detection. This ranges from random mutation in genes encoding highly variable surface proteins, such as with influenza or HIVs, to controlled variability in surface antigens of trypanosomes (Rudenko et al. 1998), directly changing in the presence of immune selection. This latter generation of variability is not in itself co-evolution, since the variation generated *de novo* is not heritable; however, the mechanism has clearly arisen from immune selection pressure.

In the long-term, given adequate selective force, mutations may occur that offer the host permanent protection. The existence of a growing catalogue of human polymorphisms, putatively linked to protection from various diseases, suggests that such mutations played a key role in the survival of human populations from the onslaught of infectious disease throughout history.

Some genetic polymorphisms, for example those in the human major histocompatibility complex (MHC) have probably been selected since before the human–chimpanzee split (Klein 1987). Other protective mutations may be thousands of years old, including deleterious mutations that persist in the population because of the protective benefits they give their carriers from the effects of widespread diseases. Among these are the diverse family of hemoglobin mutations and other polymorphisms, which are often lethal in the homozygous state but protect heterozygous carriers from malaria. These mutations probably date from the time of origin of the disease (Livingstone 1989; Nagel 1994; Tishkoff *et al.* 2001). Other protective polymorphisms may date from recent history within the past two millennia. For example Tay–Sachs, a genetic disease found at unusually high frequencies in Ashkenazi Jewish populations, may have reached its high incidence in the population as a direct consequence of disease selection in the impoverished ghettos that the community was forced to live in during the middle ages (O'Brien 1991). These conditions were ideal for the spread of a number of diseases such as TB and cholera, which place intense selection pressures on a population. Cystic fibrosis, the most frequent genetic mutation found in Caucasian populations (affecting 1 in 2000–4000 individuals, with a carrier frequency of around 1 in 30; Morral *et al.* 1994), may also be explained by its association with protection from infectious diseases. There are suggestions that it may give carriers a protective advantage against diarrheal diseases like cholera and *Escherichia coli*, although the age of the mutation indicates a very long evolutionary history predating many of the social environments associated with these diseases (Morral *et al.* 1994). Still others, possibly the result of some previous selective pressure may in the future become important protective mutations to some contemporary human diseases—the mutation in the CCR5 T-cell surface receptor for example. Although the selective agent that led to its elevated frequencies in certain populations is unknown (Stephens *et al.* 1998), the protection it offers carriers against HIV invasion is already a significant survival factor (Dean *et al.* 1996) and may be selected for in the future.

Of all the diseases malaria is certainly the most robust example of an infectious disease that has consistently selected for mutations in the human genetic makeup that offer protection. Why this is so is not so clear. It may simply be an accident of the ease of detecting malaria adaptive polymorphisms since malaria tends to be found under specific conditions of climate, altitude, and geography (often determined by the distribution of the mosquito) that lends itself to epidemiological studies (Lederberg 1999). But it is also likely that as an infectious disease malaria has been the single greatest cause of mortality and morbidity in human history (Haldane 1948). Thus the wealth of human adaptive polymorphisms is perhaps not surprising. Even today malaria remains a serious problem and either alone or in combination with other diseases, is estimated to kill between 1.1 and 2.7 million people worldwide each year, with over 2400 million globally at risk (WHO 2000). As such it continues to remain a potent selective force. The diversity of mutations that have been (putatively) selected by long-term malaria morbidity and mortality are covered extensively in the literature. Examples include G6PD (Ruwende *et al.* 1995), HbS (Allison 1954; Hill *et al.* 1991), HbC (Agarwal *et al.* 2000; Modiano *et al.* 2001), ovalocytosis (Genton *et al.* 1995), the alpha and beta thalassaemias (Flint *et al.* 1986), all of which offer some degree of protection from *P. falciparum*, the most virulent human malaria agent. The Duffy negative mutation also appears to have been recently selected (Hamblin and Di Rienzo 2000; Hamblin *et al.* 2002), perhaps by another human malaria agent, *Plasmodium vivax*. Homozygotes for Duffy negativity are completely resistant to infection by this malaria, suggesting a likely mechanism that led to the mutation's fixation in West Africa. These mutations clearly demonstrate the selective power that pathogens have placed on human populations and the dramatic way that they have shaped our genetic makeup.

6.8 Gene-for-gene co-evolution, the holy grail of host and pathogen co-evolution

Our discussion of the genetic interaction between host and pathogen has so far focused on the non-specific and multi-genetic interactions between the two, with the host evolving multiple defense mechanisms in an intimate arms race with the parasite's generation of variability avoiding immune detection. However, the holy grail of co-evolutionary studies has been to show the co-evolution of discrete and specific host and pathogen genetic loci. Unfortunately, as will imminently be clear, few studies have unquestionably shown 'gene-for-gene co-evolution' (Fig. 6.1).

Various studies have sought to investigate whether the individual host genotypes influence the infecting parasite genotype, linking parasite gene with host gene. Although malaria is the best studied of human–pathogen interactions, few loci have given promising results. For example, no association was found between the human MN blood group and genetic differentiation in the malaria parasite receptor known to bind to it (Binks et al. 2001). Conflicting associations have been found between the human sickle cell trait and variation in malaria parasite surface antigens, highlighting the dangers of trying to correlate genetic markers between host and parasite when there is no obvious functional hypothesis tying the two together. More promising has been the association between HLA class I restricted cytotoxic T-lymphocyte (CTL) response against the malaria parasite, and epitopes from liver stage parasites in a Gambian population (Gilbert et al. 1998). This demonstrates the possibility that host HLA type may influence the genotype of the malaria parasite (as defined by these particular liver stage epitopes) found in the host, although whether this association holds true across populations is not yet clear. Detailed investigations using monkey models have been more rewarding because of the potential to do controlled experiments with known primate MHC types and known virus types, something that is impossible in humans. Such studies have shown very clearly

how primate virus-specific CTL responses (influenced by the host genotype) can actively influence selection on and therefore the evolution of SIV CTL epitopes (parasite genotype) (Allen et al. 2000; Evans et al. 2000), a very clear demonstration of gene-for-gene evolution (though not co-evolution).

Clearly proof of gene-for-gene co-evolution in humans remains elusive. With a better understanding of the precise mechanisms of disease and the molecular interactions between human host and pathogens, we will be better placed to find evidence of its occurrence.

6.9 Concluding remarks: what can the study of co-evolution tell us about the future of disease?

The introduction of the large-scale use of medical treatments, both for our own diseases and for those of our domesticated livestock and crops, is probably the most significant modern factor affecting the current and future development of human disease. Domestic livestock in particular, such as cattle, goats, sheep, and chickens given their economic importance in our society, are frequently given drugs to control infection and to increase their yields. This development has dramatically changed animal exposure and susceptibly to natural pathogens, making them more vulnerable to diseases, and has placed increasing selection on their pathogens to evolve resistance.

The same can also be said of our own use of medical treatments. When antibiotics first came into common use in the 1940s it was perhaps not surprising that many felt it heralded the end of infectious disease for the human race. At the beginning of the antibiotic era virtually all Gram-positive bacterial infections were susceptible to penicillin (Garrett 1994); however, today, particularly in hospitals, the vast majority of infections (e.g. *Staphylococcus aureus*) are penicillin resistant, and large numbers are resistant to alternative drugs like methicillin (Palumbi 2001). Replacement drugs are often more expensive and medicine is rapidly running out of options. Penicillin introduced in the 1940s was soon failing and required a second antibiotic, methicillin, in the 1960s to overcome resistance. Later in the 1980s vancomycin 'the drug of last resort' (Garrett 1994) was needed to address methicillin resistance. Now linezoid and quinupristin–dalfopristin are being mobilized to tackle vancomycin resistance (Palumbi 2001). The emergence of resistance is likely to go on, and antibiotics are certainly going to be harder to find and more costly, unless we can stop the causes that lead to the emergence of resistance. The primary factor is certainly the inefficient use of our front-line drugs. The administration of suboptimal doses, the widespread failure to complete courses, and the inappropriate prescription of antibiotics for viral illness, all create ideal environments for the evolution of resistant bacteria. Over the last 50 years the number of species and strains of pathogenic and commensal bacteria resistant to their use has increased at a steady rate worldwide. For example, *Haemophilus influenzae*, one of the most common causes of respiratory infection has, during the past three decades, not only become resistant to penicillin but more recently multidrug resistant (MDR) to a host of antimicrobial agents such as ampicillin and amoxicillin. The emergence of MDR *M. tuberculosis* is another major threat, given the increasing burden that this disease lays on global populations. Furthermore, resistance can spread even in the absence of selection because bacteria are able to exchange genetic material between species, transmitting pathogenicity islands and resistance mutations (Levy 1994).

This problem is not restricted to bacteria. Malaria is also increasingly resistant to the former front-line drugs. Chloroquine, once the champion of malaria control, is now all but useless in many parts of the world following its widespread and often over-zealous usage (Payne 1988; Desowitz 1991; Newton and White 1999). Other drugs, such as mefloquine and the antifolate drugs (Newton and White 1999), are also increasingly showing signs of failing. With resistance to many insecticides prolific in the mosquito populations as well, malaria is resurgent as a leading cause of both infant mortality and population morbidity (Greenwood and Mutabingwa 2002). Influenza on its annual winter visit to temperate climates requires a reformulated vaccine each year as we run to keep apace of its evolution. HIV continually evolves away from the host immune system, so much so that many drugs are rapidly of little use in treating the infection (Palumbi 2001).

Understanding the selection pressures on diseases that lead to widespread resistance will be key factors in helping us to design long-term strategies for the control of disease. Furthermore, as we understand the way pathogens evolve we will be better equipped to judge what will happen in the future, not only for drug resistance but also for the emergence of, and virulence of pathogens that we have not seen or pathogens that we assumed were once beaten.

Predicting the evolution of diseases is increasingly a focus of attention (May and Anderson 1990; Ewald 1994; Levin 1996; Lipsitch 2001). Indeed by understanding the details of their evolution, such as the correlation between transmission and disease severity, it may be possible to coerce a pathogen to become more benign (Ewald 1994, 1998, 2000). For example effective sanitation programs may ultimately lead to the evolution of more benign water-borne pathogens (Ewald 2000) because reduced efficiency of transmission requires more extended association with the host, therefore selecting for a reduction in virulence of the pathogen. Similarly by reducing the degree to which vector-borne pathogens can be transmitted between immobile hosts we may similarly reduce the virulence of disease (Ewald 1998).

We can certainly learn lessons from the way pathogens have arisen in the past and relate them to possible future disease emergence and re-emergence. Diseases are especially likely to arise when new niches are created or when protective barriers are removed. The repeating specter of habitat destruction and the emergence of disease throughout human history must serve as a severe warning to the global community today. The increasing numbers of immunocompromised individuals, both immunosuppressed in transplant patients and immunodeficient in the expanding HIV-positive population, provides a very welcoming niche to diseases that could not previously get a foothold in the population. Xenotransplantation (Weiss 2000), biological warfare, and the explosion in global travel all serve to increase the exposure of human populations to new pathogens. These and other changes that mankind has introduced to the world are probably the most dramatic influences that serve to select for newly emergent diseases. 'Nothing in biology makes sense except in the light of evolution' (Dobzhansky 1973). Human disease is no exception, therefore like good students we must learn from evolutionary history, understand the dominant factors that led to the establishment of mankind's most formidable foes, its pathogens, and better understand ways of predicting the development and future evolution of our established diseases.

Acknowledgements

We are very grateful to B. Baum, S. J. O'Brien, and J. Slattery for helpful insights, editorial comments, and suggestions. We would especially like to thank Professor C. Greenblatt for all of his enthusiasm in the work, his encouragement in getting both of us interested in disease emergence and evolution, and for facilitating such an exciting project. J.B. is funded by a Wellcome Trust Research Studentship in Biodiversity. G.K. is funded by the Fogerty Foundation.

CHAPTER 7

The state and future of paleoepidemiology

Mark Nathan Cohen and Gillian Crane-Kramer

The most adaptable and therefore most widely distributed of today's large land animals are human beings, and this has been true of the members of the species Homo sapiens *and their hominid predecessors for a very long time—long from their point of view. Other creatures had to wait for specific genetic changes to enable them to migrate into areas radically different from those of their ancestors—had to wait for their incisors to lengthen into daggers before they could compete successfully with hyenas on the veldt, or had to wait for hair to thicken into fur before they could live in the north—but not humans nor hominids. They made not a specific but rather a generalized genetic change: they developed bigger and better brains wired for the use of language and for manipulation of tools.*

That growth of nerve tissue crammed into the treasure box of the skull began several million years ago, and as it did, the hominid became increasingly capable of 'culture'.

(Alfred W. Crosby, Ecological Imperialism, The Biological Expansion of Europe, 900–1900, 1986)

The study of prehistoric pathogens and the evolution of general health (Cohen 1989) calls attention to a long-term pattern of interaction between human behavior and pathogens which demonstrates the profound importance of human social and ecological change on disease organisms.

7.1 How we know

Contemporary knowledge of prehistoric and historic paleo-epidemiology derives from three very different types of evidence:

- the biological characteristics of known disease organisms and parasite cycles, which can be projected into the past ('retrodicted') using uniformitarian principles of natural law (i.e. the assumption that the laws of biology do not change in a serendipitous manner);
- ethnographic or contemporary observations of the presence and distribution of diseases among both human and non-human primate populations;
- evidence from the skeletons of prehistoric human populations.

Research in each area has its own problems and distortions. Uniformitarian assumptions about the unchanging inherent logic and stability of nature tend to minimize the fact that the behavior of disease organisms can change and that organisms can and have evolve(d). And it ignores the fact that the knowledge of the present from which we 'retrodict' the past is often itself incomplete or simply wrong.

The study of disease in ethnographic human populations thought to mirror patterns in prehistory (i.e. small mobile groups of hunter-gatherers, small-scale farmers, etc.) suffers from the often limited nature of the medical studies that have been done, the small number of such populations, and the fact that few if any of such extant populations are actually independent of modern civilization. We have to allow for the probability that their health patterns may have been affected by modern alterations in ecology and by secondary reintroduction of modern diseases. For example, most such societies show signs of having suffered virulent epidemic diseases or other density-dependent diseases usually considered diseases of civilization.

The study of diseases of non-human primates suffers from more limits than the study of human diseases. Their diseases are of course much less well-known than those of people. Perhaps more important, it has proved difficult in many cases to establish whether particular diseases are natural infection of primate populations and native to their environments or have been introduced to them secondarily by people or their domestic animals. Many primate diseases appear (or are observed) for the first time among captive populations, but it is not always clear whether the diseases are newly introduced by people or carried into captivity by the animals themselves. Equally puzzling is whether the contemporary distributions of diseases, even those that seem endemic in nature, reflect historical distributions or have been expanded by human activities, travel, and/or the transportation of domestic animals.

The use of skeletal studies suffers from the fact that relatively few diseases leave a mark on the skeleton. Of those that do, most leave marks on the skeleton only in a small percentage of the individuals actually infected, and very few leave marks on the skeleton distinctive enough to permit differential diagnosis. We have been unable to see signs of epidemic viral diseases in prehistory because most (except possibly smallpox) do not affect the body for long enough to leave skeletal lesions.

The representative nature of skeletal samples (whether they accurately reflect real health in the once-living populations) has been challenged (Ortner 1992; Wood et al. 1992). The criticisms relate not so much to the appearance or existence of a disease in a particular population (which is of most interest here) but to estimates of incidence or prevalence in a once-living population based on cemetery samples, the ultimate goal of all paleodemographic research. The challenge is important for comparisons of actual prevalence of a particular disease in, say, prehistoric hunter-gatherers and farmers. Major challenges include the observation that diseases are slow to develop in the skeleton so that an apparent increase in skeletal pathology might actually represent greater life expectancy, with more people living long enough for the disease to register in the skeleton, rather than an increase in actual disease. Paradoxically, a high rate of visible

pathology might therefore reflect relatively *good* health. The other major criticism is that skeletons in a cemetery are not random representations of pathology in a living population, but may be a selective sample for a variety of reasons. Hence an apparent increase in pathology might reflect only a change in patterns of selectivity in the cemetery rather than an actual change in real health. Cohen (1997) has argued, however, that when two populations are compared for *relative* frequency of disease, and when many such comparisons produce the same result [as in the comparison between hunter-gatherers and farmers at the origins of agriculture (Cohen and Armelagos 1984; Larsen 1995)], skeletons are very likely to offer a valid comparison of relative rates of pathology.

The 'proof is in the pudding': paleopathology consistently reproduces results that are supported by other lines of evidence or that accord with uniformitarian expectations. The three approaches discussed usually converge on common conclusions about prehistoric disease. For example, the three strikingly confirm one another with regard to the increase in infection rates associated with the emergence of relatively large, sedentary communities at the adoption of farming; and they correspond with one another regarding the probable and actual increase in parasite-related anemia with the adoption of sedentary farming (Cohen 1989, 1997). Sedentary farmers ought, by the nature of the parasites involved (particularly blood-consuming hookworm), to harbor heavier parasite loads than nomadic foragers and therefore suffer more anemia. Ethnographic comparisons suggest that both parasite loads and anemia *are* greater in sedentary farmers than in more mobile hunter-gatherers. And skeletons in prehistoric cemeteries invariably show an increase in porotic hyperostosis, loosely diagnostic of anemia, as populations become sedentary. Tuberculosis (TB) appears in skeletal populations in the Old World at the time of Neolithic domestication of cattle (its probable source), as would be expected. In the New World no cattle existed, so pre-Columbian TB probably came from an avian source or from herds of wild bovids. TB has been found in New World wild bovids and, though density dependent, could have occurred in herd animals such as bison (Rothschild et al. 2001).

Nonetheless, according to skeletal evidence, TB became a common and devastating disease in late, sedentary, very large urban populations in both hemispheres, as contemporary observations suggest it should (Buikstra 1981; Cohen and Armelagos 1984; Cohen 1989; Larsen 1995). Moreover, the skeletal evidence of the patterns of emergence of leprosy and its decline as tuberculosis spreads in Europe mirrors the historical documents of the periods. To interpret this as the indirect results of sampling problems, the fact that not all individuals suffering a disease show skeletal lesions, and the possibility of differential mortality, seems an extreme violation of the principle of parsimony.

Techniques of skeletal analysis described above represent the foundation of paleoepidemiological research but have been limited in their application to skeletons showing visible osseous change. However, the availability of new techniques promises to add considerably to our knowledge of cemetery populations, and of the evolution of diseases themselves, by allowing researchers to examine a multitude of new questions within the discipline. Since, however, the development of the new investigative techniques is a recent occurrence, it is too early to assess their full impact within the field of paleoepidemiology.

The recent development of PCR-related techniques for identifying ancient DNA has allowed the mapping of the complete genetic code of several pathogens with a long evolutionary history and the identification of the DNA in ancient bones. It may permit scholars to identify properties of the genome that have changed over time, allowing for possible identification of the causative factors for the genetic changes (Spigelman and Greenblatt 1998). Recent work has identified DNA of leprosy (Yoon *et al.* 1993; Jamil *et al.* 1994; Rafi *et al.* 1994; Haas *et al.* 2000b), of treponemal infection (Walker *et al.* 1991, 1995; Fraser 1998), and of tuberculosis (Eizenach *et al.* 1990; Spigelman and Lemma 1993; Salo *et al.* 1994; Molto 1995; Baron *et al.* 1996; Taylor *et al.* 1996; Ferman *et al.* 1997, 1999; Nerlich *et al.* 1997; Haas *et al.* 1999). A discussion of the advantages, disadvantages, and potential for applying molecular techniques to archeological material has been very ably presented by a number of scholars (Pääbo 1989; Spigelman 1996; Herrmann and

Hummel 1998; Spigelman and Greenblatt 1998; Kennedy *et al.* 1994; Lassen *et al.* 1994).

Recent research within the fields of immunology and microbiology has produced important new information regarding the complex relationship between invading pathogen and host response. Although this research does not focus on the identification of pathogens in bone *per se*, it furthers our understanding of the etiology and pathogenesis of organisms with a long-standing relationship with human populations. It has been suggested that ancient biomolecules such as proteins, lipids, and carbohydrates may provide unique chemical signatures for the investigation of ancient pathogens (Eglinton 1998). Several studies have concentrated on an examination of the immunological properties of *Mycobacterium leprae* and *Mycobacterium tuberculosis*.

The new techniques will enable us to differentiate more precisely between skeletons showing macro-osteological change that have, until recently, not always permitted reliable differential diagnosis, for instance, between treponemal infection (yaws/syphilis) and leprosy. The techniques allow us to evaluate more precisely the historical interactions between diseases, for example, the historical interaction of leprosy and TB mentioned above. They have also shed light on the relationship of yaws to syphilis, caused by disease organisms once undistinguishable from each other and contributing to one of the major mysteries of disease evolution. These may now be differentiated on the basis of a single DNA base (Noordhoek *et al.* 1989, 1990), suggesting that a mutation, as well as travel and social change, may be involved in the origins of venereal syphilis. The genome of *Treponema pallidum* subsp. *pallidum* also suggests possible clues to the virulence of the causative spirochete. In particular, genes have been identified that code for proteins that may assist the spirochete in attaching to and infiltrating skin, bone, heart, and other host tissues (Pennisi 1998). The sequencing of surface proteins may present the key to how the spirochete evades the human immune system.

The techniques will enable us to identify diseases in skeletal populations that no longer exist, to assess their impact on human evolution, and to assess the social and ecological factors involved in their demise.

They will permit us to identify members of a cemetery population who have a particular disease but who do not display skeletal symptoms, helping to elucidate the mathematical ratios of symptomatic to non-symptomatic individuals and thus helping to resolve issues discussed above. They may help to settle the question of whether pathology in a cemetery is the result of differential representation of pathology based on longevity raised above by Ortner (1992). (They will not otherwise, however, address the question of the possibility of selective relationship of cemetery samples of once-living populations raised above by Wood et al. 1992.)

The techniques will allow us to identify significant diseases of history and prehistory that leave no gross pathological change in bone, such as the most severe malaria, bubonic plague, and cholera, caused, respectively, by *Plasmodium falciparum*, *Yersinia pestis*, and *Vibrio cholerae*, as well as the many crowd-based viral diseases.

7.2 What we know

The known history of human pathogens can be divided into several phases:

● diseases of primate evolution and/or pre-human zoonoses in ancestral rainforests among tree-living and ground-living species, with special reference to Old World primates who are related to early *Homo sapiens* both by evolution and by interaction;
● the diseases of human evolution as we diverge from other primates, spread from the rainforest and adopt new economic practices, summarized under the rubric 'hunting and gathering';
● diseases associated with sedentism and domestication and farming;
● diseases associated with the rise of civilization and of the modern world system.

7.3 Diseases shared by non-human primates and people in a presumed evolutionary continuum or in mutual transmission

Reconstruction of the diseases of non-human primates is based almost entirely on two of the research techniques described: the combination of

epidemiological expectations and observations of contemporary primates. In other words, they are identified by their common occurrence among primates—or at least among the apes—and their life cycles, which permit prolonged and continuous transport in small primate populations or in the environments they frequent, such that they can be assigned probable or at least possible pre-human antiquity. The record of paleopathology of non-human primates is exceedingly thin.

The review that follows is based on a number of sources (Hubbert *et al.* 1975; Fiennes 1978; Acha and Szyfres 1987; Kiple 1993; Bramblet 1994; Karlen 1995; Boesch and Ackerman 2000; Rothschild, this volume, Chapter 9). There is a great deal of disagreement among these sources about primate zoonoses because of the problems discussed above. What follows is an attempt to pick our way among these sources.

1 Rainforest primates are likely to have suffered a wider variety of pathogens and potentially heavier infections in their relatively pathogen-permissive environment than do primates and people in drier environments. This assumption is based on the fact that human populations of the rainforest and of drier environments display a similar gradient of infections. On the other hand, savanna primates such as baboons are probably better indicators of early human experience.
2 Given their generally small groups and low population densities, non-human primates, like small human groups, could not have suffered on a continuous basis from rapidly spreading, lethal, or permanent-immunity-provoking diseases (discussed below), which we now refer to as diseases of civilization. Modern primate populations can of course be secondarily infected by larger populations of human beings that harbor the infections.
3 Open-air treetop environments and individual movement would have minimized the spread of respiratory infection. Moist conditions and a lack of involvement with the soil would have minimized the risk of exposure to mycoses or other soil-borne organisms, but histoplasmosis, cryptococcosis, trichosporosis, aspergillosis, and possibly botulism and tetanus in soil may occasionally afflict primates in the wild.

4 Arboreal existence should have made primates ready targets for mosquito-borne infections such as yellow fever and for nearly ubiquitous cycles involving birds and mosquitoes or ticks. Arboviruses are probably the most important category of zoonotic diseases for primates, as they are for people. Forest canopy arboreal viruses, transmitted by mosquitoes and recognized in modern primates, include not only yellow fever and, by some definitions, dengue of the flavivirus group, but also Sindbis, Semiliki Forest virus, West Nile encephalitis, Chikungunya virus, and Zika viruses. Mosquitoes also transmit protozoan infections such as malaria. Tsetse flies spread one variety of **trypanosomiasis** (sleeping sickness, the **gambiense** form) in the African rainforest both in the forest canopy and to ground-living primates on the forest fringe. In addition, mosquitoes, fleas, deerflies, **ticks**, midges, and mites spread a variety of helminthic infections. Cockroaches can spread acanthocephalus, the thorny-headed worm, but probably only, or at least primarily, in savanna-living forms. Ground living among primates poses two additional problems. In the rainforest, shaded jungle trails would not benefit from the disinfectant quality of sunlight, and ground-living species outside the rainforest, where the sun acted as a disinfectant, would have to share common stable sources of water prone to parasitic infestation and fecal contamination with diseases such as dracunculiasis (guinea worm) and possibly typhoid. The filaria causing onchocerciasis, transmitted by black flies inhabiting fast-moving streams, occurs among primates. Chimps, in fact may be a reservoir. Sand flies transmit leishmaniasis. Primates also share in a variety of zoonoses transmitted by other mammals. Rats and other rodents spread hemorrhagic fevers, rat-bite fever, and capillaria. Rodents also transmit Lassa fever in the rainforest, and it may have been transmitted to non-human primates. Wild animals spread toxoplasmosis, zoonotic salmonella, rabies, and, in drier areas, occasional plague, brucellosis, and Q fever. Leptospirosis occurs in ground-living non-human primates.

5 Largely herbivorous diets and lack of continuous interaction with most other wild species would have prevented the occurrence of many zoonotic diseases associated with carnivory. However, some primates, notably our closest relatives the chimps, do eat meat on occasion, greatly increasing the range of potential zoonotic infections. A recently discovered case in point is the possibility that chimps may be exposed to Ebola virus through eating Colobus monkeys. Trichinosis, which circulates among wild animals and spreads by ingestion, is another possibility, but only if chimps eat other carnivorous species. Nodular worms occur. Occasional ingestion of tapeworms by meat-eating primates may have been possible but it is difficult to imagine the kind of intimate, ongoing relationships with herbivores that would be necessary to maintain a regular infection. But non-human primates do maintain the cysticercarial stages of several tapeworms. Primates that eat fish, particularly macaques, are susceptible to the fish tapeworm diphyllobothrium and its larval form, sparganosis. And paragonimiasis could have been contracted from eating undercooked crustacea, the crustacea being themselves infected by feces.

6 The fur, which protects areas of skin from abrasion and open wounds, helps prevent infections such as penetration of the system by normal surface commensals (*Staphylococcus* and *Streptococcus*), soil fungi mentioned above such as *Histoplasma*, *Cryptococcus* and *Aspergillus*, as well as tetanus, which enters wounds. It would also have minimized skin infections such as screwworm from eggs laid in cuts or wounds. But the commonplace grooming of fur among primates involves close proximity and the possibility of fecal contamination from skin. It also involves the risk of ingesting the tapeworms hymenolepis from fleas and bertiella from mites because groomed fur parasites are commonly eaten.

7 Tree-living species would probably have suffered relatively small loads of fecal–oral infection because feces would not accumulate in their arboreal habitat, and because sources of water (in trees) are small in scale, isolated from one another, temporary, and isolated from fecal contamination. Ground-living apes in their wanderings (in contrast to people) do not generally return to the same spot for days or weeks. Some primates, especially ground-living baboons—but not great apes—do return to favored sleeping areas. But primates do

not ordinarily collect or gather food but rather forage for it, i.e. they move themselves to food rather than transporting food. Nor do they maintain a 'home base' (a universal feature of even the most mobile forms of human social organization discussed below), further reducing the possibility of fecal contamination. Moreover, non-human primates rarely transport or share food, so there is little chance of food being contaminated by other members of the group. But several fecal and/or oral parasites do in fact spread, particularly among ground-living non-human primates (aided perhaps by shared ground trails), including several forms of hookworm such as *Necator* and *Ancylostema*. Other parasitic worms include strongyloides, oesophogastomum, ascarids (which are rare), trichuriasis, and dracunculiasis. Schistosomiasis, in which mammalian feces or urine play a part, has been observed in ground-living baboons and by its nature could have spread among monkey groups sharing a water hole with one another and with other wild animal hosts. But baboons would not have contributed to spreading schistosomiasis very far, and it is a matter of debate whether the infection is secondary to human occupation. Primates are known to harbor a number of protozoan enteric infections including *Entamoeba, Balantidium,* and *Giardia*. Bacteria would include actinomycetes and salmonellas of animal origin (which can live for long periods in soil). *Shigella* (which must be transferred relatively rapidly from host to host) is possible but less probable.

8 Chronic diseases maintained by primates themselves include several herpes viruses, hepatitis virus, a varicella-like virus, cytomegaloviruses, possibly poliovirus, SIV, simian tumor virus, and three forms of pox virus: monkey pox, tanapox, and white pox. African primates also maintain a treponemal infection and may also support bacteria including *M. tuberculosis* or a very similar organism (although the possibility that the latter is a secondary infection has been debated), pneumococcus, staphylococci, streptococci, and meningococcus.

Recent recognition of Marburg virus, HIV, Ebola-like viruses, and other emergent diseases attacking people but found in primate reservoirs suggests that there are many more natural infections of primates—and many more zoonotic infections of people from primates—that we have still to discover.

7.4 Evolution of the human line

The following review is from Cohen 1989; Ortner 1992; Kiple 1993; Ewald 1994; Aufderheide and Rodriguez-Martin 1998; Bannister *et al.* 2000; Rothschild (this volume, Chapter 9).

Theories of disease suggest that the likely candidates for infection among hunter-gatherers are likely to have fallen in one of two groups. The first group includes chronic infections, which can live for a long time in each human host or which come from ectoparasites with similar habits, resulting in long periods of potential contact with new victims. They must also involve reliable communication directly from person to person by touch or perhaps by shared food or fecal contamination, without requisite or common intermediaries of other vectors, water, or the soil. Virulent or immunity-provoking diseases transmitted from person to person, such as many of the 'diseases of civilization', could not have survived, although they could flare up for brief periods as a result of mutation.

The second group of diseases includes those essentially exogenous to human groups, such as zoonotic diseases normally carried by one or more alternative hosts or soil-borne infections that do not depend on human populations to support them. Because they do not rely on small groups of human hosts, such diseases can be quite virulent. But, typically, they are not transmitted directly between people or, if they are, do not spread very far because of their virulence. Ebola, which we now assume to have a relatively long history in non-human primates, seems to be a case in point.

Early human beings can be assumed to have carried over a number of chronic infections, particularly enteric and commensal infections, as well as most of the arboviruses of their primate ancestors, and added a few of their own. The list would include bacteria—*Salmonella* and possibly *Shigella, Staphylococcus, Streptococcus,* pneumococcus, diplococcus, meningococcus, and probably gonococcus; protozoa—including amebae, ciliates, and flagellates; helminths—such as ascarids (now more prominent), trichuris, enterobius, and hookworm; and viruses—herpes, hepatitis, and poliovirus (the last a normal commensal of the human gut by some reports, unable to survive in small populations by others).

Three are of particular interest. The occurrence of simian forms of treponemal infection suggests that treponematosis may have been an early infection of human beings, initially in a form (like pinta) that does not produce symptoms in the skeleton and later as skeleton-scarring yaws. The apparent presence of TB in non-human primates as well as in birds and ruminants also suggests that some form of the tubercle bacillus may have been an early fellow-traveler of evolving *H. sapiens*. There is the interesting possibility that leprosy (whose origins are otherwise unknown) might also have been a fellow-traveler, even if symptoms in the skeleton began to show only in medieval times. Since all three diseases are now identifiable by their DNA in human skeletons, this is an obvious area for future research as discussed above.

Human beings would also have continued to suffer from many of the same threats of incidental zoonotic exposure, including leptospirosis, rabies, and toxoplasmosis, zoonotic salmonella infection, soil-borne mycotic infections and arboviral diseases, including possibly yellow fever and dengue as well as Sindbis and West Nile encephalitis.

Human evolution from primates changed some challenges and added some additional ones. First, group sizes larger than those of most primate groups would have increased the prevalence and intensity of many infections that respond to group size, including many fecal–oral infections such as ascarid worms.

Second, hair loss would have eventually eliminated grooming, hence limiting ingestion of ectoparasites and related infections. But it would have increased the risk of secondary wound infection.

Third, ground living and expansion into niches outside the rainforest perimeter into generally drier (and eventually cooler) climates would in certain micro-environments have altered infection problems considerably. Several fungal infections, as described above, would have increased in

importance. Limited, permanent, and centralized water supplies increased fecal–oral infections such as typhoid and zoonotic infections associated with unclean water. For example, dracunculiasis would have increased in prevalence, and leptospirosis, spread in surface water in drier areas of Africa, would have emerged or increased as a new threat. Ground living and the maintenance of more permanent locations would have greatly increased the significance of the hookworms *Necator* and *Ancylostema*, and *Strongyloides*. The abandonment of the rainforest would have tended to reduce the overall parasite load and possibly reduced the influence of mosquitoes. The risk of soil-borne infections such as mycoses and tetanus would have increased, but probably remained of lesser importance than viral, protozoan, and other bacterial infections. Vectors such as tsetse flies (trypanosomiasis) and phlebotamous flies (leishmaniasis) would have taken on new significance. Filarial infections, particularly those spread by vectors other than mosquitoes such as onchocerciasis would have continued or increased. Expansion from the rainforest would have increased the risk of bubonic plague, leishmaniasis, and trachoma in drier areas and tularemia in cooler areas. Ectoparasites (and their diseases) would have been largely those of undisturbed grass- or scrublands such as ticks and some species of 'wild' mosquitoes in dry areas (as opposed to the more 'domestic' mosquitoes that came to predominate in response to the forest-clearing activities of tropical farmers). Ticks transmit relapsing fevers.

Fourth, since potential prey species, herbivores of the savannas, commonly live in denser herds than those of the forest, the risk of herd diseases such as brucellosis and anthrax would have increased.

Fifth, the reference to ancient human beings as 'hunters and gatherers' as opposed to 'foraging' apes raises two important contrasts. The existence of 'home-base' organization (in which even in camps of brief duration, human beings collect food, return it to camp, and share it rather than searching for it as a group) would have attracted more rats and other scavengers, possibly enhancing the risk of zoonotic diseases including plague, rabies, ratbite fever, and hemorrhagic fevers. In addition human gathering of vegetables in a savanna environment typically involves finding things underground involving far more intimate interaction with the soil and vastly increasing the potential for soil-borne infection

Sixth, meat eating (although rarely more than 40–50% of the diet, despite the common appellation 'big game hunters') would greatly have increased the risk of zoonotic infection including echinococcosis, tapeworms, trichinosis, and salmonella; but infections dependent on very specific interactions between people and animals (tapeworm and echinococcosis) would be rare as long as people and animals did not continuously share a common habitat and as long as encounters between people and animals were irregular and brief. However, handling animal carcasses can introduce anthrax, brucellosis, occasional relapsing fever and hemorrhagic fevers, leptospirosis, toxoplasmosis, and salmonella. Moreover, since most anthropologists assume that scavenging meat is likely to have preceded and accompanied hunting, the risks would also have included gangrene, botulism, and tetanus. Hunter-gatherers in fact have been reported to eat meat that by Western standards has reached a revolting condition and is full of worms. On the other hand, the vast majority of visible worm infections of wild animals that do not result from human fecal contamination are likely to be species that cannot survive in, let alone colonize, the human system. In short, many may mostly have helped supply protein.

Seventh, until very recently, human hunter-gatherers lived in relatively undisturbed wild habitats and would be susceptible to diseases common to such habitats, such as tick-borne infections but not to those of habitats disturbed by human behavior. Ticks can transmit very serious diseases such as Rocky Mountain spotted fever, but typically do so as a rare infection of an isolated individual rather than resulting in person-to-person transmission.

Eighth, human hunter-gatherers commonly keep domestic dogs that could have transmitted or shared diseases with people including rabies, salmonella, dog tapeworms, echinococcosis, and paragonimiasis. Dogs also act as mechanical vectors of other diseases. As long ago as 1975 Hubbert *et al.* provided a tabulated quick summary of human diseases shared with various domesticates

listing as many as 16 diseases shared between dogs and man. (For complete lists of these and other zoonoses of domesticates see tables in Hubbert, pp. 1129–38 and updates by Acha and Szyfres 1987, 2001; Kreger 1997; Shakespeare 2001.)

If early human populations are compared with later human groups instead of with non-human primates, however, low population density, small groups, and mobility create benefits of several types. Overall, relative to other human populations, mobile hunter-gatherers enjoy low parasite loads. Even in the rainforest (which human beings apparently re-entered only in the relatively recent past), hunter-gatherers suffer lower parasite burdens than the larger and more permanent villages of farmers in the same habitats (although parasite loads for all populations are greater than in drier environments). The reasons are several.

First, most strictly human diseases are population-size and density dependent and commonly have lower prevalence and intensity in small groups with low overall population densities. Major epidemic diseases of civilization might have occurred by mutation but such diseases would rapidly have died out since small populations cannot support them (Cohen 1989). For example, work by Francis Black and colleagues (Black *et al.* 1974, 1977) suggests that among isolated Amazonian tribes epidemic diseases do die out and have to be reintroduced, suggesting that they would not have spread in prehistory before urban centers promulgated them, resulting in frequent exposure. On the other hand, once such centers do exist, being an unexposed population newly in contact with such an urban reservoir can be devastating, as evidenced by the plight of Native Americans after 1492.

Second, mobile populations leave behind feces and garbage contamination, minimizing fecal transmission and the accumulation of rodent/flea vectors. Such populations tend to leave behind parasitic infections that must mature in soil, such as hookworm. They may also move beyond the range of insect vectors that require sequential human hosts for reinfection. For example, although mosquitoes are capable of traveling long distances, it may be more difficult for them to maintain unbroken chains of infection among human hosts if human populations move any significant distance.

The same argument applies to the snail vectors of schistosomiasis.

Mobility and small group size also help protect hunter-gatherers against one particularly important class of parasites, those implicated in infant diarrhea (which is a major killer among larger and more sedentary groups).

These conclusions are drawn from all three sources of information. Epidemiological principles predict lower rates of infection in a small group. Studies of skeletal populations of hunter-gatherers (e.g. Cohen and Armelagos 1984; Larsen 1995) not only report low rates of non-specific infection (the only specific identification may be yaws), they show low rates of parasite-induced anemia, relative rarity of childhood osteoporosis, and low rates of childhood growth disruption (enamel hypoplasia of teeth, thought to reflect, *inter alia*, juvenile diarrhea), and other skeletal markers. Ethnographic comparisons of the health of hunter-gatherers and farmers commonly suggest that the former have lower parasite loads than the latter. For example, observations on! Kung San in southern Africa (Heinz 1961; Truswell and Hansen 1976) and of the Hadza of East Africa (Jelliffe *et al.* 1962) suggest low parasite rates and low rates of childhood diarrhea compared with their larger and more settled neighboring populations.

Finally, hunting and gathering populations enjoy excellent qualitative nutrition (limited mostly in calories and fats, and only in some cases)—far better than that of later farmers and, particularly, the poor of civilized and modern societies enjoy. The good nutrition can be predicted from the amount of meat and the variety of fresh foods eaten and documented both in ethnographic descriptions of contemporary populations and in skeletal remains of such populations (Cohen 1989). Since synergy between infection and malnutrition is widely reported, relatively good nutritional background would have helped individuals to fight off any infections to which they were exposed.

7.5 Sedentism and domestication

1 The beginning of sedentary farming involved increases in community size and the development of more permanent locations, accompanied by a predictable increase in overall parasite loads. Larger settled communities and permanent communal water sources would have increased particularly the prevalence of fecal–oral infections. Paleopathological evidence of the increase in parasite load includes common (but not universal) declines in stature and an almost universal increase in the frequency of skeletal pathologies indicative of non-specific infections. Infants are particularly at risk of fecal–oral infections, so rates of childhood diarrhea would have become higher. Increasing rates of enamel hypoplasia of deciduous and early-forming adult teeth in the skeletons of early farmers are thought to be related to such increases in infantile diarrhea. Sedentism also made it possible for many parasites such as hookworm to complete their life cycles in human proximity, greatly increasing the intensity of infection. In addition, it would have permitted vectors such as mosquitoes to more easily maintain chains of infection in human communities, thereby adding to the range, prevalence, and intensity of infection. The accumulation of garbage also resulted in greater fecal–oral contamination and contamination with animal products and rodent-borne diseases.

2 Background nutrition probably declined as farming was adopted, since the variety of foods, their freshness, and the accessibility of animal protein were sacrificed in favor of crops that were storable and would grow in large stands in the proximity of human settlement. As noted, all three lines of evidence indicate that anemia increased in association with hookworm or other parasitic infestation and/or with declining quality of diet (particularly a decline in heme iron from animal sources and the dependence on cereals with ingredients inhibiting iron absorption).

3 Many anthropologists assume that because of changing activity patterns and the availability of new weaning foods (cereal gruel), weaning was commonly carried out at an earlier age, exposing infants to more pathogens and more diarrhea without the support of maternal antibodies. Early weaning is thought to have increased female fertility by reducing the contraceptive effects of nursing.

4 The relatively well-constructed houses that sedentism facilitated would have provided better shelter against the elements, but would also have increased the transmission of respiratory diseases by closing circulation patterns and eliminating some of the disinfecting power of the sun.

5 Planting and care for the soil would have greatly increased the risk of soil-borne infections and of fecal–oral transmission, especially if night soil was used as fertilizer.

6 At least in temperate regions, crops are markedly seasonal and easily stored. The possibility and *necessity* of storage contributed indirectly to declining health through the attraction of parasites to stored grain. Storage also attracts rats and other mammalian and insect disease vectors, aggravating the risks of insect-related tapeworms and of rat-borne diseases. Hubbert *et al.* (1975) list 35 diseases transmitted from rats and their own ticks and fleas to people, including bubonic plague and hemorrhagic fevers. Reliance on storage also resulted in the loss of water-soluble vitamins (B and C). Storage may reduce starvation, but by many accounts the reverse may have been more common. 'Primitive' storage systems are notoriously unreliable; they are subject to human predation and, by making sedentism binding, they reduce freedom to move in search of alternative resources. There is little if any evidence to suggest that storage routinely improved the reliability of human foodstuffs.

7 Sedentism and new foods altered the means of food preparation by enabling people to use pots for cooking and boiling food. The result may have been more thorough cooking and better sterilization, although porous pots can store their own infections. Moreover, boiling results in further loss of B and C vitamins. It also produced softer foods for babies and the elderly (which were, however, generally lower in quality than the foods they replaced). The altered texture of foods had profound effects on the pathology of teeth. Tooth wear was reduced, but because tooth decay involves sticky substances adhering to teeth and density-dependent streptococcal infection, the frequency of caries increased dramatically and almost universally with the adoption of farming everywhere in the world. Whether a decline in overall calcium intake contributed to an increase in dental caries is not clear. But caries do develop in the thinned enamel of enamel hypoplasia, and an increase in the latter may have helped the frequency of caries to increase.

8 Sedentism and storage increased the vulnerability of populations to crop failures, losses in storage, and expropriation by other human communities. The vulnerability of farmers who cannot leave either their growing crops or their stored foods set the stage for the emergence of coercive ('civilized') societies.

9 Land clearance for farming in moister areas would have helped keep ticks at bay, but might have increased the prevalence of mosquitoes. In West Africa and the Mediterranean, land clearance for farming resulted in either the appearance or new virulence of falciparum malaria, with consequent evolution of a number of genetic adaptations ('abnormal' hemoglobins) creating partial immunity to malaria [see, for example, the classic reference by Livingstone (1958)], which is still reported in a largely positive manner by more recent references. Forest clearance, particularly in the New World after slavery, may have brought recently introduced yellow fever down from the treetops to ground level where human beings became infected in large numbers. In addition, in many parts of the world intensified farming resulted in the construction of irrigation ditches, which harbor mosquitoes and the snails of schistosomiasis. (It is also one of the bases on which civilization seems to have been built.)

10 The domestication of plants and animals had profound and far-reaching effects on human health. The domestication of plants *per se* probably had relatively little impact on health, since few if any diseases are transmitted from domestic crops to people. But the artificial concentration of crops, their movement into new niches, and the selective elimination of their defense mechanisms probably exacerbated their vulnerability to stressful conditions and therefore decreased the reliability of the food supply.

Domestic animals would have been a different story altogether. Diseases of domestication commonly involve several factors: introduction of animals to human habitats and their prolonged contact, increased population size of both animals and people, crowding and enclosure of both animals and people, potentially infecting both human beings and domesticates with new infections and permitting the maintenance of regular cycles of carnivore or carnivore/herbivore infections such as tapeworms and their cysticercarial forms, and trichinosis. Many of these diseases, including salmonella, leptospirosis, brucellosis, and anthrax, can be spread by several different groups of domesticates. Hubbert *et al.* (1975) provided a neatly tabulated list of 20 diseases transmitted to people and/or shared by domestic horses, particularly tetanus, rabies, encephalitis, and 36 by domestic cattle, the most important of which is probably tuberculosis, which appears for the first time in prehistoric skeletons after the emergence of farming and domestication. Other added or intensified infections include beef tapeworm, anthrax, salmonellosis, brucellosis, Q fever, and leptospirosis. Hubbert's list includes 22 diseases passed to or shared with people by pigs, including trichinosis, pork tapeworm, brucellosis, and salmonella; and 29 passed by or shared with sheep and goats, particularly echinococcosis (dependent on maintaining both dogs and herd animals in proximity to one another) and brucellosis. Domestic birds spread at least 19 diseases to people, including histoplasmosis, psittacocis, salmonellosis, botulism, TB, and encephalitis; domestic cats share 33, including toxoplasmosis (as above, see updates in Acha and Szyfres 1987, 2001; Kreger 1997, Shakespeare 2001). Fish living in permanent proximity to people, particularly in water fertilized by human feces, and then eaten by people can spread diphylobothrium tapeworms on a regular basis.

Some of these newly shared diseases are important only for their immediate effects, but several seem to have evolved strains that became the basis for the later spread of epidemic human diseases as described below.

7.6 The diseases of civilization

The evolution of 'civilization' (the 'state' in political terms, which are the most important terms) marks a third major epidemiological transition in human history. As defined by anthropologists, civilization means the following.

1 Civilization involves very high overall population densities and urban centers of far higher density than average. It is a commonplace observation that the rise of large population centers connected by transport permitted the maintenance of a number of density-dependent diseases, many of which first emerged as zoonotic infections from domestic and/or occasionally from wild animals, particularly herd-living herbivores. The recent argument that pathogens can increase in virulence when a large host population has enlarged through time instead of evolving to benign status, suggests that these diseases may be less new than suddenly much more virulent. Tuberculosis is an example of a disease that, having long been present, dramatically increased in virulence with civilization. Both skeletal and historical records of leprosy and venereal syphilis first occur in the context of recent civilizations. The other major new epidemic diseases are primarily viruses: measles, presumed to come from domestic dogs or cattle; smallpox from cattle; influenza from pigs or birds; and the common cold (rhinoviruses), probably from domestic horses. In addition, civilization marks the first appearance of epidemic cholera. All can be shown by mathematical modeling and historical record to be diseases of civilization because they 'burn out' (i.e. do not continue to spread) in any population of less than civilized size (estimated, in the case of measles, to be about 600 000).

Civilized investment in irrigation agriculture has provided a perfect home and means of cycling for the organisms of schistosomiasis. In addition, many diseases changed their mode of transmission. Yellow fever shifted from jungle zoonosis of monkeys and people to urban scourge, as did bubonic plague, which switched from an occasional infection caused by handling wild rodents to an urban disease from essentially domestic rats, and then sometimes became airborne pneumonic plague. Typhus shifted from occasional tick-borne murine disease to louse-borne epidemic human infection. Yaws was transformed from a largely endemic disease to venereal syphilis, a clear example of increased virulence in large populations.

2 Massive food storage, food transport, food processing, and, above all, food handling have increased the risk of food-borne infection from both human sources and from zoonotic infections. Civilized societies may well face higher risks of

diarrheal infections than 'primitives' despite improved sanitation.

3 Civilization involves specialists who are susceptible to diseases particular to their professions and may act as disease conduits into the larger population. It also involves varied ethnic groups (whose interests may conflict with one another and with those of the state, as opposed to a unified society of ethnically homogeneous food producers of earlier societies). The obvious outcome is a loss of both interest and ability in maintaining public health programs.

4 Class stratification means poverty, and the ability to back up stratification by force means that government can become a means of ensuring bad conditions by exploitation or neglect and monopoly of funds, reinforcing disease presence. It also means that government is at best to be distrusted and at worst hated, further reducing the efficacy of public health measures. People cannot be trusted to carry out governmental or social mandates even when they are in the interests of the population (as we are discovering in our attempts to get the poor and homeless to abide by health rules, particularly with regard to tuberculosis medication).

5 Poverty causes artificially high concentrations of population within cities, combined with filth, fecal–oral disease, and scarcity of amenities such as sanitation, services, and education. Ghettos of poverty attract rats, especially in cities without modern sanitation. Tuberculosis, plague, and Epstein–Barr virus are infections that thrive under such conditions.

6 Large-scale trade and transport played an additional role in spreading disease. The devastating introduction of Old World epidemic diseases into the New World after 1492 is a major case in point. Interestingly, the same diseases did not reach Australia because, although they could survive without dying out on transatlantic voyages, they could not survive on the far longer voyage to Australia. The faster the transport, of course, the greater the risk. The recent fear (unfounded, as it turned out) that despite its virulence and rapid mortality, Ebola might have spread from West Africa to Toronto by airplane is a case in point. At present, AIDS, a disease of civilization and not primarily of homosexuality, demonstrates the role

of transport in spreading disease, as syphilis and gonorrhea did in the past. By some theories (Baker and Armelagos 1988) yaws/syphilis was transported from the New World to the Old by the return of Columbus's ships (cf. Dutour *et al.* 1994; Mays and Crane-Kramer 2001, who argue for pre-Columbian Old World origin). Movement of potentially contaminated goods from one location to another has also helped to spread disease.

The transatlantic trade brought Old World zoonoses, including yellow fever, possibly dengue, and malaria, bubonic plague, and their vectors to the New World. It also brought domestic animal reservoirs of disease. Schistosomiasis has been spread among agricultural regions of the world by large-scale movement of agricultural workers. On a smaller scale, urbanization and trade within France are implicated in the spread of bubonic plague on a more local level. Biraben (1968) has shown that the incidence of plague in Europe is strikingly correlated with seaports as well as with large population centers. On a more local level, truck transport has been involved in spreading local strains of malaria from place to place within the African continent. Labor migration has also spread disease, the most dramatic examples being venereal diseases. Anthrax was spread from sources of animal hides and wool to those handling products for industrial use.

The contemporary world system involves subordination of whole areas of the globe to the needs of favored areas. What Pelto and Pelto (1983) have called the 'delocalization' of the diet in the modern world system has also played a significant role in spreading diseases. Foods and their diseases are transported long distances. In addition, delocalization tends to rob peripheral areas of nutritious foodstuffs in favor of the monotonous and less nutritious diet of poverty, which not only induces malnutrition but decreases the ability of the body to fight infection.

7 Organized warfare and the large-scale movement of extremely large and dense populations have created prime targets for massive epidemics and pandemics, the most notable of which are typhus and influenza during the First World War. In the 19th century, British troops were responsible for bringing cholera from India to Britain and then from Britain to its other colonies.

The use of bio-terrorism in warfare has added a new dimension of risk. Early records of bio-terrorism involve the catapulting of dead animals over the walls of castles under siege, and the devastating use of smallpox-infested blankets by the US cavalry against Native Americans on the western fringe of the USA. As we write, a new wave of bio-terrorism, this time against the United States, has resulted in a new threat of anthrax already realized, and in the more distant and far more dangerous threat of recirculation of once-supposedly-defunct smallpox.

7.7 Where we stand

Beginning in the last few centuries, in various parts of the world science has begun to overcome many diseases with inoculation, sophisticated social structures such as land clearance, plumbing, and water purification, as well as antibiotics, symptom relief, and some cures. The history of disease described above, however, demonstrates quite clearly that the number and intensity of infections suffered by people has generally increased through history. New disease threats and increasing virulence of old ones are emerging because diseases respond to the changing environments that people create through their 'progress'. One implication is that scientific advance is necessary just to keep up; a second is that health care is not a privilege wrought by progress but an obligation of modern society to treat and control diseases of its own making.

However, our ability to combat pathogens successfully has limitations, and we are falling behind. The diseases are winning because they can evolve responses to our activities, and even our medical advances, more quickly than we can develop technological responses. If one considers the trajectory described, the prognosis for the future is not good. The conclusion of most epidemiologists is that the world is more ripe for a major epidemic or pandemic than at any time before in its history.

But ours is a society built not on science but on politics, ignorance, and pseudoscientific assumptions. Society and politics probably contribute more to public health and modern disease control than does scientific advance. In *Betrayal of Trust* (2000), Laurie Garrett calls attention to the failure of economic and political will to apply even the science and technology we already have to national and international health problems. The problems stem in part from a misunderstanding of disease processes, and from a kind of contemporary popular optimism deriving from excessive faith in science to cure or eliminate disease, as it often does. Nonetheless, the popular sense that infectious diseases have largely been conquered is unfounded.

This optimism has been dimmed only in part by the emergence of AIDS and Ebola. Epidemiologists are aware of the speed with which pathogens can evolve. They know that several infectious agents (e.g. *M. tuberculosis* and *Staphylococcus*) have already evolved forms which cannot be killed by any antibiotics now in our possession. However, the public remains largely unaware of the dangers. They are ignorant of how fast germs can evolve and how many diseases are of recent origin. My students are always surprised to learn that almost all of the diseases they know are recent in origin or virulence and that those diseases respond rapidly to changes in human society and technology. A little teaching about the risk is in order.

CHAPTER 8

Anthropological perspectives on the study of ancient disease

Douglas H. Ubelaker

[Anthropology demands] the open mindedness with which one must look and listen, record in astonishment and wonder that which one would not have been able to guess. (Margaret Mead, Sex and Temperament in Three Primitive Societies, 1963)

8.1 Introduction

Recent concern about emerging infectious diseases that threaten human populations has focused new attention on the complex factors involved. Fischer and Klose's (1996) edited volume on infectious diseases stresses the importance of historical perspective in the understanding of disease emergence. Within that volume, Wilson (1996) describes factors that can be involved in the disease process such as migration, environmental shifts, changes in land use, socioeconomic factors, technological aspects, medical treatment, demographic factors, population vulnerability, formation of new habitats, and microbial evolution. These factors are not only key to our understanding of the emergence of new diseases in contemporary populations, but also to the past human disease experience. Our interpretations of past human disease depend not only on research capable of identifing the organisms involved but also our understanding of conditions such as those that contribute to the emergence and human impact of disease.

Some of the most direct evidence for the past human disease experience originates from the remains of the people themselves. Although mummified diseased soft tissue can be important in areas where exceptional long-term preservation occurs, most such evidence is found on the skeleton. The science of interpreting such evidence, usually referred to as paleopathology, has progressed dramatically in this century, but significant challenges remain. In my opinion, the six key challenges to this research are to develop:

1 an objective descriptive terminology to facilitate comparison;
2 more effective techniques of differential diagnosis;
3 improved interpretation of the individual impact of disease;
4 assessment of the disease experience at the population level;
5 improved training in interpretation of disease processes in ancient tissues;
6 effective use of new available technology.

8.2 Descriptive terminology

Any sophisticated interpretation of a past disease process, even at the regional level, usually requires comparison of data published by multiple investigators. In such comparisons, individual disease diagnoses by those involved are of less importance than the criteria used for diagnosis, and especially the basic documentation of the evidence examined. All too often in the paleopathological literature, frequencies of a condition are reported without adequate presentation of the characteristics involved. Comparison with such examples can be especially problematic.

For example, many studies of human skeletal remains offer frequency data on a condition termed 'porotic hyperostosis'. Classically this term refers to cranial lesions involving the expansion of the spongy area of the bone with thinning of the outer cortical bone and abnormal surface porosity (Angel, 1966). Usually the condition is found on the bones of the cranial vault. When present in the roof of the inner orbits, the term 'cribra orbitalia' is frequently applied. The condition continues to receive considerable attention since many workers feel it largely reflects anemia, and thus evidence of possible genetic or acquired factors that produce such effects (Stuart-Macadam 1985, 1987a,b; Stuart-Macadam and Kent 1992; Ubelaker 1992).

Comparison of data on porotic hyperostosis is problematic because definitions of the minimal expression needed to be recorded vary considerably. What extent of surface porosity represents the abnormal condition? Are workers confusing or inappropriately grouping expansion of the diploë with a periosteal bone formation appearing especially porous? Are inexperienced researchers inappropriately including examples of taphonomically produced erosion?

Not only might the orbit lesions represent different processes than vault lesions, but may themselves represent differing conditions. For example, Webb (1995) further classifies cribra orbitalia into porotic, cribrotic, and trabecular forms, although noting that they can also be grouped.

In recent years, the need for thorough, objective description has been compounded by political and legal factors limiting professional scientific access to pathological specimens (Ubelaker and Grant 1989). In the USA, sentiments have led to legislation (Goldstein 1995) mandating transfer of collections of human remains from museums and other institutions to contemporary groups with ancestral linkages to these materials. Frequently this cumulates in burial of the materials with loss of scientific access. The need for objective, thorough description has never been greater.

To facilitate comparison and other issues, especially with remains threatened with loss, a diverse group of scientists representing differing related specialities and varying academic backgrounds agreed to minimal standards of data collection from human remains (Buikstra and Ubelaker 1994). These 'standards' represent the minimal information needed to document the material adequately and to provide data for future interpretation and comparison. Regarding the pathological section of the standards, the emphasis is on documentation and objective description rather than interpretation and diagnosis. The descriptive, objective observations on bone pathology are grouped into nine categories: abnormalities of shape, abnormalities of size, bone loss, abnormal bone formation, fractures and dislocations, porotic hyperostosis/cribra orbitalia, vertebral pathology, arthritis, and miscellaneous conditions.

Within these categories, lesions are further classified according to specific characteristics. For example, lesions of abnormal bone loss are further described as to location (e.g. periosteal surface, endosteal surface, mixed), extent of involvement (e.g. percent of unit affected), organization (well-organized, diffuse, boundaries well defined but no sclerosis, etc.), and if accompanied by structural collapse. Coupled with detailed inventory, photographs, radiographs, and other documentation, the system offers information hopefully useful for future comparison and diagnosis. Revisions of this work and related efforts in paleopathology should focus on making it even more comprehensive and objective, perhaps providing more detailed visual guides to classification.

8.3 Differential diagnosis

Human remains offer limited information about the disease processes that affected them. Many diseases leave minimal imprints on bone or can cause death before enough time has elapsed for the bone to respond. Many diseases that do affect bone in noticeable ways do so in a similar manner, making it most difficult for the paleopathologist to differentiate between them.

Recent years have witnessed advances in differential diagnosis at two levels. More sophisticated techniques are available from specialists, allowing individual disease processes to be identified. Examples of this new methodology vary from Rothschild and Heathcote (1995) focusing on specific diagnostic problems to recent general treatments by experienced pathologists offering thoughtful criteria (e.g. Campillo 1993, 1994, 1996).

Advances also have been made through the recognition of more general disease conditions, which without offering information about the history of particular diseases, succeed in providing more general data on morbidity at the population level. Examples of such approaches are such broad categories as generalized periosteal reaction, osteomyelitis, and porotic hyperostosis. Frequency data on such abnormalities can be coupled with cultural information about the populations represented to allow broader interpretations of the interplay between disease and culture.

The challenge here is to (i) increase the accuracy of individual disease diagnosis, (ii) discover general categories of disease that are more meaningful, and (iii) integrate the above two approaches into a more sophisticated disease characterization that makes analysis more meaningful.

Proper interpretation of pseudopathology is needed as well, since confusion of the effect of disease with natural processes acting upon ancient tissue remains a problem. Greater appreciation of post-mortem taphonomic influences (Nawrocki 1995) that can mimic disease conditions should reduce error and enhance interpretation. Those collecting data on disease of bone need to be familiar with processes of normal bone growth and post-mortem change.

The irregular bone surface in the metaphyseal area of a growing, immature long bone marking normal osteoclasis in the remodeling process can easily be confused with a destructive disease process. The pacchionian depressions on the endosteal surface of the parietal can mimic pathological focal bone loss. Frontal grooves, metopic sutures, and many other normal anatomical variants offer potential error for the analysis of the naïve paleopathologist. For example, enlarged parietal foramina (Fig. 8.1) can potentially be confused with trephination (Fig. 8.2).

Much information about disease interpretation, especially confusing taphonomic factors, continues to be gleaned from forensic analysis (Haglund and Sorg 1997). Forensic anthropologists have the opportunity not only to observe the skeletal effects of disease and disease-mimicking taphonomic conditions, but to learn the truth when identification is made. This applied work has great potential to add needed knowledge about the effects of disease on bone, taphonomic factors, interpretation of time since death, and trauma analysis (Ubelaker 1996c). Published examples include hydrocephaly in calves that can be confused with human pathology (Ubelaker et al. 1991) and circular bone lesions produced by post-mortem chemical breakdown (Fig. 8.3) that mimic pathological focal bone loss (Fig. 8.4) (Ubelaker 1996c).

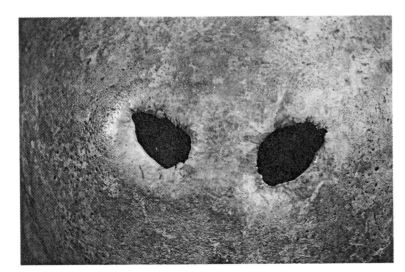

Figure 8.1 Enlarged parietal foramina, NMNH 276981. Photo by Jane Beck, Smithsonian Institution.

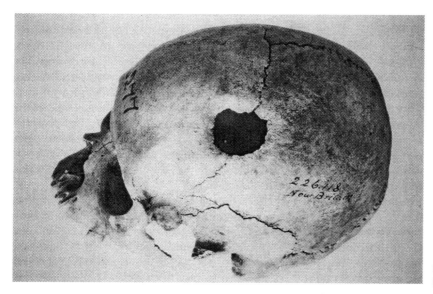

Figure 8.2 Trephined cranium from New Britain, NMNH 226118.

Figure 8.3 Perforation in cortical bone produced by post-mortem chemical alteration.

8.4 Individual impact of disease

An important challenge in assessing the past disease experience of human populations is to gauge the individual impact of the conditions discovered. Statistics can be accumulated on the frequencies of a particular problem in the samples studied or the antiquity of the appearance of a particular disease condition. An additional challenge is to assess the impact of the ailment on the host. Was the condition mortal, debilitating, or so benign that the host would not even have been aware of its existence?

When these individual assessments are compiled into data for populations, then information about impact emerges (Palkovich 1981; Saunders *et al.* 1995). The severity of the condition may be judged by the areas of the body affected, the extent of bone involvement, and the nature of the bone response. The extent of bone remodeling offers important information here about the length of time before death during which the individual suffered the problem. The assessment of 'perimortem' versus 'antemortem', so important in forensic analysis, is relevant here to differentiate a problem likely

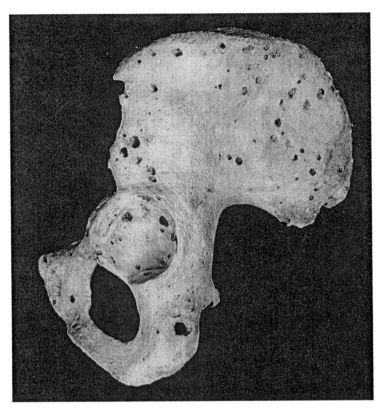

Figure 8.4 Lesions associated with multiple myeloma, NMNH Terry Collection 787. Photo by Jane Beck, Smithsonian Institution.

associated with death from one sustained much earlier in the person's life that may be unrelated to those factors leading to death.

Progress will be made in paleopathology through critical assessment of interpretations of the individual disease impact. Most disease conditions observed on archeologically recovered bone represent those sustained by the individual long before death, likely unrelated to cause of death. They provide a record of the disease experience of the individual, but largely represent problems that the individual survived. Does the occurrence of a well-remodeled periosteal lesion indicative of localized infection suggest that the person represented was less healthy than another individual lacking such lesions? Does survival of the disease process make the individual more hearty and capable of resistance to future disease exposure or is the lesion an indicator of a disease-prone lifestyle ultimately leading to death? These questions are being debated in paleopathology (Wood *et al.* 1992;

Cohen 1996) and, when resolved, may strengthen interpretation.

One recent attempt to move beyond lesion description and frequency data comparison has been discussion of the health index. An effort centered at Ohio State University seeks to sort through the various indicators of morbidity and mortality that can be observed in skeletal analysis and assess their relative impact on the individual (Steckel 1996). The result is the creation of the 'health index', a multifactorial assessment of the disease impact. The health index concept flows from medical economics, a field that considers the impact of medical efforts in terms of the quality of life measured in health terms. The attempt here is to utilize the skeletal record to make such assessments in the past. The 'health index' represents the total years lived in an individual's lifetime, adjusted for the 'quality'. Quality is assessed through evidence of bone pathology. Although agreement on the relative importance of various ingredients in the 'health index' will probably be difficult, this approach to understanding the impact of disease potentially adds an important new dimension to the study of past populations.

Coupled with conventional paleopathological analysis of bone and preserved soft tissue is paleodemography, with its focus on the analysis of mortality. The various expressions of this endeavor, life tables, mortality curves, survivorship curves, male:female ratios, adult:immature ratios, etc., all seemingly offer valuable direct information on the most extreme result of the disease experience— death. Such data potentially offer evidence of the existence of diseases that kill without leaving signatures on bone or kill sufficiently rapidly that there is no time for the bone to respond.

The accuracy of demographic analysis continues to be debated. Sources of significant error potentially include inaccurate age at death estimation and a variety of sampling problems. The impact of population growth or decline on life table reconstruction can be of such magnitude that some workers use life table data more as an indicator of fertility than mortality (Sattenspiel and Harpending 1983; Buikstra et al. 1986; Johansson and Horowitz 1986; Milner et al. 1989). If these variables can be controlled, demographic data have powerful potential to elucidate the most severe disease experience in past populations, especially when coupled with detailed lesion analysis.

8.5 Population analysis of paleopathological data

If adequate samples of human remains are available, data can be collected to examine past disease experience at the population level (Ubelaker 1996c). This approach can be traced back historically to the work of E. A. Hooton (1930) and others but has advanced dramatically in recent years. If meaningful frequency data can be acquired of relevant disease processes, they can be used to address wide-ranging problems of the past (Pérez-Pérez 1996). Correlations are possible with factors such as diet, geography, general nutrition, settlement pattern, population size and density, and even technological factors (Buikstra 1981; Walker 1996). Such associations offer tremendous potential to reach

meaningful understanding of the dynamics of the disease process and the factors that involve past emerging diseases.

The challenge of population analysis is largely twofold: to utilize a rigorous, sound, and meaningful system of disease classification, and to utilize large representative samples that can be accurately dated and correlated with appropriate cultural information. Samples are critically needed from diverse geographical areas and time periods to ferret out the likely complexities involved. Well-dated samples from different periods offer opportunities for studies of temporal change likely associated with cultural shifts that are known to have occurred as well. Geographical studies offer opportunities to explore environmental factors and physical/cultural variation in habitat disease preference.

8.6 Training

Thirty years ago, Kerley and Bass (1967) published in *Science* that the rapidly emerging science of paleopathology formed the meeting ground of many disciplines. The common interest in the history of disease brings together professionals from anthropology, evolutionary biology, history, medicine (especially pathology), literature, and many other areas. The interdisciplinary nature of this endeavor is as true today as it was in 1967. The interest in the history and impact of disease brings in such a diverse group of professionals not only because the subject is both fascinating and important, but also because no single professional field offers the type of comprehensive training needed to address all of the issues involved. The medically trained pathologists bring the clinical experience, the greatest diagnostic skills, and a strong specimen orientation, but usually need help in assessing the taphonomic and cultural factors involved. The anthropologists bring the population approach and knowledge of post-mortem change, cultural factors, etc., but are less cognizant of the complexities of disease diagnosis. The need is for broader training in all areas, or at least enhanced communication and collaboration among participants. The intellectual minefields along disciplinary borders must be defused to approach the goal of understanding the complexities of the past disease experience.

8.7 New technology

Future efforts to understand past disease will be enhanced greatly by newly emerging technology. The most exciting of these new approaches seem to be chemical in nature (Katzenberg and Pfeiffer 1995). Diet may be established not just by assessing faunal bone assemblages, or finding the elusive corncob, but through chemical analysis of trace elements or isotopic structure in bones and teeth or analysis of food residue in ceramic vessels. Disease diagnosis now flows not just from evaluation of the bony response of the host but through molecular detection of the pathogen itself. The molecular methodological breakthroughs are poised to verify the presence of the pathological organism with a level of precision not previously possible. The need remains, however, for careful evaluation of the context of this presence. This involvescomplex anthropological analysis of the social and geographical factors that explain the origin and impact of the disease.

8.8 Disease in ancient Ecuador

Since about 1973, I have collaborated with archeologists working in Ecuador, South America in the excavation and analysis of large, well-documented samples of human remains. Analyses have now been published on over 1500 skeletons from 22 archeological contexts, originating from both coastal and highland sites (Ubelaker 1996c, 1997). The samples range in antiquity from the 192 individuals in the preceramic, preagricultural Vegas complex dating between 8250 and 6600 BP (Ubelaker 1980, 1988) to a sample of 33 individuals from the central ossuary of San Francisco Church in central Quito, dating to the early part of this century (AD 1850–1940) (Ubelaker 1995).

These samples span nearly the entire temporal range of the human presence in Ecuador. Of special interest, the samples represent people living in (i) the earliest preceramic times prior to agricultural activity; (ii) the early Formative period when agricultural practices and animal domestication were initiated; (iii) the later prehistoric periods when population density increased and complex social stratification and trade blossomed, and

finally; (iv) into the historic periods, involving the introduction of Europeans and their new diseases. The many time periods and the large samples involved enable analysis directed at assessing the long-term health changes in the population and their geographical/cultural correlations.

Over the years, analysis has proceeded at many levels. Detailed inventories and descriptive data have been published documenting those aspects of mortuary customs and establishing a record for future comparative studies. Biological analysis has produced a wealth of interpretive data on physical characteristics, stature, etc. The data have also revealed much about the biological perspective on cultural development. Some of this information is quite direct (e.g. cranial deformation and tooth alteration).

Other aspects of the analysis focus on cultural explanations for patterns of biological data. An interesting example originates from the study of human remains recovered in the highland shaft tombs of La Florida on the outskirts of Quito in the Ecuadorean highlands (Ubelaker *et al.* 1995; Ubelaker 2000). Excavation revealed six very deep shaft tombs containing both high-status individuals and their likely subordinates. The tombs date to the Chaupicruz Phase (*c.* 100–450 AD) of the Regional Development Period, a time of flourishing agriculture as well as marked social stratification. To examine possible status-related dietary differences between the upper and lower classes, we analyzed stable isotopes of carbon and nitrogen from 32 individuals, nine high status and 23 low status. Stable carbon isotopes, preserved in ancient human bone, reveal whether the individuals consumed a diet composed primarily of plants involving a C_3 photosynthetic pathway and/or C_3 plant-consuming animals; or one based upon C_4 plants. In prehistoric Ecuador, the only C_4 plant available was maize; all other plants involved in the dietary chain were C_3. Abundant research has demonstrated that humans consuming C_4 plants have less negative carbon isotope ratios than those consuming other plants, although marine foods also produce less negative values.

Stable nitrogen isotope values reflect trophic level, where the numbers increase with movement up the food chain. Leguminous plants have quite low values, while herbivores are slightly higher, and carnivores higher still. Vegetarian humans will have lower numbers than those including meat in their diet (Ubelaker *et al.* 1995).

Early ethnohistorical evidence suggested that status differences would likely be reflected in the nitrogen values since the only dietary difference noted was consumption of more animal food by the elite. The historical sources also focus on the value of maize and the role of the elite in converting maize to maize beer. The elite provided this service to the community and used the distribution of the product as a means of social solidarity and reinforcing their status.

To chemically examine the dietary issues outlined above, bone collagen was extracted from each individual and analyzed for carbon and nitrogen isotopes. In addition, bone samples from five high-status and five low-status individuals were analyzed for stable carbon isotopes in bioapatite. Some have argued that carbon isotope analysis from bioapatite better indicates the lifetime diet of the individual and that such analysis from collagen reflects greater influence of dietary protein (Krueger and Sullivan 1984; Lee-Thorp and van der Merwe 1991).

In the actual analysis of the bones from La Florida, differences were found between high- and low-status individuals in the carbon values in both collagen and mineral bone components. The magnitude and direction of the difference suggested greater consumption of C_4 plants by the elite. No significant difference was found between the two groups in nitrogen values, indicating lack of status difference in trophic level reflected in the diet. Thus, the chemical isotope analysis provides no support for the ethnohistorical suggestion that the elite were eating more meat. In contrast, it suggests greater elite consumption of maize. Strong ethnohistorical evidence indicates that the form of this consumption was likely corn beer. Apparently, the elite were not only distributing their manufactured beer, but consuming a disproportionate amount of it. Supporting evidence for this interpretation takes the form of specialized ceramic vessels suitable for beer storage found buried with the elite remains.

At the more general level, population-based analysis of all of the samples enables an opportunity to examine the emerging temporal pattern of mortality

and morbidity. For example, Fig. 8.5 plots the ratios of frequencies of bones with periosteal lesions to adults in the samples, grouped into six sequential time periods, arranged in temporal sequence. The lowest ratio is found in the preceramic sample representing the earliest in the sequence. The ratio increases steadily through intermediate precontact, declines during the late precontact, and then increases substantially during the early and late historic periods. Explanations for these data are complex but focus on increasing sedentism and population density. At the earliest preceramic period, when populations were most mobile and had the lowest density, periosteal lesions were most uncommon. The subsequent increase with the settlement pattern shifts associated with agriculture is logical. The reduction during the late precontact is more difficult to understand but may reflect regionalism inherent in the data. The historic sites are all urban (highland Quito) and thus reflect not only high urban density, but also include the new pathogens introduced from Europe and the cultural factors included in that package. Whereas the precontact samples presumably are all Indians, the historic samples undoubtedly contain individuals of European ancestry. The general label 'periosteal lesions' does not differentiate among the many specific diseases that may be involved, rather it provides a general pattern that may reveal aspects of the factors involved.

Hypoplastic defects are frequently consulted by anthropologists as general indicators of morbidity (Goodman and Rose 1996). Basically, they represent defects of the tooth surface left behind when formation of the tooth was temporarily disrupted. Although the specific disease or other problem causing this disruption usually can not be pinpointed, the defect is used as a non-specific indicator of morbidity during the childhood years. The age of the individual when the problem was sustained can be estimated by observing the location of the lesion on the particular tooth. In the Ecuador data, as with periosteal lesions, the frequency of hypoplastic defects is lowest in the earliest preceramic group. The frequencies increase in the subsequent precontact periods, especially during the late precontact. This general pattern is similar to that seen with periosteal lesions and likely reflects similar factors.

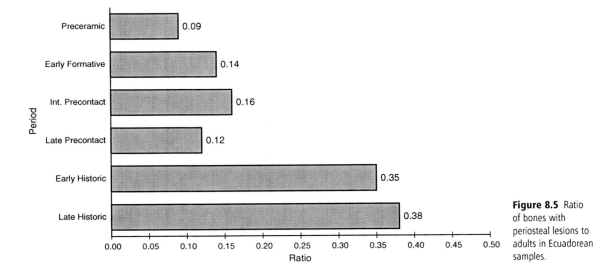

Figure 8.5 Ratio of bones with periosteal lesions to adults in Ecuadorean samples.

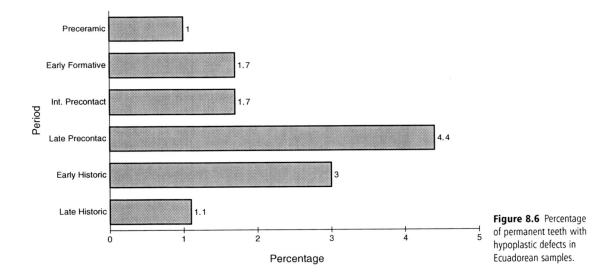

Figure 8.6 Percentage of permanent teeth with hypoplastic defects in Ecuadorean samples.

In contrast to the pattern seen with periosteal lesions, frequencies of hypoplastic permanent teeth (Fig. 8.6) decrease during the historic period (Ubelaker 1992), a pattern also reported from other regions (Larsen and Hutchinson 1992). This reduction from the late precontact sample contrasts with that seen with periosteal lesions and is difficult to reconcile with the probable increase in overall morbidity during the historic periods. A possible explanation is the relatively high rate of dental caries (Fig. 8.7) during the historic periods, probably related to diet. Hypoplastic defects on teeth represent ideal sites for oral bacteria to colonize. Thus, another health problem, dental caries, may be destroying the evidence of the dental defects either by replacing them with carious lesions (Storey 1992) or through tooth loss.

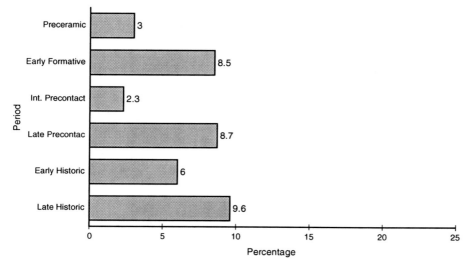

Figure 8.7 Percentage of permanent teeth with carious lesions in Ecuadorean samples.

In general, the Ecuadorean data support those derived from many other temporal studies throughout the world, suggesting increasing population morbidity with time, likely in association with the increased sedentism and population density associated with agriculture and urban living.

A current project collecting these types of data within large temporally seriated skeletal samples from northeastern Hungary seeks to elucidate whether this pattern also prevails within central Europe (Ubelaker and Pap 1997). This area was chosen for the planned research because of the availability of well-documented skeletal samples from various time periods. In addition to a complete inventory of each skeleton, the research examined such important variables as age at death, gender, dental hypoplasia, dental caries, alveolar abscess, antemortem tooth loss, cribra orbitalia, porotic hyperostosis of the cranial vault, vertebral osteophytosis, trauma, abnormal periosteal bone deposits, lines of arrested growth (radiographic observation), femoral and tibial mid-diaphyseal circumference, and estimated living stature. Comparisons between samples of the Bronze Age and the later Iron Age revealed slight temporal increases in most indicators of morbidity (Ubelaker and Pap, 1998).

8.9 Summary

Paleopathology has come a long way since Kerley and Bass noted the coming together of many disciplines back in 1967. Breakthroughs in understanding the past emergence of new diseases likely will flow when (i) objective terminology in describing lesions enables detailed, accurate comparison of data, (ii) specific diseases can be diagnosed with increased accuracy either through their effect on tissue or through molecular analysis, (iii) disease recognition can be coupled with broad-scale population analysis of general morbidity and mortality and anthropological interpretation of relevant environmental and cultural factors, (iv) research becomes truly interdisciplinary to avoid the pitfalls of interpretation in isolation, and (v) new emerging technology is fully utilized and incorporated into sophisticated problem-oriented research. The past has much to teach us about the factors contributing to the emergence of disease, if we can successfully address the challenges.

CHAPTER 9

Infectious processes around the dawn of civilization

Bruce Rothschild

A pox of this gout! or a gout of this pox! for the one or the other plagues the rogue with my great toe. (William Shakespeare, Henry IV, Part 2, Act 1, scene 2)

Patty was a modest maid
Patty was of men afraid
Patty grew her fears to lose,
And grew so brave, she lost her nose.
(Horace Walpole, in Horace Walpole's Miscellany, 1786–95)

9.1 Introduction

Being condemned to repeat the history that we ignore carries an important public health caveat. How can we anticipate and prepare for future epidemics and health care challenges without knowledge of the history of disease? Documentation of past disease character and epidemiology provides an opportunity to better understand the origins or causes of diseases whose present etiology may be unclear.

Such an approach requires that a disease be confidently recognizable. This demands as broad a diagnostic context (database) as possible. Recognition of disease in fossil or recent bone material requires that there be previous documentation of the effect of the disease on the bones of individuals diagnosed in life, and that the combination of bone findings be specific for the disease (i.e. that no other diseases routinely produce the same damage). Population studies are critical to this kind of analysis. A single individual represents only one possible set of disease damage. Disease, however, is variable, and resultant bone damage represents a spectrum. While a variety of diseases can cause a given type of damage to an isolated bone, the pattern of that damage in an individual is more specific. The distribution of disease manifestations in a population affords the strongest comparisons. Without independent corroborating evidence, isolated bones are usually not informative.

A data-based approach transcends opinion (repeatedly expressed but unverified hypotheses)-based approaches and allows a more meaningful analysis of osteological material. Availability of such databases for treponemal disease, arthritis, and tuberculosis has allowed the progress delineated below. Preconceived notions/theories can be tested against these databases. One of the clearest examples is a misguided concept that periosteal reaction is simply a stress or non-specific trauma reaction. To test that hypothesis we must first standardize comparative data, including techniques for data acquisition. Different evaluations of periosteal reaction in the same population vary from 0 to 100% involvement, indicating some failure in the criteria used, but how do we determine which method is giving useful results? In the case of periosteal reaction, independent physical chemistry-based comparisons (thermographic assessment of entropy) validated a technique giving an intermediate value (Rothschild and Rothschild 1999). This technique, because it has independent validation, should be preferred for recognition of periostitis.

9.2 Arthritis

Identification of arthritis in living individuals is based on the nature of bony features (e.g. preservation of size and configuration, presence of erosion, reactive bone formation or remodeling), as well as soft tissue alterations (Resnick and Niwayama 1988; Rothschild 1982). While bony alterations represent only a subset of those parameters, they are usually all that is available to the paleontologist. Most of the 200 varieties of arthritis are not known to produce any bony alterations (Resnick and Niwayama, 1988; Rothschild 1982) and are therefore not amenable to the archaeologic or paleontologic record (with currently applied technology). This considerably narrows the list of diagnoses recognizable in prehistoric specimens.

Arthritis is typically classified as erosive or non-erosive. One form of non-erosive arthritis is osteoarthritis. Bone remodeling occurs with spur (osteophyte) formation. The density of the subchondral bone is increased and metaphyseal cysts may form (as recognized by X-ray or cross-sectional view). With increasing severity, the articular surfaces may become grooved and eburnated. Extremely rare in dinosaurs, it is more commonly noted in Quaternary mammals, although still rare. Its analysis provides intriguing perspectives on 'lifestyle' and habitat in extinct animals and societies. The osteophytes of osteoarthritis are distinguished from those of spondylosis deformans, a vertebral centra spurring phenomenon. As the disc spaces are not diarthrodial joints, the term osteoarthritis is not appropriate for spine spurring, exclusive of the zygapophyseal joints.

9.2.1 Spondyloarthropathy as a model for recognition of disease across species lines and understanding its significance

Spondyloarthropathy and reactive arthritis are terms applied to a group of disorders characterized by a tendency to reactive new bone formation, asymmetrical pauciarticular peripheral joint erosions and ankylosis, and axial (spine and sacroiliac) joint disease (Resnick and Niwayama 1988). Both marginal and subchondral erosions occur. Population analysis is essential, as axial joint disease is not universal and peripheral arthritis is occasionally polyarticular. Zygapophyseal joint erosion or fusion and ossification within the anulus fibrosus are pathognomonic. Evidence of spondyloarthropathy abounds in the literature of human skeletal disease (Ruffer 1921; Zorab 1961), and the precedent for its non-human occurrence has been documented (Rothschild and Woods 1989).

Reactive arthritis is considered an infection-related phenomenon. *Salmonella*, *Shigella*, *Yersinia*, *Campylobacter*, enteropathic *Escherichia coli*, and *Chlamydia* are the bacteria most commonly responsible. The paleontologic/zoologic record of spondyloarthropathy is not limited to humans, nor even to primates (Rothschild and Woods 1989, 1992a,b, 1993; Rothschild *et al.* 1993, 1994; Rothschild and Rothschild 1994a). Observation of the breadth of species affected (Rothschild and Woods 1989, 1992a,b, 1993; Rothschild *et al.* 1993a, 1994; Rothschild and Rothschild 1994a) and equivalent Pleistocene frequencies (to that noted in the Holocene) suggested the possibility of very early development of this trans-mammalian phenomenon (Rothschild and Woods 1989, 1992a,b, 1993; Rothschild *et al.* 1993a, 1994; Rothschild and Rothschild 1994a). Although Ferigolo (1988) claimed signs of osteoarthrosis on the radius and ulna of the sloth *Mylodon*, this animal clearly had spondyloarthropathy involving many other areas. The term 'cave gout' (*Hohlengicht*) was used by Rudolf Virchow to describe arthritis in cave animals. It is unclear whether they had gout or actually were suffering from spondyloarthropathy.

Comprehensive evaluation has revealed spondyloarthropathy to be almost panspecific, often with quite high population penetrance with epidemic increase over time in some groups (e.g. baboons, but also the Perissodactylae—the group of horse and rhinoceros-like mammals). The frequency in baboons increased from 4% in the 1920s and 1930s to 10% in the 1970s to 30% in the 1980s (Rothschild and Rothschild 1996a,b)! In proboscidea, the frequency in the extinct mammoths (7%) is indistinguishable from that of contemporary elephants. Mastodons, however, did not develop spondyloarthropathy.

Recognition of spondyloarthropathy, on the basis of pathognomonic sacroiliitis, extends the antiquity of the disease to the Eocene [54–38 million years ago (mya) in the American west] in horned rhinoceros-like forms of the Coryphodontidae (extinct order Pantodonta) and Uintatheriidae (extinct order Dinocerata, horned forms with sharp canine teeth). The 12.5% and 25% population frequency of spondyloarthropathy in *Coryphodon* and *Uintatherium*, respectively, suggest that this was an important problem in the Eocene for those genera.

Sacroiliac joint erosions and/or fusion and syndesmophytes in odd-toed ungulates—Brontotheriidae, Chalicotheriidae, Rhinocerotidae (Subhyracodon, Peraceras, Aphelops, Teleoceras and all extant species), Tapiridae (tapirs), and Equidae are pathognomonic for spondyloarthropathy (Resnick and Niwayama 1988; Rothschild and Martin 1993; Rothschild and Rothschild 1994a; Rothschild and Woods 1989, 1992a,b, 1993; Rothschild *et al.* 1993a, 1994). Recognition of spondyloarthropathy extends to the Eocene in the Perissodactylae. It was common in extinct families such as Brontotheriidae (13% in the Eocene) and Chalicotheriidae (7% in the Oligocene, 38–23 mya) and present in 20% of extant Tapiridae. Progressive increase in frequency of spondyloarthropathy was observed through geologic time. This ranged from 9% of Oligocene (38–23 mya) rhinocerotidae to 15% in the Miocene (23–5 mya), 20% in the Pleistocene (1.8 mya—11 000 years ago) and 32% in extant animals. This compares with an increase in equid frequency from less than 1% in the Miocene to 2% in the Pliocene (5–1.8 mya), 3% in the Pleistocene, 8% in extant animals.

9.2.2 Human record

New World

North America Examination of two proximate early 19th-century cemetery populations, discordant for socioeconomic status, revealed indistinguishable character and frequency of spondyloarthropathy (Rothschild and Rothschild 1993). Two contemporary 19th-century sites—the Monroe City poorhouse (also referred to as the Highland Park site, near Rochester, New York) and Belleville (a prairie town in Ontario)—were examined for evidence of spondyloarthropathy in any member of the cemetery. This compared the Highland Park cemetery, serving the Rochester poorhouse between 1826 and 1863, and the St Thomas' Anglican Church Belleville Cemetery, used from 1821 until 1874. The cemetery samples examined included 296 individuals from the Highland Park site and 259 individuals from Belleville. Sixteen instances of spondyloarthropathy were noted in the Highland Park site, representing 5.4% of the cemetery population. Twenty-one individuals with spondyloarthropathy were identified in the Belleville site, representing 8.1% of the cemetery population, not significantly different from that noted in the Highland Park site ($\chi^2 = 1.62$). Spondyloarthropathy was present in 5.9% of skeletons identifiable as female and 8.5% of skeletons identifiable as male from the Highland Park site for a 1.4:1 male predominance. Joint involvement, number and symmetry of joint erosions and fusion and skeletal distribution were indistinguishable in the two populations. It would appear that socioeconomic status does not bias skeletal populations, at least with respect to diseases for which no effective treatment was available.

The antiquity of spondyloarthropathy has been established for early [>3000 years before present (years BP)] upstate New York (Late Archaic, Fontenac Island) and Proto-historic, Ontario sites (Kidd 1954; Rothschild and Woods 1992b). While the population frequency cannot be established for the latter ossuary site, four of 87 individuals (4.6%) from Fontenac Island had spondyloarthropathy. This was not significantly different ($\chi^2 = 0.09$) from the 5.4% found in the upstate New York (Highland Park) or from the 8.1% found in the Belleville sites examined in the present study. Given the sanitary conditions of the times and the potential for infectious agent diarrhea to cause spondyloarthropathy (Rothschild 1982; Resnick and Niwayama 1988), the increased frequency of spondyloarthropathy prior to the 20th century is perhaps not surprising. Similar environmental factors may explain the similar frequency of spondyloarthropathy in early native and immigrant populations. Reactive arthritis, of the Reiter's variety, frequently complicates infectious agent diarrhea (Rothschild 1982; Resnick and Niwayama 1988). In view of Old World sanitary conditions, Reiter's syndrome is also a reasonable diagnosis.

Findings are pertinent in 1657 additional individuals with spondyloarthropathy identified in 16 skeletal populations from Alaska to California, the Midwest, Southwest, and Florida, covering the period from 4700 to 400 years BP. The frequency of spondyloarthropathy varied in individual populations from 0.7 to 5.1%. Analyzed according to degree of antiquity (years BP), significant time-related frequency variation (1.3–2.7%) was not present ($\chi^2 =$ 2.30, n.s.). The population frequency and joint distribution of spondyloarthropathy were relatively uniform (1.3–2.5%) during the 4000-year interval studied. The frequency of spondyloarthropathy also was unaffected by geographic ($P = 0.066$, Fisher's exact test) distribution (of the site). It was within the range noted for Blackfeet, Watford, and Cree Indians (Gofton *et al.* 1972) and supports the contention that contemporary clinical studies may underestimate the frequency of spondyloarthropathy. The frequency was lower than that noted in specifically (spondyloarthropathy) predisposed Indian groups [e.g. the 9.5% reported for Haida Indians (Gofton *et al.* 1972)]. Absence of significant variation over time or in widely disparate geographic sites suggests that the endemic population frequency of spondyloarthropathy in susceptible populations ranged from 0.7 to 5.1%, at least for the time period from 3000+ to 400 years ago. Given the large number of unrelated Native American groups affected, environmental rather than genetic factors are probably of primary importance in the development of this disorder. Uniform vertebral involvement (characteristic of ankylosing spondylitis, but found in 5–10% of other varieties of spondyloarthropathy) was reported by Bass *et al.* (1974) in 0.07% of North American sites. The oldest case reported in North America may be that of Neumann (1966), dated at 5000–8000 years BP.

South America Evidence of spondyloarthropathy was found in 7% of individuals from Chile, dating back 4000 years (Arriaza 1993).

Old World

Spondyloarthropathy is well documented in the Old World (Ruffer and Rietti 1912; Zorab 1961; Kramar *et al.* 1990).

Europe Most of the literature on European cases (though some were unrecognized at the time) appear to represent isolated cases (Bruintjes and Panhuysen 1995).

Africa One hundred and thirty-eight (138) Meriotic Nubians from the Semna South Site in Northern Sudan were examined (Arizona State University). The site is on the west bank of the Nile river, 15 miles south of Wadi Halfa. It is dated at 2000–1600 years BP. Sixty-one individuals were examined from Hassi el Abiod sites in northern Mali. Fifteen individuals were dated at 7000 years BP and 46 at 4500 years BP. Three other sites (*datum incertae*), housed at the American Museum of Natural History, were examined. The largest collection, designated El Hessa or the van Luschan Collection, comprised 115 individuals. A second site, identified only as 'near Pyramids of Light', consisted of 10 individuals, and a third site, designated 'Nubian Egypt', comprised seven individuals.

Presence of subchondral, as well as marginal, erosions, characteristic of spondyloarthropathy (Rothschild and Woods 1991), was observed both in this study and that of Kilgore (1989), as clarified by Rothschild *et al.* (1999). Lack of peri-articular loss of bony density in this study and that of Kilgore (1989) is also characteristic of spondyloarthropathy (Resnick and Niwayama 1988; Rothschild 1982; Rothschild and Woods 1991). The frequency of spondyloarthropathy in the Meriotic Nubian sample was 4%, compared with 6% in the Mali sample and one of 10 Egyptians. Ruffer and Rietti (1912) had reported a 3000 years BP individual with diffuse involvement. Africans were clearly afflicted with a form of spondyloarthropathy, perhaps of the Reiter's or psoriatic variety.

Asia Goldstein *et al.* (1976) report findings (spinal fusion) and illustrate sacroiliac erosions compatible with a diagnosis of spondyloarthropathy in 200 years BP Negev Bedouin. Rothschild and Rothschild (1992) reported several examples from a 2000 years BP ossuary at Ein Gedi, Israel.

9.2.3 Implications

The wide spectrum of susceptible animals and increasing frequency through time suggest that whatever predisposes to spondyloarthropathy may have a hidden benefit. It is unclear whether this is inherent in the disease or in a factor linked to its susceptibility.

9.3 Treponemal diseases

Three patterns of peripheral osseous changes characterize the treponomal diseases (Rothschild and Rothschild 1995, 1996a,b; Rothschild *et al.* 1995a): low population frequency, pauciostotic; high population frequency, pauciostotic; and high population frequency, polyostotic (see below). These patterns are indistinguishable from those reported in clinical studies (as discussed below individually for each treponematosis), with perhaps one exception. Skeletal examination revealed unilateral tibial involvement only in syphilis.

An unanticipated observation has been the infrequency of osseous, or even dental, evidence of congenital syphilis in the archeologic record. Hypoplasia, reducing chances of tooth survival, probably explains the rarity of documentation of pathognomonic dental changes (El-Najjar *et al.* 1978; Mansilla and Pijon, personal communication). Observation of periosteal reaction in a subadult is actually much more common in yaws, and actually quite rare in syphilis (Rothschild and Rothschild 1994b,c, 1995, 1996a,b, 1997; Rothschild *et al.* 1995a). Review of the clinical literature revealed that periosteal reaction from congenital syphilis is a short-lived event, usually with complete remodeling so that it is no longer recognizable within 3 months of occurrence (McLean 1931; Levin 1970). Thus its archeological record rarity (<5% of subadults affected) is not surprising.

Treponemal infection was suspected in an extinct North American bear, *Arctodus simus*, on the basis of peripheral and axial skeleton erosions with exuberant reactive new bone formation, draining 'gumma', and a characteristic spiculated periosteal reaction (Rothschild and Turnbull 1987). Immunofluorescent analysis revealed focal clumps of treponemal antigen in the margins of the vertebral erosion. Thus, the treponemal origin of the lesions was documented. This demonstrated an early occurrence in North America. Syphilis, variously referred to as the great pox, Venetian, Neapolitan, and French disease, and initially confused with other diseases

(e.g. leprosy) (Baker and Armelagos 1988), has controversial origins (Hackett 1963). Assigning responsibility is complicated by the diagnostic vagueness of the historical written record (Baker and Armelagos 1988; Quetel 1990; Froment 1995), and our present inability to distinguish biochemically, histologically, microbiologically, immunologically, or even with sophisticated DNA techniques, between yaws, bejel (non-venereal syphilis), and venereal syphilis (Norris *et al.* 1993). Many investigators have found pre-Columbian evidence of treponemal disease in both the New and Old Worlds (Hudson 1958; Hackett 1963; Bogdan and Weaver 1992; Skinner 1995). However, confident recognition of syphilis among the treponemal diseases has until recently been problematic (Rothschild and Rothschild 1995).

Patterns identical to those previously documented as reproducible for syphilis were identified 600 years BP in Michigan and West Virginia, 700 years BP in Florida, 800 years BP in Ecuador, 1000 years BP in Wisconsin, and 1500 years BP in New Mexico, but in no pre-Columbian European, African, or Asian sites examined (Rothschild and Rothschild 1995, 1996a,b; Rothschild *et al.* 1995b). The worldwide presence of yaws, geographic separation of syphilis and yaws in the New World, and chronologic progression to recognizable syphilis (as a regional phenomenon) suggest that syphilis derived from yaws.

9.3.1 New World

North America

Baker and Armelagos (1988) speculated that venereal transmission was a post-Columbian event in North America. Others have not been so certain, instead identifying non-venereal treponemal disease (Powell 1995). Such studies have speculated (on the basis of examination of 'classic cases') that treponemal infection in North America was likely yaws or bejel. Saint-Hoyme (1969) specifically reported on yaws-like changes in the Midwest of North America. Validated criteria (Rothschild and Rothschild 1995) allowed verification of their perspective, further clarifying yaws and later syphilis in such North American sites (Rothschild and Rothschild 1994b,c, 1995; Rothschild *et al.* 1995a).

Sites examined included Windover, Ward, Carrier Mills, Frontenac Island, Lu-25, Oconto County, Palmer, and Amaknak Island (Table 9.1). Amaknak Island is in the Bering Straits, the Ward site in Kentucky, Carrier Mills in Illinois, Frontenac Island in New York State, Lu-25 in Alabama, Oconto County in Wisconsin, and Windover in Florida. The latter represents the oldest known skeletal population from North America. Six populations from the Colorado Plateau were also examined. Ninety-one adult individuals from Kuaua (dated at 650–400 years BP), 89 from Pottery Mound (dated at 650–500 years BP), 27 from Mattock (dated at 1650–850 years BP), 40 Pueblo I Anasazi (dated at 1300–1100 years BP), 25 Basketmaker III (dated at 1500–1800 ypb), and five adult Basketmaker II from White Dog Cave (dated at 2000 years BP) (Emslie 1981) were examined. Skeletons from four additional sites located south of the Mogollan Rim in Arizona were also examined. Eleven adults from Palo Pardo (dated at 1250–1450 ypb), 15 from the Chinchera site (dated at 1300–1100 byp), 173 from the Grasshopper site (dated at 1250–1450 years BP), and 139 individuals from the Turkey Creek site (dated at 1200–1400 years BP) (Emslie 1981) were examined.

Skeletal remains from the earliest sites (769 individuals) were examined (Table 9.1). Periostitis was identified in the Windover, Ward, Carrier Mills, Frontenac Island, Lu-25, Oconto County, Palmer, and Amaknak Island sites. The population frequency, subadult affliction frequency, polyostotic nature of disease, incompleteness of sabre shin remodeling, frequency of hand and foot involvement, and absence of unilateral tibial disease were characteristic of yaws.

As a result of these studies, the history of treponemal disease in the New World can be definitely traced almost 8000 years. It is suggested that yaws migrated to the New World with the first Asian immigrants.

Two clearly separable patterns of treponemal disease were noted in the southwest. Anasazi who died less than 1800 years BP and Mogollan who died less than 1100 years BP had a low population frequency (<13%) of treponemal disease, pauciostotic (less than two bone groups affected) in distribution, with occasional unilateral tibial involvement, and sabre shin remodeling. The latter was frequently so complete as to efface all surface indications of periosteal reaction. The hands and feet were spared. This pattern was characteristic of syphilis and quite distinct from that noted for yaws and bejel (Rothschild and Heathcote 1993; Rothschild and Rothschild 1994b,c, 1995; Hershkovitz et al. 1995; Rothschild et al. 1995a). Consideration of subadult sparing further substantiates this conclusion. Anasazi who died more 1800 years BP and Mogollan who died more than 1100 years BP had a very different pattern. Both had high population frequency of periosteal reaction, with frequent subadult affliction (especially notable related to the relatively low number of juveniles available—13 individuals—for examination). Disease was polyostotic. The extreme of distribution noted in the White Dog site is representative of the population variation expected in yaws (Rothschild and Heathcote 1993; Rothschild and Rothschild 1994b,c, 1995; Rothschild et al. 1995a). Hands and feet were frequently affected. Unilateral tibial involvement was not noted and sabre shin remodeling was always limited (such that surface periosteal reaction was always recognizable). While both groups (Anasazi and Mogollan) experienced replacement of yaws by syphilis, the time course of replacement clearly differed. It is of interest that the transition periods (from yaws to syphilis) above and below the Mogollan rim both represented time of onset of drought cycles.

Review of published literature on treponemal disease revealed Nebraska and Iowa sites (Schermer et al. 1994) with sufficient documentation for application of criteria for distinguishing among the treponematoses. Those sites still have the characteristics of yaws (Rothschild and Heathcote 1993; Rothschild and Rothschild 1994b,c, 1995; Rothschild et al. 1995a) and suggest a transition time later than that noted on the Colorado plateau.

Syphilis thus appears to be a North American disease that transmuted from yaws c. 2000 years BP on the Colorado plateau. The time course of mutation (disappearance of yaws and first observation of syphilis) appears to vary within the North American continent (Rothschild and Rothschild 1994b,c, 1995; Rothschild et al. 1995a).

Character and distribution of disease west of the Sierras Skeletal populations from the western

Table 9.1 Characteristics of Bering Strait and earliest North American treponemal disease contrasted with documented syphilis and yaws

	Syphilis[a]	Yaws[a]	Windover	Ward	Carrier Mills	Frontenac Island	Amaknak Island	Lu-25	Oconto WI
Site age (years BP)[b]	70	500	7900	4300	6300	2000	2000	4300	3250
Population size, total	2906	40	112	203	159	63	16	89	37
Youth affected > 5%	No	Yes	Yes	Yes	Yes	Yes	c	Yes	Yes
Sabre shin without periostitis	Yes	No	No	No	No	No	No	No	No
Unilateral tibial involvement	Yes	No	No	No	No	No	No	No	No
Average number of bone groups affected ≥ 3	No	Yes	Yes	Yes	Yes	Yes	Yes	Yes	Yes
Hand or foot commonly affected	No	Yes	Yes	Yes	Yes	d	Yes	Yes	Yes
Frequency perspectives									
Percentage at risk of population affected	5	33	27	36	31	38	27	40	35
Percentage of youth affected	<1	14	12	20	16	13	c	17	18
Average number of bone groups affected	1.9	4.0	3.4	3.1	3.1	3.1	3.3	3.0	3.2

[a]Standard criteria for disease recognition, derived from Rothschild and Heathcote (1993); Rothschild and Rothschild (1995).

[b]Rothschild et al. (1992).

[c]Not present, but inadequate numbers (five subadults) to rule out.

[d]Hands and feet were absent from burials, precluding analysis.

coast of North America were clearly afflicted with a treponemal disease very different from that previously documented elsewhere in North America. Six populations from west of the Sierra Cascades were compared with five sites east of the Cascades, one from the Aleutians, and one from Alaska. Eighty-three adult individuals from Los Palmos (Baja peninsula, dated at 500–800 years BP); 88 from CCo295 (San Francisco Bay area, dated at 2250 years BP); 13 from Baldwin, six from Parizeau Point, and 26 from Lachane (all from Prince Rupert Harbor, British Columbia, dated at 3200 years BP); and 27 from Crescent Beach (coastal British Columbia, dated at 3000–4000 years BP) were examined. Additionally, 500 individuals from Humboldt and Carson Sink, Nevada (9000–1000 years BP), Point Hope (Alaska, 300–600 years BP), and Amaknak Island (Aleutians, 1900 years BP) were examined. A high population frequency (both in adults and subadults) of pauciostotic periostitis was noted in the six western skeletal populations, identical to those reported previously with bejel in Negev Bedouin, Sudanese Nubians, and the Kish site from Iraq.

The pattern of treponemal disease on the west coast of North America was markedly different from either the early (yaws) or later pattern (syphilis) noted east of the Cascades. The west coastal region from Canada to Mexico is characterized by a high population frequency (21–35%) of treponemal disease (both in adults and subadults), pauciostotic in distribution, and invariably bilateral in its tibial distribution (Table 9.2). Sabre shin remodeling was always incomplete, with surface evidence of periosteal reaction easily recognized. This pattern was characteristic for bejel and quite distinct from that noted for yaws and syphilis (Rothschild and Heathcote 1993, Rothschild and Rothschild 1994b,c, 1995; Hershkovitz et al. 1995; Rothschild et al. 1995a).

While yaws was clearly the original treponemal disease to enter central North America through Beringia, and was clearly present east of the Cascades (Rothschild and Rothschild 1994b,c, 1995, 1996a,b; Rothschild et al. 1995a), the western coast of North America was just as clearly host to a very different treponemal disease, bejel. Replacement by syphilis was noted in the Los Palmos, Baja region

only after 1700 (Molto 1995), significantly later than observation east of the Cascades, and corresponding to the timing of European contact.

Mexico Six sites were chosen to assess the population frequency, nature, extent, and character of periosteal and other osseous reaction. These included Tlatilco IV, a pre-Classic site dated at 3100 years BP, two Marismas sites (Tecualilla and Chalpa Nay), dated 1150–1300 Common Era (CE), and excavations of three subway tunnels, dated 1300–1521 CE (Table 9.2).

This study confirms Goodman's work (Goodman et al. 1988) indicating that periosteal reaction is very common in such pre-Columbian skeletal populations. Their interpretation (that this is simply an indicator of stress) is clearly rebutted by comparisons with first millennium and post-Columbian English populations, which lack evidence of significant periostitis. This (periosteal reaction) is clearly a treponemal phenomenon, indistinguishable from populations documented as afflicted with treponemal disease (Rothschild and Heathcote 1993; Rothschild and Rothschild 1995; Rothschild et al. 1995a).

The population pattern of periosteal disease clearly demonstrates the presence of three sequential phenomena (Table 9.2), each quite different from the preceding phenomenon and from one another. The character of periosteal reaction clearly was uniformly represented in Tlatilco and Marismos sites, spanning 2000 years. It is only in more recent sites that significant changes were noted, clearly reflecting the appearance of one and then another new disease. While all three disorders were recognized as treponemal in nature, there clearly was a transition between the three diseases. The initial disease had all the characteristics of bejel (Hershkovitz et al. 1995; Rothschild and Rothschild 1995a). Recognition of bone involvement in 32–48% of the Tlatilco and Marismos skeletal populations suggests that essentially the entire population was afflicted, a phenomenon typical of bejel (Hershkovitz 1995).

The time period characterized by tunnel #2 witnessed replacement of the initial disease, bejel, with a new disease (Table 9.2), with all the findings previously documented as characteristic of yaws (Rothschild and Heathcote 1993; Rothschild and Rothschild 1995). This disease is easily distinguished

Table 9.2 Skeletal manifestations of treponemal disease in early Mexican populations

Population Provenience	Marismas			Subway		
	Tlatilco 1100 BCE	Tecualilla 1150–1300 CE	Chalpa Nay	Tunnel #2 1300	Tunnel #1	Tunnel #3 1521 CE
Population no.	125	32	25	64	96	50
Percentage afflicted	32	47	48	27	10	12
Subadult no.	28	5	2	36	44	18
Percentage afflicted	14	40	[a]	14	0	6
Tibial						
Unilateral	No	No	No	No	Yes	Yes
Sabre without periostitis	No	No	No	No	[a]	[a]
Hand/foot affected (%)	1	0	0	16	0	0
Average no. of affected bones	2.4	2.1	2.6	4.5	1.7	1.5
Diagnosis[b]	Bejel	Bejel	Bejel	Yaws	Syphilis	Syphilis

[a] Inadequate number present to assess.
[b] Derived according to criteria documented in Rothschild and Rothschild (1995a,b, in press).
CE, Common Era; BCE, before Common Era.

from the more pauciostotic syphilis ($\chi^2 = 3.973$, $P < 0.05$), in which hand and foot, and subadult, affliction are rarely observed in skeletal populations (Rothschild and Rothschild 1994b, 1995; Rothschild *et al.* 1995a). Other evidence for syphilis (in the form of complete sabre shin surface remodeling and unilateral tibial disease) was also lacking. Disease in the tunnel #2 population was also easily distinguished from the more pauciostotic bejel, which infrequently affects hands and feet ($\chi^2 = 13.91$, $P < 0.001$) (Hershkovitz *et al.* 1995; Rothschild and Rothschild 1995a).

Another transition clearly occurred between the tunnel #2 and tunnels #1 and #3 sites. The polyostotic pattern became pauciostotic ($\chi^2 = 6.088$, $P < 0.01$), with infrequent involvement of hands and feet, but with new observation of unilateral tibial affliction. The pattern had changed from that of yaws to that of syphilis (Rothschild and Rothschild 1994b, 1995; Rothschild *et al.* 1995a).

The character of treponemal disease remained stable for over 2000 years, until the advent of the Aztecs. Migrating from northern Mexico, they apparently populated the area below the Mogollan plateau, in which yaws was endemically represented. The juxtaposition of timing of documentation of yaws in the Tenochtitlan–Tlatilco region (capital of the Aztec civilization) and the advent of Aztec conquest of the region, suggest the culpability of the Aztec invaders. If Aztecs were infected with yaws in their original northern Mexico habitat, it would not be surprising that the disease was transported to the central Mexican plateau. As the Aztec culture permeated the area, displacing the endogenous populations, one disease (bejel) was replaced by a second disease (yaws), already endemic in Aztecs. If it is assumed that the Aztecs were responsible for the introduction of yaws, perhaps tunnels #1 and #3 represent victims of another contacting culture, that of Spain. If Spanish mainland contamination by syphilis was as epidemic as suggested (Quetel 1990), it would not be surprising that Spanish sailors transmitted it back to sites of Spanish conquest in the New World.

Additional literature perspectives Baker and Armelagos (1988) speculated that venereal transmission was a post-Columbian event in North America. Others have not been so certain, instead identifying non-venereal treponemal disease (Schermer *et al.* 1994; Powell 1995). Such studies have speculated (on the basis of examination of 'classic cases') that treponemal infection in North America was likely yaws or bejel. Saint-Hoyme (1969) specifically reported on yaws-like changes in the Midwest of North America. Validated criteria (Rothschild and Rothschild 1995) allowed verification of their perspective, further clarifying yaws and later syphilis in such North American sites (Rothschild and Rothschild 1994b,c, 1995; Rothschild *et al.* 1995a).

South America and the Caribbean

Burgos *et al.* (1994) identified antigens of *Treponema* in a 5000 years BP South American. While treponematosis was clearly present, the specific disease was not identified. While yaws was identified in early Patagonia and syphilis in 700 years BP Ecuador (Rothschild *et al.* 1995a), most sites have not yet been sufficiently evaluated to determine the nature of the underlying disease.

Denouement

Treponematoses were apparently transported to the New World by way of at least two migrations, one bringing yaws, the other bejel. The population with bejel likely derived from a population different from that with yaws. Given the absence of treponemal disease variation in the very wide spectrum of environments represented by the bejel-afflicted populations, it is clear that environment is not the factor determining disease character. This study expands on animal studies documenting that the individual treponematoses are separate diseases and not simply climate-induced 'variations on a theme'.

9.3.2 Old World

Africa

The first recognized polyostotic periosteal disease was in KNM-ER 1808, dated at 1.6 million years BP (Rothschild *et al.* 1995b). Diaphyseal periosteal reaction of up to 7 mm thickness was noted in humeri, radii, ulnae, femora, tibia, and in a partial fibula. Similar periosteal reaction in the isolated humerus KNM-ER 737 was also noted. Although originally thought to be a hypervitaminosis A-related disorder, new understanding of the nature of hypervitaminosis A made that diagnosis untenable for KNM-ER 1808 and 737 (Rothschild *et al.* 1995b). Among the treponemal disorders, only yaws is frequently polyostotic and only yaws frequently affects the humerus. Thus the antiquity of treponemal disease is established with the origins of humans in Africa. Further, the first form of treponemal disease appears to have been yaws. While Cockburn (1963) suggested (apparently on the basis of blood serology) that yaws may have had its origins in Pleistocene apes, no non-human primates have yet been confirmed with bone lesions of the yaws variety. Wright (1971) reported an alleged occurrence of syphilis in Neandertals. Subsequent information suggests that it was actually yaws.

Bejel was clearly present in northeast Africa 15 000 years BP. The bejel in Wadi Halfa is indistinguishable in character and population frequency from that noted 13 000 years later in Meriotic Sudanese, in Iraq 4000 years BP, and the Negev 2000 years BP. Study of North African populations revealed bejel in the Sudan (Rothschild and Rothschild 1996b), while yaws was prominent in southern Africa.

Europe

It is suggested that yaws was the resident treponemal disease in Europe, replaced by syphilis in the late fifteenth or early sixteenth centuries. Interpretation of the historical record suggested that the non-venereal treponemal disease—yaws—was present in the British Isles and replaced by syphilis sometime between the late fifteenth and early sixteenth centuries (Cockburn 1995). Thus, sibbens in Scotland appears to represent what is today recognized as yaws. That assessment is compatible (on a time line basis) with the Columbian hypothesis of New World origins of syphilis. Three sites from the pre-Columbian millennium and a post-Columbian British site were chosen to assess the population frequency, nature, extent, and character of periosteal and other osseous reaction. These included Poundbury (a 4th- to 5th-century Romano-British cemetery), Cannington (a 6th- to 10th-century British cemetery), Winchester (an 11th- to 14th-century site consisting of individual and ossuary burials), and Spitalfields (an 18th- to 19th-century—i.e. post-Columbian—cemetery site).

The time course of population health between the Cannington (6th–10th century) and Winchester (11th–14th century) sites is clearly reflected by the appearance of a new polyostotic disease (Table 9.3). That new disease has all the characteristics of the treponemal disease yaws. Recognition of bone involvement in one-fifth to one-third of the skeletal population suggests that essentially the entire population was afflicted, a phenomenon typical for yaws (Moss and Bigelow 1922). This disease is easily distinguished from the more pauciostotic syphilis ($\chi^2 = 3.973$, $P < 0.05$), in which hand and foot and subadult affliction are so rarely observed in skeletal populations (Rothschild and Rothschild 1994b, 1995; Rothschild *et al.* 1995). Other evidence for syphilis (in the form of complete sabre shin surface remodeling and unilateral tibial disease) was also lacking. Disease in the Winchester cemetery was also easily distinguished from the more pauciostotic bejel, which infrequently affects hands and feet (Hershkovitz *et al.* 1995; Rothschild and Rothschild 1995).

While the reports by Hurley *et al.* (1994) of a 14th-century (period III) individual aged 12–13 suggested syphilis, examination of the actual skeleton revealed changes unlike those of treponemal disease and actually representative of histiocytosis.

The observations in 13th-century England confirm those of Goodman (Goodman *et al.* 1988), that periosteal reaction is very common in such pre-Columbian skeletal populations. Their interpretation (that this is simply an indicator of stress) is clearly rebutted by comparisons with first millennium and post-Columbian English populations. Stirland (1995) reported polyostotic disease in the St Margaret Fyebridgegate cemetery (usage 1254–1468 AD). Her description (in selected skeletons) fits that of yaws and appears indistinguishable from the findings reported above for the same time period.

Another transition clearly occurred between the 11th- to 14th-century (Winchester) and 18th- to 19th-century (Spitalfields) epochs. The initially

Table 9.3 Skeletal manifestations of treponemal disease in early British populations

Population	Poundbury	Cannington	Winchester	Spitalfields
Century	4th–5th	6th–10th	11th–14th	18th–19th
Population no.	210	112	34	75
Percentage afflicted	0	0	24	9
Subadult no.	19	12	5	3
Percentage afflicted	0	0	20	0
Tibial				
Bilateral (%)	–	–	100	71
Sabre shin (%)	–	–	23	43
Sabre without periostitis (%)	–	–	0	100
Hand/foot affected	–	–	Yes	No
Average number of affected bones	–	–	3	1.9

polyostotic pattern became pauciostotic, with infrequent involvement of hands and feet, but with unilateral tibial affliction and a new degree of sabre shin remodeling. The latter had become so extensive that in the Spitalfields site, surface evidence of periosteal reaction was often no longer detectable. The pattern had changed from that of yaws to that of syphilis (Rothschild and Rothschild 1994b, 1995; Rothschild *et al.* 1995). These observations provide further support for the onset of syphilis in England as a post-Columbian event. A mystery remains: treponemal disease was clearly absent in England, Germany, and Hungary from the 5th and perhaps through the 11th century.

Asia

Study of Asian populations revealed bejel in 19th-century Arabian bedouins and in Iraq, 4000 years ago (Rothschild and Rothschild 1996b). While treponematoses were clearly present in the Asian Pacific (Rothschild and Heathcote 1993), only Guam has had a modern analysis (Rothschild and Heathcote 1993).

Distributional data suggest that syphilis and bejel represent mutations from a parent disorder, yaws. While some might suggest recurrent mutation, the epidemiologic pattern suggests a single mutation, with subsequent spread. In all areas where yaws or bejel were replaced by syphilis, there is no evidence of regression (to the former variety). The pattern of first identification of syphilis in a region is also geographically consistent. Further

supporting the replacement hypothesis is *de novo* recognition of syphilis in areas wherein there was no evidence of treponemal disease (e.g. periosteal reaction and sabre shin changes). The time course of recognition of syphilis invariably represents contact with populations in which syphilis was previously documented. The Northwest Territories of Canada and pre-Russian-contact Alaska were apparently free of treponemal disease. So, too, was eastern Canada. There is no evidence of treponemal disease in the Archaic Maritimes and at least the eastern portion of Ontario. This contrasts with an earlier presence of yaws in Manitoba, Minnesota, Ohio, and New York.

9.4 Tuberculosis

9.4.1 Tuberculosis as a model for recognition of disease across species lines and understanding its significance

Isolated cases of vertebral body destruction with gibbus formation abound, although most not validated (Verano and Ubelaker 1992). Some cases in mummies have been validated by recognition of the causative organism on microscopic examination (Zimmerman 1979) and by detection of *Mycobacterium tuberculosis* DNA (Salo *et al.* 1994).

As tuberculosis appears to be predominantly a population-dependent disease, its origins in Chile seem to coincide with new immigrations (e.g. Cabuza-Tikwanaku people) around 1500 years BP

(Arriaza *et al.* 1995). They suggested a population frequency of bone disease of 2%, confirming Garcia-Frias's (1994) recognition of acid-fast bacilli in Inca mummies. Rib periostitis, reported in 18% of 306 individuals from Chichester, Sussex, England, was attributed by Roberts *et al.* (1994) to tuberculosis (or at least to a pneumonic process). Tuberculosis is widely distributed in humans during the Holocene.

Tuberculosis has now been traced back to the Pleistocene (Rothschild *et al.* 2001). Natural Trap Cave, in the foothills of Little Big Horn Mountain, is simply a hole in the middle of a game trail. The hole is not visible until one is almost on top of it. This lack of visibility allowed unbiased sampling of passing herds and their accompaniment. Thus bears, lions, cheetahs, wolves, wolverines, martens, marmots, and voles were found, as well as antelope, horses, camels, sheep, bison, musk oxen, mammoths, pika, lemmings, and rabbits. Among the animals that fell into Natural Trap Cave, tuberculosis-like pathology was limited to three genera, showing marginal erosions, undermining articular surfaces. They were found in the metacarpal phalangeal joints of *Ovis* (big-horn sheep), *Bison*, and *Bootherium* (musk ox), affecting 15–25% (Rothschild and Martin 1993). Granulomas extending from marginal areas of joints, undermining subchondral bone, are characteristic of tuberculosis and perhaps of other granulomatous disorders (Resnick and Niwayama 1988). However, extracted DNA amplified by PCR revealed unequivocal tuberculosis (Rothschild *et al.* 2001).

Vertebral destruction with partial collapse, minimal reactive bone formation, and fusion of the affected vertebrae was noted in Queensland F178989 (Rothschild *et al.* 1997). The latter was a set of fused proximal caudal macropod vertebrae from the Pliocene of Chinchilla, Australia. The gross and radiologic appearance was that of partial collapse of one vertebra, angulated 90 degrees in juxtaposition to a second, and associated with extensive new bone formation. There is a complete posterior dislocation of a vertebral body. Three-dimensional computerized tomography (CT) reconstruction revealed the classic collapse and fusion with no significant new bone formation. That represents the earliest identified case.

The concept that tuberculosis is a result of domestication and derived from a bovine variety now seems unlikely. *M. tuberculosis* is the parent, not the child, of *M. bovis*. To understand this new concept it is important to understand what has been identified previously. *M. tuberculosis* is a specific member of what has been termed the *M. tuberculosis* complex. The latter includes *M. tuberculosis*, *M. bovis*, *Mycobacterium microti*, and *Mycobacterium africanum*. Until recently, tests on PCR-amplified DNA did not distinguish between these bacteria. Ancient DNA is by its nature fragmented, such that simple sequencing of a tiny fragment of the tuberculosis genome would not allow the required specificity. Polysegmental analysis by spoligotyping, however, simultaneously analyzes a large number of DNA fragments. Among this group are four that are present in *M. tuberculosis*, but never in *M. bovis*. Presence of these four spoligotype groups in the fossil from Natural Trap Cave eliminates the possibility of *M. bovis*. The pattern was that of *M. tuberculosis* and is far earlier than the domestication of cattle. Similar metacarpal lesions were common in late Pleistocene (20 000 years BP) to Holocene *Bison*, *Ovis*, and *Bootherium*, showing a broad distribution in bovids that might have served as a source of tuberculosis transmission to humans.

Given the number of cases in the Amerindian skeletal record, tuberculosis may have been endemic in Amerindians of North America. The high frequency of tuberculosis in North American bovids and their position in the late Pleistocene food chain make it reasonable to suggest that the original Amerindian tuberculosis derived from bison exposure.

As the human osseous changes of *M. tuberculosis* and *M. bovis* cannot be distinguished, it is only the PCR-amplified material that will reveal the time course of *M. bovis* evolution, possibly from the artificial influence of domestication—but the latter is only a suggestion for future investigation.

The derivation of the *M. tuberculosis* documented in bovids is an intriguing question. Spoligotyping suggests a relationship with *M. africanum*. The latter causes tuberculosis in modern Africa. As the organism afflicting bison appears ancestoral to modern-day tuberculosis and related to *M. africanum*, the roots of modern day tuberculosis may be found in Pleistocene Africa.

9.5 Denouement

Preconceived notions often compromise scientific advancement. Anthropologic concepts of tuberculosis, treponemal disease, and arthritis are prime examples, wherein new technologies permit transcendence of stereotypic paradigms. Contrary to disregarding observations when they refute the popular paradigm, these new techniques provide rigorous independent assessment and the opportunity to surplant promulgated and engrained perspectives/theories.

CHAPTER 10

Evolution and ancient diseases: the roles of genes, germs, and transmission modes

Paul W. Ewald

Evolution is the law of policies: Darwin said it, Socrates endorsed it, Cuvier proved it and established it for all time in his paper on 'The Survival of the Fittest'. These are illustrious names, this is a mighty doctrine: nothing can ever remove it from its firm base, nothing dissolve it, but evolution. (Attributed to Mark Twain in Three Thousand Years Among the Microbes, John S. Tuckey, 1967)

Evolution through natural selection is the fundamental organizing principle of the life sciences. It explains not only how the diversity of life came to be but also provides a framework for understanding what is possible and what is not. Considering that the health sciences is largely a subset of the life sciences we can expect that principles of evolutionary biology will be able to provide some general guiding principles into any particular focus of study within the health sciences. Because the study of ancient infectious diseases involves time periods over which humans and especially pathogens can evolve in response to each other, an evolutionary perspective would seem especially appropriate. However, apart from the use of phylogenetic analyses, inquiry into ancient diseases has proceeded largely without reliance on evolutionary principles. Now that powerful techniques for analyzing the molecular makeup of humans and pathogens in ancient tissue are becoming available, it seems an especially appropriate time for incorporating evolutionary insights into decisions about what to look for and where to look. This chapter addresses two topics that are currently being developed in evolutionary medicine that also bear on the study of ancient diseases: the evolution of virulence and the scope of infectious causation.

10.1 The evolution of virulence

10.1.1 General theory

For most of the 20th century the prevailing dogma held that disease organisms should eventually evolve toward benign coexistence with their hosts. A corollary is that harmful diseases represent a transitory state of maladaptation (Dubos 1965; Burnet and White 1972). This traditional viewpoint is flawed because it fails to cast the problem in the context of natural selection. Rather than asking whether harmful or mild variants would win in competition with each other over the short run, the focus was on what was stable over the long run. Natural selection is powerless to favor long term stability if the variants that win over the short term destabilize the system. Natural selection may thus favor the evolution of extreme harmfulness if the exploitation that damages the host enhances the ability of the harmful variant to compete with a more benign pathogen, even if this harmfulness destabilizes the association in the long run. If predator-like variants of a pathogen population out-produce and out-transmit benign pathogens, then peaceful coexistence and long-term stability may be precluded much as it is often precluded in predator–prey systems. A few authors expressed reservations about the traditional dogma during the middle of the 20th century (Ball 1943; Cockburn 1963; Coatney *et al.* 1971) but they were largely ignored. Only during the last two decades of the 20th century, was a theoretical framework developed that is consistent with the principles of natural selection.

Paleopathology offers an opportunity to evaluate these ideas by determining whether the pathogens that cause severe modern diseases have been present in humans for long periods of time. Because infectious agents can evolve over very short time periods, the past several thousand years provides ample opportunity for disease organisms to evolve increased or decreased virulence. If harmful agents have been present in humans for several thousand years, this presence offers direct evidence that their current virulence cannot be attributed simply to the newness of their association with humans.

Reciprocally, paleopathology can benefit from this theoretical framework because the framework provides a sense of whether one can expect to find particular pathogens in ancient tissues. Modern theory about the evolution of virulence suggests not only that some modern virulent pathogens may be well adapted to human hosts, but that they may have been maintained as virulent, well-adapted pathogens throughout human existence. We may therefore find them in ancient tissue. The theoretical framework identifies the kinds of pathogens that could be maintained indefinitely in a highly virulent state according to modes of transmission.

10.1.2 Host mobility and virulence

Pathogens that can be readily transmitted from severely ill patients should be molded by natural selection to exploit hosts severely and thus be highly virulent (Ewald 1994a). Several categories of human pathogens fall into this category (summarized in Table 10.1). Biting arthropods can transmit pathogens effectively from immobilized hosts and should therefore favor evolution to higher levels of virulence. Accordingly such vector-borne pathogens are more lethal than directly transmitted pathogens (Ewald 1983).

Plasmodium falciparum malaria is noteworthy because it is one of the most virulent of all pathogens regularly transmitted among humans. The traditional viewpoint would ascribe its high virulence to it recent entry into humans. This viewpoint could have a detrimental effect on scientific inquiry by discouraging attempts to search for *P. falciparum* in ancient remains of humans or mosquitoes. The current view, in contrast, encourages such searches regardless of the antiquity of such samples. A similar argument can be made for *Trypanosoma cruzi* in New World remains. The finding of *P. falciparum* and *T. cruzi* in some of the oldest materials tested thus supports the validity of the current theory about the evolution of virulence (Miller *et al.* 1994; Guhl *et al.* 1997; Cerutti *et al.* 1999).

The traditional argument that high virulence in humans results from a recent entry into humans has been used to explain the harmfulness of *Trypanosoma brucei* in humans (Herbert and Parratt 1979). It might also be feasible for *T. cruzi*, which appears to move frequently between humans and other vertebrates hosts via kissing bugs. Although the argument has been invoked for *P. falciparum* (Waters *et al.* 1991; Rennie 1992; Wills 1996), it is not tenable because the evidence

Table 10.1 Associations between lethality of infections and the dependence of pathogens on host mobility for transmission

Characteristic allowing transmission from immobile hosts	Association with lethality	Reference
Arthropod-borne	Lethality higher among arthropod-borne pathogens than among directly transmitted pathogens	Ewald (1983)
Water-borne	Lethality of diarrheal bacteria correlated with tendencies for water-borne transmission	Ewald (1991a); Ewald *et al.* (1998)
Attendant-borne	Lethality of *Escherichia coli* correlated with duration of attendant-borne cycling	Ewald (1991b)
Durability in the external environment	Lethality of respiratory tract pathogens correlated with durability	Walther and Ewald (manuscript in preparation)

suggests that *P. falciparum* has been transmitted exclusively in human–mosquito cycles for a minimum of about 10 000 years and perhaps for much longer. This time scale represents vastly more time than would be necessary for evolutionary changes in virulence.

The modern evolutionary perspective does not dismiss the possibility that harmful pathogens may be harmful in part because they are recent and have therefore not yet lost through evolution negative effects that provide no compensating competitive benefits. It emphasizes, however, that although sources of virulent human diseases may often be recent and zoonotic, virulent vector-borne pathogens may be maintained indefinitely in human populations. The modern theory therefore encourages studies that attempt to seek evidence of virulent vector-borne pathogens in ancient populations. Reciprocally, as is the case with *P. falciparum*, the finding of such ancient associations lends credibility to the modern view that vector-borne pathogens can maintain high levels of virulence indefinitely.

Like vector-borne pathogens, water-borne pathogens should evolve to relatively high levels of virulence because they can be transmitted from immobilized people. Reliance on the mobility of infected hosts is low for water-borne pathogens because the waste disposal activities of attendants and the movement of water can contaminate sources of drinking water. Accordingly, the lethality of diarrheal bacteria is positively correlated with the extent to which they are water-borne (Ewald 1991a). Similarly, the prevalence of harmful species of *Shigella* relative to mild species is greater in areas where the potential for water-borne transmission is greater, and has declined wherever water supplies have become protected (Ewald 1991a).

The recent outbreak of cholera in Latin America provides an opportunity to determine whether a specific pathogen can evolve changes in virulence over a much shorter time period. The causative bacterium, the El Tor biotype of *Vibrio cholerae*, is particularly amenable to such study because its virulence can be quantified *in vitro* by measuring toxin production. The bacterium was first isolated in Peru in January 1991. It had spread to neighboring countries within 2 months and throughout most of Latin America within a year. In Chile, where access to safe water supplies is relatively good, the toxigenicity of isolates declined as the decade progressed, whereas in countries with poorer access to safe water no decline occurred (Ewald *et al.* 1998, and unpublished data).

Searching for harmful water-borne pathogens such as *Shigella dysenteriae* type 1, *Salmonella typhi*, and *V. cholerae* in ancient human remains is therefore justified by current theory. Water-borne pathogens would be difficult to isolate from human remains, because they tend to cause acute infections of short duration or are present in low densities in the bodies if they cause persistent infections. Searching for traces of them in waste disposal sites may be more fruitful because they can be shed in vast quantities in fecal material; moreover, the damage they cause may have led to the abandonment of the sites and hence left them as the last layer deposited at the site.

The Harrapan civilization, which existed from 3000 to 1800 BC along the Indus river valley, offers an illustration. Harappan cities were characterized by elaborate sewerage systems. Waste water flowed through brick-lined drainage canals just below the roads, and emptied into soak pits, from which it would gradually seep into the ground (Mackay 1931). In houses, wastewater flowed from interior channels through outside walls into the road drainage system, a pit or a large pot (Mackay 1931; Wheeler 1968). Waste water, collected in the pots, would seep into the ground through holes in the bottom of the pot, or, in pots without holes, would be carried and emptied into the street drainage system or allowed to overflow (Mackay 1931). This drainage system formed a network around the open wells (Gokhale 1959), the rims of which were typically no more than a few inches high and were often situated next to drain openings (Mackay 1931). The water supplies in this system must have been highly prone to contamination, especially because extensive flooding was known to occur in Harappan cities. Water-borne transmission must therefore have been common, and evolutionary effects of water-borne transmission on virulence would be expected.

Harappan cities were abandoned during the first few centuries of the second millennium BC (Ghosh 1982). Hypotheses based on warfare, earthquakes, deforestation, climatic changes, and reduced trade have been advanced, but most are inconsistent with available evidence (Dales 1964; Ghosh 1982) and none seems sufficient to explain the asynchronous but persistent de-urbanization within a century or so (Ewald 1991b). The association between water-borne transmission and the evolution of virulence offers a hypothesis that is consistent with available evidence: the infrastructure facilitated water-borne transmission, which led to the evolution of lethal diarrheal disease and the abandonment of urban areas.

Cholera seems the most likely culprit because of its lethality, the clinical and epidemiological suddenness of its onset, and its persistence in South Asia. When cholera is transmitted by sewerage-contaminated water it can spread explosively through urban populations and kill a high proportion of infected individuals, some within a few hours of the first signs of illness (Snow, 1855). As Charles Rosenberg (1962) put it, 'It was not easy for survivors to forget a cholera epidemic. The symptoms of cholera are spectacular; they could not be ignored. The victim often "felt no premonition of cholera at all ... until he pitched forward in the street, 'as if knocked down with an axe'."' The sudden deaths of hundreds of city dwellers and the absence of such mortality in rural areas could generate a strong motivation to abandon urban areas, especially before people were aware of the biological causes of infection and protective measures. Indeed abandonment of urban hot-spots in response to cholera outbreaks occurred regularly during the 19th century prior to the establishment of the germ theory (Rosenberg 1962).

An illness indistinguishable from cholera can be traced back to the second or third millennium BC from Sanskrit Vedic literature (W. B. Greenough, personal communication). Techniques for DNA extraction offer a chance to test directly whether cholera or some other virulent modern pathogen was present in urban centers at the end of the Harappan civilization. Heaps of solid waste have been uncovered as they were left in the streets by sanitary squads who periodically cleaned this drainage system (Marshall 1931; Wheeler 1966, 1968; Rao 1973). The microbes can be searched for in these haystacks using the polymerase chain reaction (PCR). Residues in the soak pits offer a similar opportunity, which may be even more informative because of sedimentary stratification. If a severe diarrheal pathogen was responsible for the deurbanization, it should be present particularly in the upper layers of any undisturbed sediment in the soak pits. *V. cholerae* would be the most likely target, for the reasons mentioned above. *S. dysenteriae* (the most lethal agent of dysentery) and *S. typhi* (the agent of typhoid fever) are also sufficiently water-borne and virulent (Ewald 1991a) to warrant testing; they might contribute to deurbanization separately from or in conjunction with *V. cholerae*.

Pathogens that are durable in the external environment can also be transmitted from very ill people because they can reach susceptible individuals by relying on the mobility of susceptibles rather than on the mobility of the infecteds. Searching for durable pathogens such as smallpox virus, *Mycobacterium tuberculosis*, and *Corynebacterium diphtheriae* in ancient tissues is therefore consistent with current theory even though it would have been a tenuous venture according to the traditional view. The report of *M. tuberculosis* in South American material dated as far back as 4000 years BP (Salo *et al.* 1994; Arriaza *et al.* 1995) is particularly interesting in this regard because it is often assumed that *M. tuberculosis* first became established in humans after the onset of agriculture when humans and ungulates were in close contact. There is an inclination among experts to assume that these infections were largely zoonotic and the specificity of the testing for *M. tuberculosis* as opposed to *Mycobacterium bovis* in the South American samples has been questioned (e.g. Diamond 1997). The high durability of *M. tuberculosis* in the external environment (Mitcherlich and Marth 1984), however, makes a longer history of *M. tuberculosis* as a virulent, fully human pathogen feasible. A longer history of human *M. tuberculosis* is also consistent with the time period of 15 000–20 000 estimated for nucleotide divergence of *M. tuberculosis* isolates from geographically diverse human populations (Kapur *et al.* 1994). Humans could have become stably infected with virulent *M. tuberculosis* long before the onset of agriculture and brought *M. tuberculosis* with them when they migrated from Asia across the Bering strait. Alternatively, ancient humans in the New World might have acquired

M. tuberculosis independently from infected New World ungulates (Rothschild *et al.* 2001). More extensive testing of DNA sequences will be needed to reconstruct the phylogenetic affinities of the various isolates and thus resolve the matter.

10.1.3 Virulence of sexually transmitted diseases

For sexually transmitted diseases the evolutionary tradeoffs associated with virulence are shaped by the timing of opportunities for sexual transmission. Specifically, the success of sexually transmitted pathogens depends on their ability to persist in and be transmissible from a host over time periods that span successive changes in sexual partnerships. Sexually transmitted pathogens may therefore need cell and tissue tropisms that allow them to persist for long periods in the face of the immune system.

Increased potential for sexual transmission should favor those pathogen variants that reproduce more extensively sooner after the onset of infection. Such an evolutionary increase in host exploitation when coupled with persistence within the host should tend to increase the probability that negative side effects of infection will occur at a later time—a large wrench left in a machine is likely to cause greater damage than a small wrench. The kind of damage depends on the cell tropisms and activities of the pathogen. For HIV, which replicates and buds virions from leukocytes such as macrophages and helper T cells, the damage manifests itself as a decimation of the immune system. For human papillomavirus (HPV) and human T lymphotropic virus (HTLV)-1, which sabotage the cellular control of replication, the infected cells are pushed closer to a cancerous state.

This framework predicts that virulence of any particular sexually transmitted pathogens should be positively correlated with the potential for sexual transmission. This prediction accords with differences in virulence across the spectrum of sexually transmitted pathogens. It accords with differences between the two major phylogenetic categories of HIV, which are designated HIV-1 and HIV-2, and with comparisons within each category (Ewald 1991b, 1994a, 1999) and with geographic differences in the time between the onset of infection

and the onset of leukemia among HTLV-1 infections (Ewald 1994b). Similarly, the more virulent serotypes of HPV are significantly associated with increased sexual partner change rates but benign serotypes are not (Franco *et al.* 1994; Ewald 1999). Isolates of human herpes simplex virus 2 (HSV-2) from Japan, where the potential for sexual transmission has been low, were more virulent in a mouse assay than HSV-2 from Thailand, where the potential for sexual transmission has been high (Sunagawa *et al.* 1995). *Chlamydia trachomatis* tends to be particularly damaging in areas with high potential for sexual transmission, where it is characterized by the presence of strains that cause lymphogranuloma venereum.

10.2 Scope of infectious causation

10.2.1 Genetics and germs in chronic diseases

During the second half of the 20th century, experts were concluding that the scope of infectious causation had been largely worked out, that infectious diseases could be largely controlled wherever sufficient economic resources could be applied, and that the future of the health sciences would involve a shift to studying other categories of disease. William H. Stewart, the US Surgeon-General during the late 1960s, characterized the prevailing mindset when he announced that it was time to close the book on infectious diseases and concentrate on chronic diseases. The obvious short-sightedness in his proposal was his failure to recognize the looming problems posed by antibiotic resistance and by pathogens that might emerge or resurge in the future. A less obvious short-sightedness was his juxtaposition of infectious diseases with chronic diseases. His assumption, like those of most of his contemporaries, was that chronic diseases were not caused by infection. We now know that several of the chronic diseases that were major problems then are largely caused by infection, diseases such as peptic ulcers, stomach cancer, and liver cancer. But there is also now evidence that many more, perhaps most of the common severe chronic diseases, are caused by infection: stroke, atherosclerosis, multiple sclerosis, Alzheimer's disease, prostate cancer, schizophrenia, and breast cancer are just a few of the chronic diseases that are being linked to infectious agents. This development of theory and evidence implicating infectious causation is providing new areas of research for studies of ancient tissues. Such studies may even help resolve current controversies by distinguishing between alternative hypotheses.

Current discussions about the causes of chronic diseases can be cast in the context of the three general categories of disease causation: genetic, parasitic, and non-parasitic environmental (the parasitic category being broadly defined to include infectious causes). Although genetic causation seems to be the default explanation, evolutionary considerations severely limit its potential scope, because the negative effects of disease on the passing on of a genetic cause of the disease will tend to weed out causal alleles to a frequency that can be maintained by mutation. If an allele provides some compensating benefit (as is the case with the allele for sickle cell anemia) it can be maintained at higher frequency, but few of the common, harmful diseases with unknown causes have characteristics that are consistent with this model.

The evidence of chronic diseases from ancient material raises concerns about the validity of genetic causation and non-parasitic environmental causation because this evidence suggests that ample time would have elapsed for purging of any genetic cause or genetic susceptibility to an environmental cause. Atherosclerosis, for example, has been found in Egyptian mummies and the Tyrolean Ice Man. If atherosclerosis were caused by bad alleles one would expect those alleles to be weeded out by natural selection. Thus the paleopathological evidence when viewed from an evolutionary perspective suggests that research needs to look beyond the concept of inferior alleles to understand the causes of atherosclerosis.

Still human alleles are known to be associated with atherosclerosis. The most important of these is the epsilon 4 (e4) allele of the apolipoprotein E gene. The e4 allele is also associated with increased risk of stroke, Alzheimer's disease, and multiple sclerosis. The negative effects of e4 appear far too great to allow it to be maintained simply by mutation. Yet e4 is maintained at frequencies that range from about 5% to 50% in different human populations.

It is lowest in populations that have been living in high densities for the past few thousand years. It is higher in populations that have been relatively small and isolated during this time, and highest of all in peoples who have been hunter-gatherers into the 20th century. This pattern holds within every continental region including the Americas. The e4 frequency is as low among Mayans is it is among the Greeks, moderately high in the Inuits and Scandinavians, and high in the Papuans and the Kyoi San. e4 is the primary allelic form of the apolipoprotein E gene in other primates and therefore cannot be considered a defective allele.

If e4 is not a defective allele, what is it? One possibility is that it increases vulnerability to some infectious agent that is tracking humans through evolutionary time by making use of the e4 protein, for example, to enter cells or garner nutrients. Several pathogens are candidates for infectious causation of atherosclerosis: *Chlamydia pneumoniae*, the oral pathogen *Porphyromonas gingivalis* (and the associated oral pathogens *Actinobacillus actinomycetocomitans* and *Bacillus forsythus*), and cytomegalovirus. Each of these pathogens has been found in atherosclerotic plaques, and risk of atherosclerotic disease has been positively correlated with evidence of infection by *C. pneumoniae*, *P. gingivalis*, and cytomegalovirus. Vulnerability to infection has been assessed relative to apolipoprotein alleles for only one of these pathogens: *C. pneumoniae*. Individuals who were infected with *C. pneumoniae* were about four times as likely to have the e4 allele as individuals from the general population (Gerard *et al.* 1999).

These findings raise several new questions for molecular paleopathology. What pathogens are in the atherosclerotic plaques of ancient mummies? Has the e4 allele been declining over the past 4000 years, wherever human populations have increased in size and mobility and thus increased the exposure to *C. pneumoniae*? Human evolution tends to occur much more slowly than pathogen evolution. Yet once the genetic basis of susceptibility to infectious disease is understood, frequencies of alleles that confer resistance or susceptibility can be generated from remains to assess whether allele frequencies have changed over time. The people who inhabited the Tarim basin in northwest China two to four millennia ago (Mallory and Mair 2000), for example, were most closely related to present-day Scandinavians, who have e4 allele frequencies of about 0.20. If the tracking of human populations by *C. pneumoniae* is at least partly responsible for suppressing e4 frequencies to this value, we would expect the e4 frequency of the mummies from the Tarim basin to be greater. Similarly the e4 frequencies of Egyptian mummies and South American mummies from this period are predicted to be greater than those of their most closely related peoples of today.

The e4 allele does not comprise a majority of the apolipoprotein E alleles in any population that has yet been studied, including hunter-gatherer populations. Whatever was responsible for the decline in e4 frequency in humans relative to other primates, therefore probably acted to a substantial degree before the onset of civilization. *C. pneumoniae* generates long-lived infections in humans that may have been able to perpetuate themselves in small groups. If *C. pneumoniae* is present in the South American mummies in addition to those of Egypt and the Tarim basin, then it probably was widespread in the human populations prior to the onset of civilization and could have contributed to a steady loss of e4 throughout human populations.

Mental illnesses in ancient populations

Paleopathology has not made much progress in studying mental illnesses of ancient populations, largely because mental illnesses do not leave distinct traces in ancient body remains. Or, if they do, research has not yet identified these clues. Descriptions in ancient literature could be useful if they were sufficiently accurate and detailed, but even diseases that are less protean than mental illnesses are difficult to judge from early descriptions. Medical historians are, for example, still uncertain about the identity of the plague of Athens. Even with current access to patients and abundant time for intensive study, psychiatrists are still disagreeing about categorizations of mental illnesses. The reality of the boundary between schizophrenia and bipolar disorder, for example, is still being debated by experts on these diseases (e.g. Jameson 1993; Torrey 1999). Experts disagree over whether schizophrenia with its distinctive symptoms, was present before the 19th century much as it is today or

virtually non-existent (Torrey 1980; Stone 1997). These disagreements indicate how unrealistic it would be to expect consensus on the occurrence of schizophrenia in more ancient periods on the basis of the ancient literature. If, however, some of these illnesses are caused by pathogens, studies of ancient tissue could offer a fresh approach to the problem.

Infectious causation of mental illness has been recognized for over a century, syphilitic insanity and rabies being the best known examples. Although less well known, the evidence favoring infectious causation for other mental illnesses is becoming increasingly persuasive. Schizophrenia, for example, has been strongly associated with *Toxoplasma gondii* (Yolken *et al.* 2001). The elevated IgM seroreactivity for *T. gondii* among mothers of schizophrenics when the patients were fetuses *in utero* (S. Buka, personal communication) suggests that schizophrenia may result from infections with *T. gondii* during pregnancy. This hypothesis accords with the firmly established season-of-birth associations (Torrey *et al.* 1997), because house cats, which are a primary source of *T. gondii* infections in humans, are more likely to be indoors during the winter months, as are humans. Pregnant women are thus more likely to be infected with cysts from cat feces during winter months. The associations with *T. gondii* therefore suggest that schizophrenia may be a delayed manifestation of a congenital *Toxoplasma* syndrome, which may also encompass miscarriages, ocular damage, and birth defects.

Just as pneumonia and hepatitis are not single diseases but rather collections of diseases caused by different pathogens, the diverse array of ailments currently grouped under the category of schizophrenia could be a variety of different diseases caused by different pathogens. Other pathogens have been strongly linked to schizophrenia. Studies have found that mothers of schizophrenics are significantly more likely to be infected with human HSV type 2 (Buka *et al.* 2001), and schizophrenics are more likely to be infected with borna disease virus (Iwahashi *et al.* 1998; Kirkpatrick *et al.* 2001) and more likely to have activated human endogenous retroviruses (Yolken *et al.* 2001). Although experts continue to believe that a strong genetic basis for schizophrenia has been demonstrated, the evidence consistent with genetic causation can also be explained by prenatal infectious causation, but not all of the evidence for prenatal infections can be explained by genetic causation. The evidence as a whole therefore supports infectious causation of schizophrenia more strongly than genetic causation, though the two need not be mutually exclusive.

Identification of causal pathogens has provided some of the strongest evidence yet of the presence of diseases such as falciparum malaria and Chagas' disease in ancient human populations. Identification of pathogens that cause mental illnesses will provide similar opportunities to confirm the presence of specific mental illnesses in ancient populations, but because the mental illnesses are generally more difficult to identify in ancient populations by other sources, the relative importance of identifying causal pathogens in ancient tissues will tend to be greater for mental illnesses than for diseases with distinct tissue abnormalities in human remains (e.g. *T. cruzi*) or unambiguous depictions or descriptions in ancient literature or artifacts (e.g. poliomyelitis).

If future studies bear out the link between *T. gondii* and schizophrenia, for example, researchers could assess whether *T. gondii* is present in ancient tissue using tests similar to those used for its distant cousin, *P. falciparum*. Considering that *T. gondii* may be most damaging when it is transmitted to humans from cats (as opposed to undercooked meat), mummified cat tissue could also be tested to assess whether cat-to-human transmission was probable in ancient Egypt, thus providing insights into the existence in ancient populations of *Toxoplasma*-induced mental illness. With little or no direct exposure to cat feces, Native Americans, in contrast, may not have been infected with *T. gondii*. If *T. gondii* can be found in Egyptian but not in Native American mummies, this difference could provide insight into the comparative mental health of ancient Americans relative to ancient Egyptians. Data from the Caucasian mummies of China's Tarim basin would be similarly informative.

10.3 Conclusion

At the close of the 19th century the germ theory had generated a new understanding of the causes of acute infectious diseases and revealed new directions for study. Researchers used pathology and eventually molecular genetics to better understand the role of parasitism in past millennia. At the end of the 20th century the second stage of this disciplinary development is occurring. The old germ theory is being expanded into a new germ theory, which, by integrating evolutionary principles with the health sciences, is finally integrating the full spectrum of biological insights. This new germ theory is emphasizing how environments and human activities influence the characteristics of infectious agents and the broad role of infection as a cause of chronic diseases. In so doing it is revealing new directions for study that may help us understand better not only the health problems we face now but also those faced throughout human history.

Part III

CHAPTER 11

The molecular taphonomy of biological molecules and biomarkers of disease

C. D. Matheson and D. Brian

the Lord carried me out in a spirit, and set me down in the midst of the valley, and it was full of bones; and He caused me to pass by them round about, and, behold, there were very many in the open valley; and, lo, they were very dry. And he said unto me: 'Son of man, can these bones live?' (Ezekiel, Chapter 37, v 1–3 Jewish Publication Society of America translation, 1970)

Taphonomy is the study of the processes that affect biological material after death. Changes over time and how they alter the patterns of deposition, within the archaeological or palaeontological record, is the study of taphonomy (Reitz and Wing 1999). However, the broader use of this term now commonly refers to any post-mortem process that occurs, both biotic and abiotic. There are a number of taphonomic models (Davis 1987; Noe-Nygaard 1988) that all focus on the natural and anthropogenic factors leading to loss of information. This occurs firstly by natural alteration before, during, and after burial (first-order process); and secondly, during the excavation, sampling, storage, analysis, and presentation of the results (second-order process).

Despite its deployment in the archaeological arena, molecular palaeopathology can be seen to draw much of its impetus from the medical and biological sciences, with both recent analogue pathology and the study of ancient DNA (aDNA) being major influences. It is nevertheless vital to consider the role of taphonomic factors. However, what remains to be done is to build these initial observations into a comprehensive taphonomic framework of specific relevance to molecular palaeopathology, one that draws on existing models within forensic taphonomy, archaeology, and palaeontology, but which also holds especial relevance for the study of ancient disease. In particular, the reinvention of the taphonomic paradigm that is required is its recalibration to the level of the *molecule*.

11.1 Toward a taphonomic model for molecular palaeopathology

It is common for taphonomic models to take into account only those processes that occur after death, and in archaeology at least, for the concept to be further narrowed so that it is virtually synonymous with 'postdepositional processes'. In attempting to devise a model to encompass all the taphonomic processes relevant to molecular palaeopathology, we need to cover events preceding deposition and those that follow recovery. In fact, here we divide the taphonomic 'pathway', or taphonomic history, of palaeopathological remains into seven sequential phases, which in turn can be placed into three quite distinct temporal stages (Fig. 11.1).

Stage I takes into account ante-, peri-, and immediate post-mortem processes, a particularly intense period of taphonomic activity, and one which, moreover, absorbs the interest of most researchers,

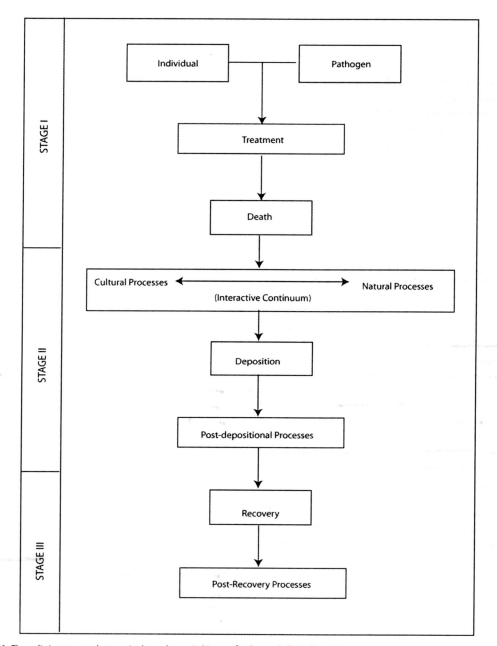

Figure 11.1 Three distinct temporal stages in the taphonomic history of palaeopathological remains.

containing as it does the key events of contraction of a disease, treatment and death. Stage II begins with deposition—the transition of the remains from the biosphere to the lithosphere in palaeontological terms, or, for archaeologists, the transition from the 'systemic' or cultural context to the archaeological context. It is this stage, consisting of depositional and post-depositional processes, that is more popularly perceived as the province of taphonomic studies. Stage III, from archaeological recovery of the remains onwards, is a stage commonly overlooked for its taphonomic ramifications, yet processes of a taphonomic nature continue to affect remains throughout their archaeological investigation, and beyond.

What is called for is a view of the multiple scales of taphonomic action as constituting a set of nested microenvironments, interactive, but differentiated from one another in terms of the kinds of processes and effects experienced at each scale (Table 11.1). For example, in the past, it has not been uncommon to propose taphonomic covering laws such as inferences made by Kumar *et al.* (2000) that 'tropical conditions (in particular exposure to heat and moisture)' is a discouraging prospect for the preservation of DNA throughout India, but it is now recognized that such statements are far too general. We believe the significance of the depositional *microenvironment* needs be highlighted by much more detailed case studies. This distinction is brought very much to the fore by the consideration of the taphonomic context of molecular palaeopathological studies, in that small-scale molecular residues may be found to survive quite well in a context where larger regional-, site- or individual-scale processes may be adverse to preservation of the remains.

Related to these issues of scale is the necessity for a dual-purpose taphonomic model for molecular palaeopathology. We must take into account taphonomic processes acting on two subjects: the remains of the host organism, and the remains of the pathogen. Despite having endured broadly similar depositional processes, their taphonomic pathways actually diverge significantly. Biomolecular residues may remain where there is poor skeletal preservation, and vice versa. We should however note that factors affecting preservation or destruction at each of these two levels are highly interdependent, and it will be useful to develop our taphonomic understanding of each in tandem with the other (Table 11.2).

The archaeological, palaeontological, geological, and molecular records are all subjected to the process of taphonomy. Biosynthetic remains of living organisms, comprised of biomolecules, biomarkers, and chemical fossils constitute the molecular record. Biomarkers are the known breakdown products of biological molecules, and suggest the presence of the original molecule.

Bone lesions often indicate the presence of disease in the archaeological or palaeontological record. Molecular palaeopathology studies have contributed greatly to the identification of disease in the past by examining the biomolecules and biomarkers associated with disease. The analysis of these disease markers requires a unique understanding of taphonomic mechanisms and their impact on disease-related biomolecules, a molecular level of understanding taphonomy.

The complex nature and interaction of taphonomic mechanisms results in different post-mortem effects on different biological substrates, and often, taphonomic variation within the skeletal remains of a single organism. Molecular taphonomy has focused on research of a few processes and interactions. The apparent magnitude of taphonomic change is astounding, as a continuous process with variation from geographical site to site, within one site, and at each step of the death, burial, excavation, and laboratory analysis (Davis 1987; Lyman 1994; Sandweiss 1996).

11.2 First-order taphonomy

11.2.1 Post-mortem mechanisms

Death marks the beginning of the study into taphonomic processes. The loss of information begins at this point, as behavioural and physiological responses cease (Martin and Rothschild 1998). At death repair mechanisms stop and the biodecay of organisms begins immediately. At the molecular level, each cell switches into programmed cell death and apoptosis (Grossman 1991), where the biomolecules are bombarded with lytic and degradatory enzymes and chemicals which will eventually destroy the cells and tissues. The enzymes released by the liver, the stomach acid, the colon bacteria all increase the rate of degradation of biological material. Dehydration reduces the rate of this autolytic decay (Eglinton 1998). The enzymic degradation process provides a nourishing aqueous environment for microorganisms already present in the body. Enhanced by the state of terminal bacteraemia where gut commensals enter the bloodstream prior to and at the point of death ensuring the rapid spread of microorganisms throughout the body.

Table 11.1 The effect of different taphonomic agents over multiple scales

Scale/agent	Region	Site	Body	Molecule
Cultural	Cultural groups resident in region at given time; diversity in cultural practices, particularly regarding illness and death	Choice of burial environment—e.g. stone cyst burial, riverbank cemetery	Medical practices such as surgery, use of drugs and dressings	Use of preservatives in preparation of dead for burial; cremation
		Curation of burial environment, e.g. cemetery landscape modification; re-use of cemeteries or tombs	Preparation for burial, e.g. embalming, mummification, cremation, defleshing, removal of organs or fluids, use of shrouds or coffins	Use of shrouds, coffins, or other grave furniture that may favour the preservation of molecular residues
			Burial practices, e.g. primary and secondary burial of individuals or groups	
Biological	Regional fauna and flora, including disease organisms	Bioturbation of the deposit by vertebrate and invertebrate organisms Infestation of the burial medium by invertebrate and microbial organisms	Vertebrate, invertebrate, and microbial scavengers	Autolysis by the body's own enzymes and endogenous flora
Physical	Landscape scale geomorphic and taphonomic processes, e.g. volcanic activity, mass wasting	Range and fluctuation of temperature, solar radiation	Disarticulation of the body by the combined effects of bioturbation, trampling, scavenging, and gravity	
	Temperature and precipitation regimes	Site-scale hydrology	Fragmentation of remains by above processes and by exposure to fire, water movement, freeze–thaw, wet–dry cycles	
			Formation of adipocere	
Chemical	Underlying chemical composition of soils and water in region	Acidity (pH) of burial medium	Formation of adipocere	Denaturation of proteins by endogenous and exogenous enzymes
		Presence of water, oxygen, and trace elements in sediments		Leaching of organic material into surrounding sediments

Table 11.2 Taphonomic processes acting on host and pathogen remains

	Host	Pathogen
Infection	Host acquires pathogen and develops corresponding disease condition; may be symptomatic or asymptomatic	Disease encounters host and begins to multiply and spread thoughout host's tissues
Severity of illness	Progression of disease may prove debilitating to host, influencing various body systems; host organism may produce antibodies and make other alterations in an effort to destroy the pathogen	Pathogen counters immune response with further reproduction, colonization of more tissue, perhaps to the point of causing bone lesions
Treatment	Medical or spiritual treatments of varying efficacy may be applied, and may diminish symptoms, have no effect, or in extreme circumstances, cause the death of the host	Treatment with chemical substances, or by other means, may destroy some or all of the pathogenic organisms
Death	Death, whether related to the disease or not, ends the above cycle of processes and begins a new set of processes of decomposition	Death of the host may destroy the pathogen, although particularly hardy pathogens may survive for much, if not all, of the taphonomic cycle of the host's remains. The processes responsible for decay of the host's tissues may likewise destroy or alter the pathogen
Deposition	The host body is transferred to the depositional environment where it continues to decompose	As the host body decomposes, the pathogen, or its remains, may be deposited in the sediments of the host's burial environment, onto other objects associated with the burial, or within the remains of the host itself
Post-deposition	The host body continues to decompose further, most often until there is nothing but skeletal tissue surviving, which in turn is subject to processes of fragmentation and diagenesis	Pathogenic remains in the sediments will most likely be destroyed or leached away, while those bonded to grave goods or surviving tissue may resist such processes
Recovery	Should the remains survive until this time in recognizable form, they may be excavated and collected for analysis. This implies exposure to the outside environment, and potential damage or alteration through excavation, screening, storage, and transport	Should the archeological signature of the depositional environment be sufficiently intact for recognition of the burial, samples of surviving tissues, grave goods, and sediments may be collected, processed, stored, and transported to a laboratory for further analysis
Post-recovery	Sampling, analysis, storage, and curation continue to impact the host's remains in various ways, from relatively benign alterations to the depositional environment, to more destructive analytical techniques	Sampling, analysis, storage, and curation also impact on the pathogenic remains; there is the chance too, in the case of some extremely hardy pathogens, that they will have the opportunity to reactivate at this stage, and commence the cycle all over again

The lack of living defences allows for microbial attack and invasion, accompanied by fungal, invertebrate, and vertebrate scavengers. The microorganisms proliferate rapidly in this nutrient rich environment and as Janaway (1985) found the organic material in an organism breaks down into proteoses, peptones, and polypeptides, encouraging a greater range of microorganisms. This microbial action can be active for a long while after death (Eglinton 1998). Anthropogenic mummification is an attempt to prevent these events from occurring. However, Lewin (1967) reported the presence of bacteria and saprophytic spores in thin sections of mummified tissue taken from Egyptian mummies,

indicating that the infestation begins rapidly even after relatively immediate intervention. The proliferation of microorganisms can be reduced but not stopped by the removal of water, which occurs in the natural mummification of organisms, by desiccation.

The cause of death of the organism may also affect its degradation. Like terminal bacteraemia and septicaemia, other diseases will enhance the degradation of the organism. Post-mortem modifications or treatments can also affect an organism's subsequent degradation. Butchery or the intentional removal of the flesh enhances the preservation of the skeletal material and the biomolecules associated with these elements due to the removal of the lytic, enzymic, acidic, and bacterial attack on the skeletal material, and increase the rate of dehydration of the remains. However, the information from the soft tissue and disease manifestations in the tissue is lost.

Loss of integrity of the skeleton and the anatomical skeletal structure of the organism, often by disarticulation of the skeleton before or during deposition, has little effect on the biomolecules within these elements. Disarticulation may increase environmental variation in taphonomy generated by physical separation of the skeletal elements.

11.2.2 Depositional mechanisms

The process of deposition is also a taphonomic mechanism, continuously manipulating the biological material into burial or a resting state. Biostratinomy is a term given to the loss of elements, often distal skeletal elements, during the process of deposition (Reitz and Wing 1999). Biological material that is not deposited in a cultural context as a complete skeleton in a formal burial, remains at the mercy of the environment, which preserves only a fraction of organisms by chance. Loss of information through the process of deposition ultimately depends on the environment and the medium of the deposit. Incorporation of biological remains into the depositional environment is focused on complete equilibrium (moisture, microorganism complement, pH, etc.) with the surrounding environment, achieved through taphonomy.

Trace remains are evidence of an organisms interaction with its surrounding, such as foot prints, scale/skin impressions, or a biological product including gall stones, kidney stones or coprolite. Coprolites are the faecal remains of animals or humans and hold a wealth of biomolecular information.

The processes of deposition can be examined as an integration of biological and non-biological or environmental factors. Scavenging animals are one factor that disarticulates and scatters skeletal remains, occasionally increasing the chance of preservation if elements are buried for later consumption or taken to a lair. Fire is a factor that may enhance preservation or destroy biomolecules depending on the temperature. A low-heat fire may create a carbon cast, the outer surface area of the remains are destroyed (Shipman et al. 1984; Lyman 1994), while the uncharred internal biological remains are preserved by drying and flame sterilizing. The burning generally fragments the elements (Coy 1975), although the biomolecules in low-temperature burnt bone exhibit good preservation (Brown et al. 1995) due to the lack of microbial activity. Calcined bone is produced by extremely high-temperature fires, which cause oxidation of the carbon, and thus destruction of biomolecules, producing friable and white or even light blue skeletal remains. Rapid desiccation of biological remains by low heat can aid enormously in preservation. Unique circumstances: some volcanic burials, such as Pompeii, would also be expected to produce good preservation due to the heat sterilization and desiccation of remains in an anaerobic burial environment.

Non-biological effects include wind dispersion of biological elements, wind-generated desiccation of the material exposed, or wind-mediated burial. However, wind combined with moisture can also aid the destruction of the material by introducing airborne fungi and destruction of the elements. Hydrological processes can relocate and disarticulate skeletal material, mediate deposition, and/or degrade biological molecules including water-soluble proteins and nucleic acids, which also accelerate structural degradation of the remains. Keratin, cellulose, lectins, lipids, cholesterol, and some proteins are resilient hydrophobic biomolecules that can aid the preservation of vegetable textiles, hair, and occasionally, soft tissue. Preservation in an aqueous environment is greatly determined by the pH; an acidic environment causes the dissolution of bone and preserves skin and soft tissue; in contrast to alkaline, which increases the decay of tissues and preservation of bone.

Rapid desiccation is integral for the survival of nucleic acids and many proteins. Salt effectively desiccates tissues, aiding molecular preservation, although extremely saline solutions can destroy biomolecules, specifically DNA in bone while preserving it in soft tissues. Unique combinations of environmental factors and accidental events often produce the most remarkable preservation. Dying of starvation or dehydration in the desert is a unique often well-preserved occurrence in the archaeological record.

Biological materials exposed to the elements in a terrestrial environment are weathered by wind, rain, and prolonged exposure to solar radiation. Weathered bone is demineralized and friable and may exhibit longitudinal bone fractures (Behrensmeyer 1978) along the collagen fibres or bundles in the organic component of bone (Shipman 1981). A transverse fracture of these fibres and bundles requires much greater force, by a different taphonomic process (Shipman 1981). Small, distal skeletal elements, including tarsals, carpals, and phalanges, weather more slowly than larger skeletal elements (Behrensmeyer 1978). Solar radiation exposure causes biolmolecular degradation. Photo-irradiation, particularly exposure to ultraviolet (UV) light will degrade most organic molecules through the absorption of light energy. Photo-oxidation is a major degradation pathway for exposed organic material in sunlight (Eglinton 1998). UV damage is reduced if the organism is rapidly buried or covered to protect it. The depth of the buried material also affects its rate of decay (Perry et al. 1988). Fresh breaks in bone expose the internal cavities, which can increase the rate of osteological degradation.

Organisms can be buried alive, and thus death occurs in the process of deposition. Transposition directly to a depositional environment before death including freezing, drowning, being buried alive or trapped in resins (like amber or copal), relies heavily on the characteristics of the environment for preservation (Eglinton 1998). The direct transfer is usually associated with an anoxic environment, which reduces the activity of microbial attack or fungal attack, and minimizes the degradation caused by autolytic processes. In most cases, with the exception of deposition in lava or hot desert sands, this involves a cooler temperature and aids in the preservation process. These anoxic environments involve a depositional medium that greatly limits oxygen and water penetration, and minimal microbial activity, which results in reduced formation of oxygen free radicals (Eglinton 1998) by anoxic microorganisms. Fossilization is most favourable with rapid burial in water-lain sediments, or wind-blown silts (loess) that provide fossilizing conditions in upland environments (Martin and Rothschild 1998).

Biological material trapped in a well may be physically protected by the structure of the walls, although without optimal pH and water chemistry, this deposit may not preserve entombed organisms. Skeletal remains and organic offerings placed into a burial chamber are vulnerable to the microclimate including water flow through the chamber, substrate chemistry, humidity within the chamber, temperature fluctuations, oxygen content, and the presence of microbes and fungi (Reitz and Wing 1999). High humidity may prolong enzymic activity and increase the rate of degradation in tombs or burial chambers (Perry et al. 1988). Skeletal and macroscopic material is generally preserved well in protected environments such as caves, crypts, and coffins; or within a stable environment including still water or dry deserts; or with rapid burial. These types of sites exhibit limited physical disturbance, minimal bacterial activity, and a generally stable environment; the molecular preservation will depend on the microenvironment within these structures or deposits (Purdy 1988).

Age alone is not the sole important factor in preservation. The condition of burial, diagenesis, and the burial environment are much more important. Optimal environments are those which protect the sample from water fluctuations and chemical exchange. The faster the transfer of the organic material to a depositional environment (like ice, amber, or tar traps) the better the preservation of material. Generally water and wind-blown sediments seal at varying rates. In other media, silicates can be formed quite rapidly depending on the chemistry, and the preservation can be quite good. Good preservation is due to low temperature, restriction of access due to close packing of sediment, absence of reactive agents such as free radicals, lack of nutrients, and lack of water fluctuations (Eglinton 1998). Fossils are preserved because they are protected from the changing outside environment. Isolated in their own depositional environment it is the characteristics of this environment that determines how well biomolecules and tissues are preserved. However, tropical environments are not usually good for molecular preservation due to high humidity, acidity of soil, high temperatures, and fluctuations of water (Eglinton 1998).

11.2.3 Post-depositional mechanisms

Temperature

Temperature increases tend to increase the rate of chemical reactions, and temperature fluctuations, together with fluctuations in moisture availability, can produce physical processes such as wet–dry or freeze–thaw cycles. Wet–dry cycles promote chemical weathering of bone, while freeze–thaw cycles act to contract and expand spaces within the bone, causing cracks to form and fragmentation to take place. Generally the lower the temperature, the better the preservation. Frozen biological material exhibits extremely good gross anatomical and molecular preservation, whereas high temperatures cause desiccation, which aids preservation, but the addition of water causes degradation by denaturation and increases many forms of damage including hydrolytic, enzymic, microbial, and oxidative damage.

Unstable environment

Instability of the depositional environment affects anatomical preservation as well as molecular preservation. The freeze–thaw cycles and wet–dry fluctuations affect biological material by the dramatic changes in water content and molecular structure. Bone histology in sites of water fluctuation is least preserved (Nielsen-Marsh and Hedges 2000), as it can cause cracking and splitting; however, in biomolecules the effects of water fluctuation rapidly speed up degradation. Generally speaking, environmental instabilities characterized by fluctuations in temperature, moisture, or other such variables, create poor conditions for the preservation of bone and its molecular organic content.

pH

The pH of both the soil and surrounding water has enormous effects on the preservation of biological material. In soil the preservation of bone hydroxyapatite is best at pH 7.8–7.9 but this destroys protein, while acidic conditions cause the bone mineral to dissolve (Gordon and Buikstra 1981; Linse 1992). Infant bones, which have little mineral content, are the first to dissolve and the most resistant are teeth, due to their enamel. The combination of soil chemistry and water content affects the pH of the depositional environment; some hydrated clays can generate a pH as low as 4.0–5.5, which does not provide the right conditions for preservation of biomolecules.

Soil chemistry

Chemical diagenesis is the loss of chemical or biochemical elements from biological material. Biological molecules can be converted to mineral depending on the soil chemistry. As the mineral can be dissolved likewise it can be recrystallized and deposited and this recrystallization depends on the composition of the aqueous solution within the deposit; pH, ion composition, redox potentials, and protein content (Nielsen-Marsh and Hedges 2000). A chemical equilibrium forms between the biological material and its surrounding environment, which affects bone and molecular preservation by the transfer of inorganic molecules into the biological material, disruption, or replacement (Weiner *et al.* 1993). However, to identify the chemical equilibrium and transfer in archaeological material is difficult because the original chemical composition at the time the assemblage was deposited may have been quite different (Karkanas *et al.* 1999). Mineralization destroys information within the material but the organic content can sometimes be preserved well when stabilized with minerals. Karkanas *et al.* (1999) found good organic preservation in depositional layers with silica and silicaceous minerals. However, breakdown products of biological material (fulvic and humic acids) can also bind to minerals in bone and silica in the purification process, generating a problem with the recovery of other biomolecules, like DNA. The presence of the minerals including dahllite, crandallite, and montgomeryite contributes to good preservation in the biomolecules (Weiner *et al.* 1993). Dahllite is a carbonated apatite, similar to the hydoxyapatite mineral in bones (Weiner *et al.* 1993). Karkanas *et al.* (1999) found poor preservation if phosphatic minerals containing calcium, aluminium, and magnesium (crandallite and montgomeryite) are formed.

Moisture

Water intervenes at all stages of the taphonomic cycle. It acts under the influence of temperature fluctuations, to produce cyclic weathering of bone and along with temperature, moisture content is a key variable in the progress of microbial decay of soft and hard tissues and in the rate and extent of chemical reactions. Additionally, water content in terrestrial depositional contexts, in the form of rainwater or groundwater percolation, will not only participate in hydrolytic reactions with the remains, but will act to leach away the products of such reactions, thus dispersing the remains throughout the deposit, at the molecular level. The chemical composition, including pH, and oxygen content of the water, are of paramount importance in determining the actions of biological and chemical taphonomic agents.

Macroscopically, bone histology is best retained in fully waterlogged sites, for example, peat bogs or dry sites; the fluctuation of water results in poor preservation (Nielsen-Marsh and Hedges 2000). This bone histology correlates roughly with the protein content of the bone; however, protein can be lost through microbial action, which also destroys the histology, or through basic pH, which would keep the histology intact (Nielsen-Marsh and Hedges 2000). Protein loss also correlates with porosity of the bone and the amount of water passing through the bone. The porosity and hydrological environment of the biological material is most important for macroscopic and microscopic preservation (Nielsen-Marsh and Hedges 2000). Intermolecular bonding such as hydrogen bonding and ionic bonding is important in the preservation of molecules. These bonds can form ionic and molecular reactions in solutions and the presence of water increases these reaction rates. One of these reactions is hydrolysis, where a water molecule or hydrated protons attack a susceptible bond and it is cleaved like the peptide bond in proteins. The conformation of the molecules is extremely important for this type of damage and the types of bonds present that can be hydrolysed. Degradation occurs with the presence of hydration layers but depends on the chemistry of the associated molecules (Eglinton 1998). Many other factors can affect the chemical interaction in the microenvironment of the molecule itself like the presence of oxygen, light, radiation, mineral particles, molecular mobility, temperature, and pressure.

Trace metals

Metals present in the soil can aid or hinder the preservation of biological molecules. Some organic material has been preserved due to its association with metals like copper, zinc, and lead (Chen *et al.* 1998). The presence of these metal ions aids anatomical preservation usually by ionic substitution of the ions into the biological material (Lambert *et al.* 1984; Pate *et al.* 1989). It is possible that copper generates a reducing environment that protects the organic material from oxidative damage (Cano 1998). However, the presence of these metal ions severely hinders the analysis of biomolecules like DNA; this is due to methodology and the inability to separate the biomolecules from the ions in the recovery process, where the ions if not separated will inhibit the polymerase chain reaction (PCR). Sometimes the metals in high enough concentration may act as a biocide, preventing further microbial degradation (Chen *et al.* 1998).

Solar radiation

Sunlight has a specific effect on the remains at the molecular level. Increased radiation in all environments damages the biomolecules but in some soils radiation can be present as radioactive trace elements. Solar radiation only affects surface material. This form of radiation can cause oxidation and the formation of free radicals from the biological material and the surrounding water. It is these generated agents that cause damage to the biomolecules.

Oxygen

An anoxic environment aids the preservation by reducing the biotic processes and limiting any activity to anaerobic organisms. In its molecular state oxygen is unreactive but when oxygen free radicals are formed they attack molecular bonds relatively indiscriminately in gas, liquid, or solid phases independent of temperature. Oxygen-induced oxidative degradation is another source of damage to biomolecules like DNA as it attacks the nitrogen bases of DNA (Eglinton and Logan 1991). Protection afforded by the bone matrix or amber does not prevent this sort of damage occurring. In living organisms mechanisms have evolved to repair such damage, for instance carotenoids, which capture the free radicals (Grossman 1991).

Biotic factors

Biotic mechanisms include all alterations induced by biological organisms such as bioturbation and biodegradation. Bioturbation is the disruption of the depositional environment by mixing caused by snails, burrowing animals, and tree roots. Microbial degradation is the most common form of biodegradation caused by the many soil microorganisms. These can be fungal, protozoal, or bacterial. Fungi and animals cause increased anatomical damage but at the molecular level are no different from bacteria. The bacteria colonize and degrade biological material rapidly except in certain aqueous or other anaerobic environments and can be placed in the category of primary invaders and secondary invaders of the biological material. Most aerobic bacteria will increase the rate of degradation by digestion and destructive secretions. The most destructive secretion generated by bacteria for DNA consists of bacterial peroxidases. Rarely, the presence of other organisms enhances preservation. Mollusc shells change the soil chemistry, reducing pH change by generating a carbonate buffer, preventing the detriments of extreme pH that would degrade biological material (Scudder 1993).

11.3 Second-order taphonomy

Second-order taphonomy involves information loss due to the methods of excavation and curation. These include survey strategies, excavation location, excavation methods, screening methods, sample size, sample selection, storage, analysis, recording information, and publishing data (Reitz and Wing 1999). Throughout the excavation material is lost, usually the microscopic material including biomolecules, small skeletal elements, and associated artefacts. Poor excavation procedures or recovery methods only increase the loss of materials. Incorrect storage of biological remains increases the loss of information by loss of the elements and increase degradation by exposure to air borne microorganisms, moisture and environmental conditions that would be detrimental to the preservation. Sometimes the best preservation for archaeological material is in the conditions it was found, still buried. The methods chosen for analysis affect the type of information recovered, from genetic analysis to elemental analysis of hair. Some methods are destructive, preventing any more information being gained, while others reduce the effectiveness of further analysis. These decisions all affect the recovery of information.

11.4 Biological molecules and biomarkers

The condition of burial and diagenesis are much more important than age in preservation. Remains are isolated in their own depositional environment and the conditions of this environment determine how well biomarkers and biomolecules are preserved. There is no correlation between gross anatomical preservation and biomolecular preservation.

11.4.1 DNA

The double-stranded helical DNA molecule is more resistant to damage than the single-stranded RNA molecule. The phosphate ester bond is susceptible to hydrolytic cleavage. The pyrimidine bases, especially thymine, are sensitive to oxidative damage, as are the sugar moieties. Crosslinking of the molecule may occur (Eglinton and Logan 1991). Conversion of bases through hydrolytic deamination (guanine changes to xanthine, cytosine to uracil or derivatives) and depurination (removal of guanine and adenine from the sugar–phosphate backbone) can occur (Eglinton and Logan 1991).

Correlation of DNA preservation with colour (Shipman *et al.* 1984), detection of trace elements (Grupe and Herrmann 1988), crystal structure (Shipman *et al.* 1984), collagen content (Tuross 1993), protein preservation (Poinar and Stankiewicz 1999), morphology (Shipman *et al.* 1984), histology (Nielsen- Marsh and Hedges 2000), and amino acid racemization (Poinar *et al.* 1996) have all been studied. These studies have indicated that DNA preservation does not correlate with age but none show an adequate correlation, so DNA may be relatively stable through geological time. The taphonomy has the greatest importance in the survival of DNA.

There are many factors that enhance the preservation of DNA. Death of the organism and rapid burial reduces damage from UV radiation, while cellulose in the cell wall seems to help to preserve plant DNA, and collagen seems to preserve the

DNA in mummified tissues (Spigelman and Greenblatt 1998). Hydroxyapatite binds DNA in bone and appears to reduce damage to DNA as it slows the rate of hydrolytic depurination (Lindahl 1993; Tuross 1994), reduced oxidizing agents and reduced microbial decay. High osmolality (Cano 1994) reduces depurination 10-fold, while some modifications, like charring in a low-temperature fire, aid preservation (Eglinton 1998). The differential preservation of skeletal elements is also important in the preservation of DNA. While temperature, humidity, pH, or exposure to seawater would have significant detrimental effects on bone, Schwartz *et al.* (1991) identify no effect on forensic teeth samples. However, no study has looked at the effects of taphonomy on pathologically modified bone.

Hydrolytic attack within the DNA molecule occurs in a number of places but it is the cleavage of the *N*-glycosyl bond between the sugar and the base that causes deamination (O'Rourke *et al.* 2000; Hofreiter *et al.* 2001). Deamination is found to be 20 times more common in the purines than the pyrimidines, although this rate is pH and temperature dependent (O'Rourke *et al.* 2000). This generates errors in the PCR where a T or an A is inserted instead of a C or a G, respectively (Hofreiter *et al.* 2001). Where the base is lost the chain is weakened and eventually cleaved or broken (Eglinton and Logan 1991; Lindahl 1993). Pääbo (1989) has shown with electron microscopy that soft tissue 13 000 years old displayed fragmentation (40–500 BP) and crosslinking of double-stranded DNA. However this is not age related as a 5000-year-old mummy had less damage than 4-year-old pork (Hänni *et al.* 1990; Hagelberg *et al.* 1991). Pääbo (1989) clarifies that it is the rate of desiccation and reduction of hydrolytic damage that determines the degree of fragmentation. Taphonomy of the sample then determines whether more fragmentation will occur over time. It is possible that different tissues exert a different degree of fragmentation depending on the speed and mode of autolysis and cell death.

Exposure of biological material to free radicals or directly to ionizing radiation causes the generation of oxidative damage to DNA resulting in modified bases (Eglinton and Logan 1991; Höss *et al.* 1996b). Oxidation is believed to be the major damage that renders DNA non-replicable (Eglinton 1998;

Lindahl 1993). Oxygen when it is involved in forming free radicals attacks the nitrogen bases (Pääbo 1989). The photochemical reaction between water and ionizing radiation causes the formation of hydroxyl or superoxide radicals, which damage the DNA (O'Rourke *et al.* 2000). Pääbo *et al.* (1990) report that large proportions of oxidative damage occur in the pyrimidines, generating alkali-sensitive sites or damaged sugar moieties. Apurinic and apyrimidinic (AP) sites were shown to be 1 in 20, which represented half the alkali-sensitive sites; the other half were attributed to the modified sugar moieties. Oxidative modification to the DNA, like oxidized pyrimidines (hydantoins), affect the recovery and analysis of the DNA while not necessarily generating strand breaks (O'Rourke *et al.* 2000). It is possible that a chelating agent like copper or other metal ions may generate a reducing environment and protect from this sort of damage. Oxidation affects primarily the mitochondria as that is the site for oxygen metabolism during the organism's lifetime (O'Rourke *et al.* 2000).

Temperature can aid preservation. Höss *et al.* (1996b) found that a 20°C reduction in temperature reduced base degradation 10- to 25-fold, therefore demonstrating the commonly found inverse correlation of temperature with the recovery of DNA (Höss *et al.* 1996b; O'Rourke *et al.* 2000). Cattaneo *et al.* (1999) studied the effects of temperature and found that burnt bone at 800–1200°C did not amplify mtDNA although gross morphology was maintained and staining was possible, suggesting DNA is more temperature sensitive than microscopic structure. The study also suggested that mitochondrial (mt) DNA should survive only 5 days at 100°C and 3 days if denatured.

Both the matrix that biological material is buried in and the matrix for which the material is stored or preserved can have effects on the recovery of DNA. Mineralization of the biological material, depending on the mineral provides poor qualities for preservation of DNA in the burial matrix (Eglinton 1998), while the storage or preservative is extremely important for the recovery of medical museum specimens stored in solutions or on slides. Ethanol is good for preservation of anatomical structure but has been found to be bad for preservation of DNA (Barnes *et al.* 2000).

Waterlogged samples are different, the anaerobic environment being good for macroscopic preservation of organic material especially skin and hair. While attempts at recovering DNA from waterlogged bone have failed, it has been successful on soft tissues (Doran *et al.* 1986; Pääbo *et al..* 1988; Graw *et al.* 2000).

Pathogen

The identification for the presence of pathogens causing disease has been facilitated by the identification of pathogen-specific sequences of DNA. *Mycobacterium tuberculosis* was first identified (Spigelman and Lemma 1993) due to the resilient nature of the bacteria. *M. tuberculosis* has continued to be identified by a variety of specific sequences in a number of different tissues (Salo *et al.* 1994; Arriaza *et al.* 1995; Baron *et al.* 1996; Taylor *et al.* 1996, 1999; Faerman *et al.* 1997; Braun *et al.* 1998; Donoghue *et al.* 1998; Norula 1999; Haas *et al.* 2000a; Mays *et al.* 2001; Zink *et al.* 2001). *Mycobacterium leprae* was the next pathogen to be identified (Rafi *et al.* 1994; Taylor *et al.* 2000; Haas *et al.* 2000b; Donoghue *et al.* 2001), then *Clostridium* (Ubaldi *et al.* 1998), *Trypanosoma cruzi* (Guhl *et al.* 1999), *Plasmodium* (Taylor *et al.* 1997), *Treponema pallidum* (Kolman *et al.* 1999), and *Yersinia pestis* (Drancourt *et al.* 1998). Taubenburger *et al.* (1997) identified the virus causing the 1918 flu pandemic, sequencing the whole genome and identifying virulence factors and viral proteins of this influenza. This systematic approach has led the way for ancient pathogen studies assisting modern medicine.

Retrieving DNA

The polymerase chain reaction (PCR) (Mullis and Faloona 1987) is integral to ancient or degraded DNA research. The cloning of ancient DNA is difficult due to post-mortem chemical modification or damage (Pääbo 1989). There are two obstacles in the recovery of ancient or degraded DNA: inhibitors and biochemical damage to the DNA molecule. Inhibitors that co-purify with the DNA affect the recovery of ancient or degraded DNA. The nature of inhibitors has been largely ignored as the failure to recover DNA has been blamed on the damage through diagenetic processes. However, the recovery methods are often at fault due to a lack of understanding of the forms of inhibition that can affect any type of tissue. These inhibitors can be native to the biological organisms like collagen in bone or polyphenolics in grapes, which prevent molecular amplification. Much of these other forms of inhibition are due to the nature of the biological material in the depositional environment; this is sometimes called kerogen, a heteropolymeric material found in the geological record. It is the mass of biological matter as it accumulates and breaks down together forming organic acids like humic and fulvic acids and tannins (Tuross 1994).

Another form of inhibition is generated by complex crosslinked structures such as maillard condensation reaction products (alkylpyrazines, furanones, and furaldehydes) that affect molecular methodologies (Evershed *et al.* 1997; Pääbo 1989; Poinar *et al.* 1998). Reducing sugars react with proteins by non-enzymic glycosylation or glycation. This modified sugar can bind to the amines of DNA and over time these adducts become advanced glycation end-products (AGEs) and become irreversibly bound (Vasan *et al.* 1996). These forms of inhibition can be found in any biological material; however, some, like tanins as inhibitors in bog bodies, are specific to the type of remains (Spigelman and Greenblatt 1998). Large amounts of divalent cations (magnesium and iron) can inhibit the amplification of DNA by causing failure of the PCR. However, removal of these ions can be quite difficult especially when they form conjugate bonds with other macromolecules.

There are many methodology errors that can affect the accurate recovery of aDNA. Even PCR, which produces an exponential amplification of target DNA to be manipulated and analysed by standard molecular biological techniques like cloning, restriction fragment length polymorphism (RFLP) analysis, and sequencing, has many modes of error that have to be understood when applied to ancient or degraded DNA. From amplification of depurination and abasic sites through to misincorporation of nucleotides. Adenine (A) is common in damaged DNA when amplified and sequenced and can be found quite commonly at the point where a PCR jumping event has occurred due to the inherent placing of an A at the termination of a strand by the *Taq* in a PCR (Pääbo 1989). However, the correct understanding of these results can be used to confirm the authenticity of the material being studied. Robbins *et al.* (1979) demonstrated this when some of the differences (10/12) found in an extinct Quagga were found to be deleterious in the amino acid sequence, which suggested to them that it was more than likely to be authentic and not postmortem damage. However, some errors can be inherent in the sample being used. The many amber studies have been disregarded due to the irreproducibility of the recovery of DNA from this material (Austin *et al.* 1997), which is therefore an unreliable source for biological and evolutionary studies.

The success of the recovery of DNA depends on refined methods for its extraction; it can not be a method designed and optimized for modern DNA extraction, but needs to be a method designed specifically for the extraction of small fragmented, damaged, and degraded DNA that has inhibitors present and may have undergone some form of diagenesis (Cano 1998).

Bacterial DNA: a special case of preservation

Some organisms have other methods that aid their molecular survival: plants produce seeds, nematodes produce massive amounts of sugar alcohols, brine shrimp encyst, and some bacteria produce spores. These mechanisms protect their DNA and other biomolecules during times of stress such as starvation. By these mechanisms they can stay dormant for very long periods of time (Roubik 1989) withstanding starvation. While these bacteria are in a dormant state they can die or be fossilized. However, with these mechanisms in place their biomolecules will not only withstand attack from their own cellular death, but also from other bacteria in the archaeological record, and in the case of pathogenic bacteria from the death and autolysis of the host.

A number of these protective mechanisms are commonly found in bacteria. Biocrystallization is the crystallization of the DNA into a structure that makes it resistant to attack, changing the DNA into a more stable compound that acts like an inorganic crystal (Grant *et al.* 1998). The major factor in biomolecular protection is the use of small acid-soluble proteins (SASPs). SASPs binding to DNA alter its chemistry, enzymic reactivity, and UV

photochemistry, which is why it becomes resistant to UV damage, depurination, hydroxyl radicals, and oxidizing agents (Potts 1994; Setlow 1994). Similarly, some hyperthermophilic bacteria employ chaperone proteins that protect their DNA and cellular machinery (Eglinton 1998). Sporulation protects the DNA, as the cortex and spore coat, and the removal of water prevent anything from entering, reducing the potential for damage (Potts 1994; Setlow 1995). The spore wall consists of peptidoglycans and dipicolinic acid, which reduce its permeability to oxidizing agents, hydroxyl radicals, and water (Felsenstein 1989; Eglinton 1998). UV and gamma irradiation, instead of forming thymine dimers, produce a spore photo-product (SP), which can be repaired much more easily than the dimers and therefore does not damage the DNA (Setlow 1995).

Racemization and DNA preservation

There have been many indicators proposed to correlate with the preservation of DNA, and most controversial is amino acid racemization. Racemization is the conversion of an amino acid from the L to the D enantiomer. This conversion continues over time until equilibrium is reached. However, the rate of this isomerization varies between amino acids and can depend on pH, temperature, the presence of water and metal ions, time of exposure to certain temperatures, pressure, and microbial action (Child *et al.* 1993; Poinar *et al.* 1996; Collins *et al.* 1999). Poinar *et al.* (1996) claim that no DNA could be recovered if the experimentally determined D/L ratio was >0.08 and continued to state that under normal conditions all amino acids are almost fully racemized in 5000-year-old bone. Due to many taphonomic factors affecting the conformation of collagen, racemization in archaeological bone is unpredictable (Collins *et al.* 1999). Similarly, different species have been shown to racemize at different rates with the racemization in rat dentine being 10 times the rate in humans (Ohtani *et al.* 1995).

11.4.2 Other biomolecules

Proteins

Proteins breakdown rapidly after death, first, by autolytic processes, second, by the actions of other organisms, and finally, by the biotic and abiotic processes in the depositional environment. There are many processes that can have effects on proteins: denaturation (Joly 1965), deamidation or transamidation (Cha 1989), glucosylation (Lea et al. 1950), hydrolysis (Asquith and Leon 1977), oxidation (Asquith and Leon 1977), and racemization (Bada 1985).

Molecular preservation may extend to tertiary structure. Retention of the helical structure of collagen and susceptibility to enzymic digestion by collagenase has been noted in 10 000 year old remains; however, generally, proteins are denatured and chemically modified (Rowley et al. 1986). Denaturation can be caused by temperature, mechanical shearing of fibrous proteins, high pressure, ultrasonic waves, nature of the substrate, surface interaction forces, pH, ionic changes, radiation, and chemical and enzymic action, which all affect the rate and degree of this denaturation (Joly 1965). The rate of denaturation is decreased by reducing the water content (Joly 1965). Structural changes from denaturation generate changes in solubility, activity, and reactivity with stains, dyes, and antibodies.

Deamination can change the net charge of the protein, influencing the isoelectric point (pI). Cha (1989) reports that molecules of deaminated bovine calbindin appeared to be homogeneous by sodium dodecy sulphate—polyacrylamide electrophoresis (SDS–PAGE) molecular weight analysis but isoelectric focusing revealed a mixture of proteins differing in net charge. This would generate differences in old proteins making accurate identification difficult.

Transamination, the conversion of one amino acid to another by chemical rearrangement, has been evident by thermally induced changes in amino acids documented geochemically, with conversions of threonine, serine, and valine to glycine (Katz and Man 1980), also making accurate identification difficult.

Glycosylation is a form of protein damage that involves a chemical reaction and crosslinking to other biomolecules. Lea et al. (1950) demonstrated that glycosylation occurs between the reducing sugars and free amino groups in serum plasma, and that the rate is dependent on water content. Day et al. (1979) found that about 6% of albumin is normally in a glucosylated state and the reaction occurs between glucose and lysine without the necessity of enzymic interaction. Glycosylation will change the conformation and electrophoretic properties of protein but only affects a few amino acids.

The removal of water enhances protein preservation (Sensabaugh et al. 1971). Water causes hydrolytic damage but also facilitates more complex condensation reactions involving amino acids and sugars in the surrounding environment (Collins et al. 1992). There is also evidence that close association with a mineral matrix of bone, ceramic, or stone may slow down the process of degradation (Newman and Julig 1989; Cattaneo et al. 1990; Evershed and Tuross 1996).

There are many other agents that can cause degradation of proteins (UV radiation, radioactive decay, ionic reaction, metal ions) causing damage to proteins primarily by inactivation, photochemical reactions, chemical modifications, and complex binding with other molecules, all changing their properties, which would make identification difficult. There are studies of amino acid preservation and bone (Hare et al. 1980; Weiner and Lowenstein 1980; Bada 1985). The oldest bone analysed for amino acids is some 150 million years old (Miller and Wyckoff 1963).

The study of disease using proteins involves detection of unique proteins. Unique proteins require identification to correlate to allow future disease like scrapies protein and recently prion diseases including mad cow disease. The use of proteins may be more valuable than the use of RNA in older materials and thus aid the detection of viral diseases through specific viral proteins.

Collagen Taphonomy, both in the depositional environment and in second-order taphonomy like museum storage, affect the preservation of collagen and its relative conformation just as it affects DNA. The different DNA recovery found from museum storage facilities is due to storage conditions like temperature, humidity, microbial activity, preserving materials, and the degree of climate control exercised on the storage facility holding the samples, which all affect the preservation.

Evidence for the preservation of protein from archaeological material began with the recovery of collagen and its use for dating and palaeodietary determination. The survival of collagen in calcified tissue is enhanced by the protein's relative insolubility in aqueous solvents (Gilmcher and Katz 1965). Collagen is the major protein found in bone, 18% in dry weight, composed of a triple helix and its content declines over time but does not correlate directly with age, but with length of time exposed to high temperatures and diagenetic processes including oxidation, biodegradation, and hydrolysis (Collins *et al*. 1999). Collagen preservation in bone and teeth is due to close association with apatite, which encloses the collagen fibres. Other minerals can aid this preservation when the empty spaces in the bone are filled with silica or calcium carbonate. Studies have shown that even in fossilized bone, some collagen survives (Tuross *et al*. 1980).

Osteocalcin This is a small acidic protein, consisting of only 47–50 amino acid residues depending on species, found in bones and teeth, which contains an unusual amino acid (carboxyglutamic acid). The molecule is extremely resilient to temperatures, which may explain its good preservation (Ulrich *et al*. 1987).

Albumin Albumin has been used to identify the presence of blood proteins from the archaeological record and has been identified in ancient human bone (Cattaneo *et al*. 1990). Loy and Hardy (1992), using high performance liquid chromatography (HPLC), stated that most proteins break down immediately after death. By comparing blood proteins of fresh blood, 6-year-old blood, and a 90 000-year-old blood residue, Loy and Hardy (1992) identified the recovery of serum albumin was 14% for the 6-year-old blood and 13% for the residue.

Enzymes The activity of some proteins as enzymes has been shown to survive. Liquid blood stored for 44 years in sealed vessels at room temperature had activities comparable to fresh blood enzymes (Keilin and Wang 1947). Another study showed dried blood maintained a 40% recovery of enzymic activity (Sensabaugh *et al*. 1971).

Antibodies Antibodies, which target small regions of molecules, generate the immune response. The preservation of immunological activity in biological material was demonstrated in human tissue and bone between 500 and 2000 years old by detection of ABO and HLA antigens (Stast 1974; Allison *et al*. 1978). Antibody detection has identified immunological activity in fossil molusc shells (Westbrook *et al*. 1979) and has been demonstrated as far back as the Cretaceous, while immunological tests have been used to identify immunoglobulins (Yohe *et al*. 1991). Rothschild and Turnbull (1987) used antibody detection systems to identify yaws in an extinct bear and IgG antibodies to treponemal antigens have been identified from human remains from 1240 BP (Ortner *et al*. 1992). As long as the specific target of the antibody regions survive then they can be identified (Curry 1987).

Toxins Toxins are another possible source as biomarkers. These are usually species specific to pathogens, e.g. cholera toxin, and could be extremely informative.

Haemoglobin Forensic investigators detect haemoglobin using tetramethyl benzidine-based reactions along with the phenylphthalein tests (Cox 1991). Haemoglobin has been identified in extracts of prehistoric human bone and other material (Cattaneo *et al*. 1990; Williamson 2000).

Lipids

The most common biomarkers are lipids. This is because they are hydrophobic molecules, which evade the damaging effects of water (Evershed 1993). Lipids have been detected by thin layer chromatography, HPLC (Passi *et al*. 1981), infrared spectroscopy, gas chromatography (Evershed 1990; Skibo 1992), combined gas chromatography and mass spectroscopy (Evershed 1990; Copley *et al*. 2001), and isotope ratio monitoring GC/MS (Evershed *et al*. 1994).

Mycolic acids Mycolic acids have been used as a biomarker for *M. tuberculosis* and tuberculosis disease in 1000-year-old human bone (Gernaey *et al*. 2001). Mycolic acids are long-chain lipids and those found in mycobacterium have 3-hydroxy fatty acids substituted at the 2 position by an aliphatic side chain. The length of the side chain (C_{70}–C_{90}), the degree of saturation, oxygen function, and methyl branches can identify each mycobacterial mycolic acid (Gernaey *et al*. 2001).

Cholesterol Cholesterol is a biomarker that has been traced back hundreds of millions of years, but because it is so widely synthesized in all eukaryotes the information it carries is minimal. However, the breakdown products can be used as biomarkers in themselves and can indicate the diagenetic and catagenetic processes that may have acted on the surrounding material (Eglington 1998).

RNA

Taubenberger *et al.* (1997) have extracted the influenza virus, an RNA virus, from archival samples of the 1918 'Spanish' influenza pandemic. RNA was also extracted by Rollo (1985) from seed remains using molecular hybridization methods (see Chapter 16).

Trace elements and miscellaneous compounds

Detection of trace elements in bone can be informative for diseases like lead poisoning. However, the levels detected in bone may not indicate the endogenous levels in the bone at death or toxic levels that may have caused the death of the individual (Grupe 1987).

Porphyrins Porphyrins are found in biomarkers such as chloroplast, haemoglobin, haemosiderin, and hemozoin. The later two, haemosiderin and haemozoin are of interest in disease studies as the molecules produced in malarial infection (Aikawa *et al.* 1980; Taramelli *et al.* 1999). Both molecules differ depending on the species of malaria. Detection of haemosiderin is usually by reaction with Prussian blue (Aikawa *et al.* 1980), which could serve as a method for analysis.

Chitin Chitin is the second most abundant biopolymer on earth and is very well preserved. When bound to proteins the protein–chitin complex is likewise well preserved. Chitin was analysed from invertebrate cuticles using pyrolysis GC/MS and found to be resistant to bacterial degradation (Eglinton 1998). Stankiewicz *et al.* (1997) used chitin analysis of a beetle found in 25 million-year-old sediments.

'Drugs' Other chemicals, such as the breakdown products of cocaine, hashish, and nicotine, have all been found in tissues both soft and hard of Egyptian mummies using radioimmunoassay and GC/MS, and may indicate medicinal or social use.

11.5 Conclusion

Molecular palaeopathology is a new field of research where molecules are being used to confirm the presence of disease in the past. It is this field that will contribute to our study of the history of diseases, their frequency over time, and the evolution of the disease, the pathogen, and the host. It is this information that provides the temporal framework lacking in modern medicine, and the study of ancient disease is the only way to provide this information. However, by relying on the biomolecules of the past one must rely on the preservation of these key biomolecules and an accurate understanding of taphonomy is required for this endeavour.

It is critical to achieve a better understanding of the process of taphonomy and how it affects molecules. Understanding the processes involved in preservation may lead to more effective and successful recovery of molecules from the archaeological and palaeontological record. We know biomolecules degrade under certain conditions over varying lengths of time, but through what processes do they degrade and through what processes can we explain their survival? Obviously much more research is needed to increase our understanding of these processes, and research designed specifically to investigate each process and the complexity of the combination of these processes. Replicating and analysing taphonomy is difficult because to isolate each taphonomic process and study it will not provide an accurate picture. In the ground all the taphonomic processes are occurring at once and therefore the complex interrelationship of each process must be fully understood. It is difficult to simulate or study taphonomic processes as isolated parts. This chapter is an attempt to address a molecular taphonomic framework for preservation, one that draws on existing models within forensic taphonomy, archaeology, and palaeontology, but which also holds special relevance for the study of ancient disease. In particular, the reinvention of the taphonomic paradigm that is required is its recalibration to the level of the molecule.

CHAPTER 12

Ancient DNA can identify disease elements

Bernd Herrmann and Susanne Hummel

Causa latet, vis est notissima. (The cause is hidden, but the effect is obvious.) (Ovid, Metamorphoses, Book IV)

12.1 Introduction

Ancient DNA (aDNA) analyses may provide more information about endemic infection rates of ancient populations than morphognostic inspection. This chapter focuses on the advantages and disadvantages of molecular approaches to diseases of the past, and presents arguments in favor of intensive promotion of such endeavors.

There are no models for estimating the ratios of infected to uninfected people in past populations. While aDNA analyses may also provide insight into the genomic co-evolution of hosts by tracing inherited diseases beneficial for heterozygotes, the crucial issue remains research and development of new technologies to investigate host and pathogen DNA. Of necessity, these will be based in good part on the polymerase chain reaction (PCR) or PCR-like techniques. Presently these tests are performed by examining one gene sequence at a time, but in the near future multiplex-PCR-based screening of ancient remains will be carried out, in which a number of different DNA sequences will be amplified simultaneously.

'Human diseases elicit a great deal of interest, as they reveal basic environmental and selective living conditions of the past and thus contribute to our understanding of the past'. Statements like this can be read in almost all grant applications—what the German scientific community cynically refers to as 'application prose or poetry'. To respond to such ambitious demands, one would have expected epistemological patterns to be established. Surprisingly enough, however, since the days of Marc Armand Ruffer or Roy Moodie, to mention only these two founders of what is usually called 'paleopathology', no such epistemological essentials have been formulated.

In the introduction to his famous *Handbook of Historical–Geographical Pathology*, August Hirsch (1881) pointed out the purpose of epistemologically oriented access to diseases of the past, which seemed to be congruent with the rationale behind his handbook: '... a presentation of the existence and behavior of diseases in individual historical periods and at individual sites on the surface of the earth, to show whether and which changes of form these have undergone in time and space, which causal relationships there are between disease factors influencing specific times and specific places on the one hand and the existence and form of the individual diseases on the other, and how these relate to each other in their spatial and temporal prevalence'.

The ultimate benefit derived from historical studies lies in hope for the future. In terms of environmental history, we regard the past as a long-term experiment under natural parameters, following Judson Herrick s principle that 'The most important experiments are those that nature has already done for us'. Certainly, any approach to the past must begin from evidence. And if we are looking for evidence of diseases in the past, there are almost exclusively two kinds of source material: (i) historical material: written sources, tools for medical treatment, pharmacological materials indicating the presence of diseases (sometimes even specific diseases); and (ii) biological materials that are the substrates for diseases: remains of humans or other organisms bearing signs of diseases, sometimes enabling specific diagnosis.

12.2 Historical sources

Bleker (1984) contended that history was a unique panorama for the study of patterns of diseases under changing conditions of life and environment on a theoretical level only. Practically all available sources are unsuitable, as they do not contain the relevant data for that problem. Sources may also be biased, as shown in Lanphear's study of a set of disease data from a 19th-century US poorhouse (Lanphear 1988). Pfister's example is also most instructive: 'According to church registers of the 18th and 19th centuries, Lutheran areas in Switzerland seemed to have higher perinatal mortality than Roman Catholic areas. Such patterns can be explained consistently on a sophisticated level in terms of sociobiology as differential paternal investment in children according to religious tradition. It might also be explained consistently by religious features influencing regional epidemiological factors related to mortality, taking perinatal mortality as a proxidatum for religion-based customs affecting mortality in a similar manner as parental investment' (Pfister 1986).

In fact, the mortality pattern was actually the same for both Lutherans and the Roman Catholics, but the two faiths had different traditions concerning stillborn or perinatal death. Thus the time between birth and death was, in practical terms, longer for Roman Catholics than for Protestants, incorrectly suggesting differential parental investment.

12.3 Disease classification

In addition to this empirio-critical evaluation of sources, another issue to be considered is the problem of nosological compatibility, a problem addressed by Michel Foucault (1972) in the history of the hospital as 'the archeology of the medical look', as to the empirical scarcity of diseases in the archaeological record (i.e. of human remains—see Table 12.1).

Table 12.1 cannot be taken solely as an example of insufficient scientific professionalism, as one would have expected many more affected skeletal elements. Many important diseases leave no traces in bone. Structural alterations may at times be seen in microscopic sections, but there is no general tradition at present for routine microscopic approaches to ancient bones. Tables such as this indicate clearly that empirical knowledge of ancient diseases derived from bone inspection is far less

Table 12.1 Frequency of diagnosed infectious diseases in 25 medieval skeletal series from Europe, 500–1500 AD (adapted from Padberg 1992)

Series	Osteomyelitis	Periostitis	Tuberculosis
Rohnstedt	1	1	–
Espenfeld	–	–	4
Schleswig/Nikol.	1	–	–
Bonaduz	4	–	–
Libice	1	–	2
Berlin/Petri.	–	1	–
Leobersdorf	–	–	1
Mittelsoemmern	1	–	–
Unterthuerheim	3	–	–
StaubingKleinlangheim	11	4	1

Minimal number of individuals: 40; total: 4463; published between 1957 and 1987, mean year of publication: 1978. The series selected can be considered to form the sample of those skeletal populations that have been studied by physical anthropologists to the best of their professional ability.

than sound knowledge, or even a reasonable estimate. Interestingly enough, we do have better knowledge for less important diseases (Table 12.2).

These tables might be useful for intra-population and inter-population comparison, but they concern minor diseases. What we consider a 'major disease' is any disease with short-term impact on human populations, influencing both the number of reproducing individuals and the genetic composition of the population by the selection of certain features. We refer, of course, to events like the European Black Death of 1348–51, or the 1875 measles infection in the Fiji Islands, in which 40 000 people out of a total population of 150 000 died after the king of Fiji visited the city of Sydney (Zinsser 1949). Such events are familiar to epidemiologists. Of course, due to the resorting of genetic material in the pathogens (e.g. the influenza virus), the selecting principle for a population is altered, but not randomly (cf. Hill 1996).

12.4 Evidence for infectious diseases

Evidence for infectious diseases in ancient populations is problematic. The absence of evidence of a pathogen in any given human remains is not evidence of its absence. The vast majority of infectious diseases in ancient populations cannot be diagnosed directly because they are undetectable: most of them are not bone-seeking diseases, and even bone-seeking infectious diseases cause characteristic tissue alterations only after a more or less long-term chronic infection.

Thus, studies that concentrate on aDNA from pathogens could certainly advance the field. In contrast to morphognostic studies of bone surfaces, the basic benefit of screening bones for pathogen aDNA is that the search for aDNA permits diagnosis even when there is no sign of disease in the source. Diseases that lend themselves especially to such research are those whose organ specificity or acute development is not manifested in bone tissue. Furthermore, aDNA analysis requires no more than a few molecules of DNA or RNA in the parasite remains.

The ratio of pathogen DNA to volume unit of bone during lifetime most likely does influence the preservation of intact pathogen target aDNA in a sample. Thus, the predictability of successful screening for infectious diseases is greatly reduced, and a ratio formula (DNA per bone volume and disease) would be necessary for each disease. This is not only a practical impossibility, but also devoid of practical value. Even if the ratio could be calculated, it is definitely not decisive, as detectability depends on the likelihood of DNA preservation in the given environment of the decaying bone.

The only advice that can be given presently is to learn by doing and to encourage scholars in the search for pathogens. However, they should rely on reasonable assumptions and experimental design. Two features must be considered and under

Table 12.2 Frequency and intensity of spondylosis (%) found in medieval skeletal series (adapted from Padberg 1992)

Series	Frequency			Intensity		
	Females	Males	Total	Females	Males	Total
Anderten	29.0	44.0	40.0	–	–	–
Duster-Reckahn	–	–	30.0	–	–	2.7
Espenfeld	60.0	72.0	65.0	27.0	43.0	34.0
Sindelsdorf	–	–	–	69.0	80.0	77.0
Endingen	–	–	–	78.0	79.0	79.0
Luebeck/Dom	73.0	76.0	75.0	–	–	–
Grossschwab-hausen	–	–	76.0	–	–	–
Mittelsoemmern	92.0	83.0	87.0	60.0	66.0	63.0
Unterthuerheim	24.0	31.0	28.0	–	–	–
Staubing	40.0	57.0	52.0	–	–	–

Frequency: relative frequency of individuals with at least one spondylotic sign. Intensity: relative frequency of vertebrae with spondylotic signs.

no circumstances overlooked: the number of cycles required for PCR (Rameckers *et al.* 1997), and the possibility of pre-contaminated disposable items used in the course of the research (Schmidt *et al.* 1995). Table 12.3 presents some results of aDNA analyses.

One would expect typhoid fever, cholera, meningitis, and smallpox to be among the next pathogens detected in ancient remains by aDNA techniques. But what about those diseases that have not been identified thus far by medical historians? Some diseases that had a significant impact on past human populations seem to be unknown to us today. Furthermore, there are no reports of screening remains from ancient pets and domesticated or wild animals that act as zoonotic reservoirs or vectors for human diseases.

Another consideration is indirect evidence of diseases related to inherited genetic disorders, for example, heterozygotes in the ΔF508 mutation of cystic fibrosis (the most common molecular defect in 20–25% of Europeans), which is thought to benefit affected individuals in areas where tuberculosis is prevalent (Meindl 1987). A presumed tuberculosis resistance is also a matter of discussion in Tay–Sachs heterozygotes (Petersen *et al.* 1993). Screening in human remains for such genetic disorders would reveal evidence of the mutation at the molecular level, but not that the defect was expressed. (For molecular techniques: we are yet technically far from competitive PCR in ancient materials, which would be absolutely necessary to discriminate between homozygotes and heterozygotes.)

Another disadvantage for the study of genetic disorders is their absolute scarcity. It would require at least 20–25 individuals today to find one ΔF508 mutation; screening 100 individuals would disclose only four cases. But how many medieval skeletons must be screened for the same number? Those who are interested in relationships between culture and genetics may also speculate about correlations between infectious diseases and the selection of genetically triggered human behavior. Selection may favor or handicap coupled features, such as levels of transmitters or related biomolecules that influence behavior (e.g. the relationship between the monoamine oxidase level and the readiness of male bearers to expose themselves to high-risk

tasks). Table 12.4 presents the maximal likelihood for detectibility of genetic disorders in ancient remains by molecular probing. Empirical testing would best start on a prehistoric sample of stillborn infants and perinatal deaths because of the increased possibility of fatal homozygotes. However, frequencies of inherited diseases are the main obstacles in detecting such cases. For instance, a sample of 10 000–20 000 skeletons would be required to detect one case of phenylketonuria (PKU).

Given the incomplete archaeological record, speculation about the reliability of data from molecular screening is open, especially for inherited disorders. However, this is the only way to raise proxidata in that area. Of course, the main obstacles remain the small number of individuals and the effect of unknown sequences in ancient individuals, to name but two of the central issues that cause epistemological disadvantages.

Table 12.5 is a schematic presentation of the pros and cons in aDNA research on ancient pathogens. The most important obstacle seems to be that such analyses are most likely restricted to those infectious diseases that are spread via the bloodstream, thus making it rather unlikely to obtain evidence of cholera from skeletal remains. In such cases sources other than bone might be helpful, for example, sediments from sewage or cesspits or cloacae (cf. Herrmann 1985). Furthermore, even aDNA analysis cannot indicate the ultimate cause of death. When it comes to the technical level, there are no routine protocols available for PCR techniques to detect pathogens from ancient materials. Thus, no ratio of effort to results is predictable, making such research not only challenging but also risky and expensive.

It might be argued that ancient pathogens may exhibit slightly or extensively differing DNA sequences or unpredictable sequence alterations and thus cannot successfully be probed for. This certainly makes the search for such sequences complicated, even impossible, as such sequences are difficult to recognize. But this argument can be somewhat disregarded, as the co-evolution of host and pathogen requires at least some highly conserved sequences of the pathogen through time. Thus aDNA analyses should preferably start by targeting those sequences, since at least a reasonable

Table 12.3 Evidence of pathogens by aDNA techniques

Disease Pathogen	Sample	Source	Reference
Tuberculosis *Mycobacterium tuberculosis* complex	Mummified tissue	Thebes-West and Abydos, Egypt 500–3000 BC	Zink *et al.* (2001)
	Bone	Medieval	Spigelman and Lemma (1993)
	Mummified lung tissue	Peru 1000 years BP	Salo *et al.* 1994
	Bone	Pre-Columbian northern Chile	Arriaza *et al.* (1995)
	Compact bone	19th-century pathology collection	Baron *et al.* (1996)
	Mummified tissue	Egypt	Nerlich *et al.* (1997)
	Bone	Medieval UK	Taylor *et al.* (1996)
	Bone	Pre-Columbian, North American	Braun *et al.* (1998)
	Calcified pleura	Byzantine Israel 1400 years BP	Donoghue *et al.* (1998)
	Bone	Sweden 17th century	Nuorala (1999)
	Bone	Hungary	Haas *et al.* (2000a)
	Rib	50 yrs BP USA	Ubelaker *et al.* (2000)
	Bone	Medieval Lithuania	Faerman *et al.* (1997)
	Bone	Medieval	Taylor *et al.* (1999)
	Bone	Medieval UK	Mays *et al.* (2001)
	Bone	17 000 years BP USA	Rothschild *et al.* (2001)
	Rib	Medieval	Gearney *et al.* (2001)
	Bone	Pre-Columbian USA	Braun *et al.* (1998)
	Lung tissue	Pre-Columbian	Salo *et al.* (1994)
	Rib	Medieval	Dixon *et al.* (1995)
Leprosy *Mycobacterium leprae*	Bone	Medieval	Rafi *et al.* (1994)
	Bone	Medieval 10th century	Donoghue *et al.* (2002)
	Bone	Medieval 10th century	Haas *et al.* (2000b)
Bubonic plague *Yersinia pestis*	Compact bone	Medieval	Hummel *et al.* (1994)
	Dental pulp	France 16th–18th century	Drancourt *et al.* (1998)
	Bone	Medieval	Raoult *et al* (2000)
Venereal syphilis *Treponema pallidum* subsp. *pallidum*	Textiles	Renaissance	Marota *et al.* (1995)
	Bone femur	Easter Island 200 years BP	Kolman *et al.* (1999)
Malaria *Plasmodium falciparum*	Rib and mummified tissue	Egypt	Taylor *et al.* (1997)
	Bone	18th century	Kolman *et al.* (2000)
	Bone, infant	Roman	Abbott (2001)
Chagas' disease *Trypanosoma cruzi*	Mummified tissue and bone	Pre-Columbian	Guhl *et al.* (1997)
	Mummified remains	Chile 1200 AD	Ferreira *et al.* (2000)
Chagas' disease *T. cruzi*	Mummified soft tissue	Andean	Madden *et al.* (2001)
Retrovirus (RNA)	Skin tissue	5300 years BP	Goudsmit *et al.* (1993)
Escherichia coli	Bowel content	1200 BC	Fricker *et al.* (1997)
T-cell lymphotrophic virus type I	Mummified material	Andean	Li *et al.* (1999); Sonada *et al.* (2000)
Influenza orthomyxovirus	Lung tissue	84 years BP	Taubenberger and Reid (this volume, Ch. 16)
Ascariasis *Ascaris* spp.	Eggs from coprolites	Middle ages	Loreille *et al.* (2001)
Schistosomiasis *Schistosoma* spp.	Mummified remains	Egypt New Kingdom	Matheson *et al.* (personal communication)
Corynebacterium	Bone	Egypt	Zinc *et al.* (2001)
Irish potato famine pathogen *Phytophthora infestans*	Potato leaves	Ireland	Willman (2001)
Various (DNA and RNA)	Archival biopsies	90–50 years BP	Taubenberger and Reid (this volume, Ch. 16)

Table 12.4 Inherited diseases suitable for PCR detection in ancient materials (i.e. indicative DNA sequences do not exceed 200–300 years BP; features checked and compiled by Menzel 1994)

X-linked diseases
- muscular dystrophy, type Duchenne, type Becker
- hemophilia A
- ornithine transcarbamylase deficiency

Autosomal dominant diseases
- polycystic kidney insufficiency, type Potter III
- Huntington's chorea
- neurofibromatosis, von Recklinghausen
- hypercholesterinemia

Autosomal recessive diseases
- 21 hydrolase deficiency-based adrenigenital syndrome
- cystic fibrosis
- PKU
- sickle cell anemia
- hemochromatosis
- Tay–Sachs disease

Table 12.5 Pros and cons of screening skeletal material for pathogens by aDNA technology

Pros
- almost ultimate certainty of diagnosis
- screening reveals percentage of infected population
- indirect evidence for inherited diseases
- essential sequence alterations are not likely to occur compared with modern pathogens

Cons
- restricted to bloodstream-dependent pathogens
- no routine PCR technique available
- no reasonable ratio of effort to result
- sequence alterations are not predictable and thus highly conserved sequences are preferred

Required benchmark
- protocols for multiplex PCRs for sets of common pathogens

amount of sequences are likely to occur both in ancient and modern pathogens.

Another consideration is the lack of any models to predict, from the number of evident cases, the actual number of infected individuals in a given ancient population. Demographers have developed formulae by which they can calculate the number of males, females, infants, individuals under 15 and over 65, etc., even if the archaeological record is incomplete. They have looked for population structures in different places and economies and have modeled populations under the assumption of a stable, constant structure. To the best of our knowledge, nothing like this has been developed for the calculation of infected persons in past populations,

although we have even tried to initiate studies of this sort.

The greatest benefit of aDNA analysis is the ultimate (or at least almost ultimate) certainty of diagnosis. Given time, money, and infrastructure, screening for pathogen DNA would also reveal the rate of endemic infections in a given population (i.e. the Durchseuchungsrate), which can by no means be provided by morphognostic approaches.

The most sophisticated approach, however, would be to look for balanced polymorphisms in a past population by means of aDNA techniques, e.g. blood groups, which are known to be selected by infectious and other diseases. If significant fluctuations in ratios are found, the search for pathogens by aDNA tools seems promising. The advantage is that it offers general information about the epidemiological burden of a population, independent of successful detection of evidence of the pathogen (Vogel and Motulsky 1997).

Acknowledgement

Parts of this work have been supported by the Bundesminister für Forschung und Technologie (Schwerpunkt NTG).

CHAPTER 13

Epidemiology of infectious diseases in the past: Yersin, Koch, and the skeletons

Olivier Dutour, Yann Ardagna, Marta Maczel, and Michel Signoli

This disease makes little discrimination, and not a few fearing God are cut off amongst the rest. They die of the same distemper with the most profane, they are buried in the same grave, and there sleep together till the morning of the resurrection. (Thomas Vincent, God's Terible Voice in the City of London, Wherein You Have the Narration of the Two Late Dreadful Judgements of Plague and Fire, Inflicted by the Lord upon That City, the Former in the Year 1665, the Latter in the Year 1666, Cambridge: Samuel Green 1667, quoted in The Wages of Sin, Peter Lewis Allen)

13.1 Introduction

Paleoepidemiology is a recent approach to the study of diseases in the past. This requires well-documented osteoarcheological series in order to minimize the numerous systematic biases represented by sample size, the material's state of preservation, the 'demographic' structure of the sample, and its chronological coherence. Very few skeletal series are compatible with this set of conditions. The plague epidemics in Europe have accounted for a million deaths over the past five centuries. This acute disease, which is classically not selective, is characterized by an age-at-death distribution similar to that of the living population. Plague epidemics are characterized by their great celerity and rapidity of death. For these reasons, the material originating from plague mass grave excavations is especially suitable for paleopidemiological studies. Besides improving our knowledge of plague in the past, these series can provide consistent data on the paleoepidemiology of other specific infectious diseases. In this approach, molecular biology offers the unique opportunity to confirm both the cause of massive death and the diagnosis of the specific infections studied by amplifying specific sequences of ancient DNA (aDNA). In this way, these new techniques allow us to develop an archaeological approach to infectious diseases and could efficiently resolve some historical outbreaks as well as explore the past history of re-emerging human infections such as plague and tuberculosis (TB).

The goal of this chapter is to present the new advances in molecular identification of *Yersinia pestis* and *Mycobacterium tuberculosis* in ancient human skeletal remains and possible ways of minimizing the different biases in epidemiology of past infectious diseases, taking as an example, the prevalence of TB in skeletal series due to plague epidemics, by studying three different samples representing 350 skeletons, from 16th- to 18th-century plague epidemics identified by polymerase chain reaction (PCR).

The re-emergence of past diseases has presented new challenges for microbiologists who have been led to examine the history of human infections in order to understand their present-day evolution over a longer time scale. Past diseases can be investigated with the aid of historical sources (written documents, paintings, engravings, etc.) or biological sources (skeletons, mummies); since the end of the 19th century, the latter science has been known as 'paleopathology.'

The analysis of ancient biological material discloses information on the history of disease without recourse to historical documents which, before the 19th century (i.e. before the development of modern concepts in biology and medicine), were too difficult to interpret medically. The relative advantage of biological sources in enabling an investigation into the far-distant past is mitigated by the fact that it is by its nature limited to the skeletal system. Paleopathology has long dealt with isolated cases, but recently the need has been noted for such investigation of past diseases on the level of an entire population, creating a new field of inquiry: paleoepidemiology (Bérato *et al.* 1990; Pàlfi 1993; Waldron 1994, 1999). The epidemiology of skeletal populations has been greatly assisted by recent advances in molecular biological techniques such as PCR, which have created new and fascinating opportunities especially in regard to infections, and are now more commonly used routinely, as attested by the increasing number of results on this topic (Salo *et al.* 1994; Spigelman and Lemma 1993; Taylor *et al.* 1996; Nerlich *et al.* 1997; Crubezy *et al.* 1998; Drancourt *et al.* 1998; Dutour *et al.* 1999; Haas *et al.* 2000a,b; Raoult *et al.* 2000; Boldsen 2001; Donoghue *et al.* 2001; Mays *et al.* 2001; Zink *et al.* 2001). These techniques may be used either to confirm the paleopathological diagnosis especially in atypical forms, or more recently, in order to investigate the prevalence of infectious diseases in a skeletal series.

Surprisingly, however, even in so-called 'paleoepidemiological' research, the nature and structure of the 'population' studied are often neglected. We believe that one of the present challenges of aDNA investigation is to obtain precise knowledge concerning the material on which these techniques are developed; as this material is archaeological rather than biological, it is submitted to a set of different parameters in the preservation processes (Waldron 1987), called taphonomy (Mays 1992). If we really intend to develop a relevant approach in paleoepidemiology, we need a better understanding of taphonomic processes at the molecular level (so that they can eventually become routine techniques), as well as at the level of the skeletal population. This concept of 'molecular taphonomy' (Dutour *et al.* 1998) is led by the recent introduction

of molecular techniques in the field of archaeology. Biologists are realizing that ancient biological material is far different from the true biological samples they are currently using. If 'molecular taphonomy' is to be extensively developed by molecular biologists [some authors have begun report to precise comprehensive knowledge of the alteration processes of aDNA (Höss *et al.* 1996a,b; Poinar and Stankiewicz 1999)], then biological anthropologists and other bone specialists must define precisely the taphonomy of skeletal populations used in ancient DNA paleoepidemiological reconstruction.

The goal of this chapter is to present briefly the main biases in the constitution of skeletal populations and possible ways of minimizing them in paleoepidemiological studies, taking, as an example, the study of the prevalence of tubercular infection on a skeletal series from plague epidemics, with a brief exploration of the new perspectives brought by molecular biology.

13.2 Paleoepidemiology

Although it may seem to be a contradiction in terms, paleoepidemiology has nothing to do with the epidemiology of past populations. Past populations are indeed mainly represented by skeletal populations—which constitute the subject matter of paleoepidemiology—but skeletal populations do not really represent past populations as they existed while alive. This is a major problem for paleopathologists and other specialists using skeletal records to reconstruct the history and evolution of disease; in fact, a skeletal series is the worst sample for a 'classical' epidemiologist. Theoretically, human skeletal remains are representative of a variable part of a dead population, which itself derives from the living population. This constitutes an immediate distinction from epidemiology, as paleoepidemiologists study disease in a community of the dead.

13.2.1 Epidemiology in a community of the dead

The dead population differs in gender and age distribution from the living population. In the less developed part of the world—which we presume to be closer to that obtaining in the past than more developed societies—the demographic structure of the mortality curve is the reverse of that for the living population. Over 40% of the living population is 15 years old or younger (Fig. 13.1). The mortality curve, on the other hand, shows high mortality of the 0- to 5-year-old cohort, relative stability between the ages of 5 and 35, a progressive increase to age 55, and a dramatic increase after age 55: a typical U-shaped profile.

It would be ideal to find the entire dead part of a given population preserved in a single cemetery, enabling us to reconstruct the structure of the living population from the age distribution of the skeletal population. However, the relationships between the two curves depend on other parameters as well. An improved economy, for example, will modify the demographic pattern, making it older and reducing the mortality of its youngest members. If the application of the demographic pattern of an undeveloped society to past populations is probably adequate in many cases (Fig. 13.2), we must none the less consider other patterns, especially those in developing countries.

For paleoepidemiologists, it is ideal when the dead population has the same structure as the living one, when a non-selected part of the population—or the population in its entirety-suddenly disappears (such as in the case of Pompeii).

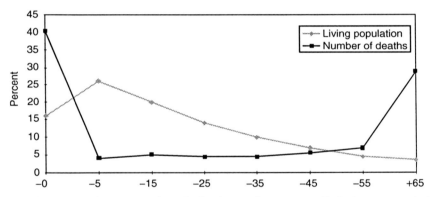

Figure 13.1 Living population and deaths in a present-day undeveloped country (percentage age distribution, age categories in years). The two curves are reversed. The curve of deaths is typically U-shaped.

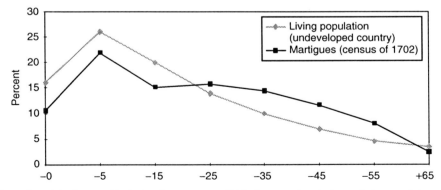

Figure 13.2 Comparison of the demographic structure (percentage age distribution, age categories in years) of a present-day undeveloped population with the structure of a historical population (France, beginning of the 18th century). The similarity of the two structures would allow the use of some demographic aspects of undeveloped countries to reconstruct the demographic patterns of past populations.

13.3 From the dead part of the population to a skeletal series

Any skeletal population studied by paleoepidemiologists is a sample of another sample (i.e. the dead population) of an ancient population. Of course, the sampling is not randomized. The main factors contributing to the constitution of an osteoarchaeological series are the burial assemblage (which is influenced by cultural practices), its duration (which can at times extend over several centuries), the taphonomic processes employed (chemical or biological), and the conditions of the archaeological excavations.

13.3.1 Burial assemblage

Let us consider a theoretical example: a medieval or modern cemetery, comprising a total of 2000 burials. If only one part of this cemetery, 200 graves, is excavated, and if this part happens to be the burial site for a local monastery, then the skeletal population may very well be comprised mainly of old males, i.e. the monastery's monks. In this case, the burial assemblage is indeed dependent on the cultural practice that prescribed that monks be buried on a particular site. If no archaeological records are available (e.g. if the monastery has been destroyed), this selected sample might even be mistakenly considered representative of the entire population. In paleoepidemiology the sample structure will have an effect on the prevalence of diseases. In our example the prevalence of a disease such as diffuse idiopathic skeletal hyperostosis (which is more frequent in old people and in males) will be very high in this skeletal population (Waldron 1985), just as we might expect the prevalence of venereal treponemal diseases to be non-existent. In another example, in studying the past prevalence of a disease involving mainly the youngest part of the population (e.g. TB), it will be of interest to know whether the youngest individuals are not missing or underrepresented in the skeletal series; at the time of the Roman Empire, for instance, the children were often buried separately.

The precise knowledge of the burial assemblage must be known in order to define the skeletal sample, and the best sample is a non-selected one.

13.3.2 Time effect

Let us suppose that our fictitious cemetery were occupied from the 10th to the 14th century. The skeletal population will be the sum of the dead portions of the successive living populations (it is usually difficult to date the burials archaeologically, and we cannot presume that over a period of 500 years the population remained static in structure and its origin, except in the uncommon case of a very closed, genetically isolated, community with a stable economic status). The reconstructed prevalence of diseases, which can be single or recurrent events, will be minimized, tending to a mean prevalence for the total period. In our theoretical cemetery of 2000 graves, we presume theoretical cohorts of 400 skeletons per century. We suppose very theoretically equality of distribution of skeletons during each century. We suppose one acute epidemic with a mortality rate of 50% (which we will have detected on the skeletons) to have occurred 40 times (i.e. 40 years) in one century, 30 times in a second century, and, finally, 10 times in a third. During these periods (representing for instance 80 years on 300 years), we consider (if we do not consider over-mortality) that half of the skeletons are provided by this acute disease (i.e. 160). Thus, in this example, the estimated prevalence of this acute epidemic will be only 13% for the three centuries in which epidemics occurred, and 8% for the entire sample, constituted over five centuries. If we suppose a more chronic disease with a prevalence of 30% occurring over two centuries, the prevalence in the total sample is 12% (Fig. 13.3). Time does have an effect on reconstituted prevalence; it must be taken into consideration in studying the epidemiology of skeletal series even if 'time effect' cannot really be predicted. The shorter the period involved in the constitution of the sample, the better the sample.

13.3.3 Taphonomy

The effect of taphonomy on paleoepidemiology is twofold. On a general level, a poor state of preservation of a skeletal sample will reduce its interest for paleoepidemiology. Preservation indexes (Nemeskeri 1963; Dutour 1989; Bello 2000) can quantify the overall preservation score of a skeletal series and provide information on the intensity of taphonomic processes. Clearly, the number of individuals alone is insufficient to prescribe the material available for paleoepidemiological studies; each skeletal population presents its own general preservation profile (Fig. 13.4). At a more detailed level, differences in preservation can occur in the same skeletal population, depending on gender or age: female and juvenile skeletons (Fig. 13.5) seem to be frailer and more often destroyed than the male, adult skeletons (Masset 1973; Dutour 1989). Anatomy is another

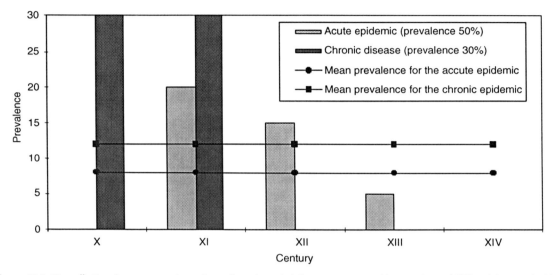

Figure 13.3 'Time effect' on the reconstructed prevalence of two theoretical diseases, one acute with a prevalence of 50% and the second with a prevalence of 30% (see the text). The prevalence estimated on all the skeletons on five centuries is a mean prevalence with low rates: 'time effect' is reducing the prevalence of acute or sporadic phenomena.

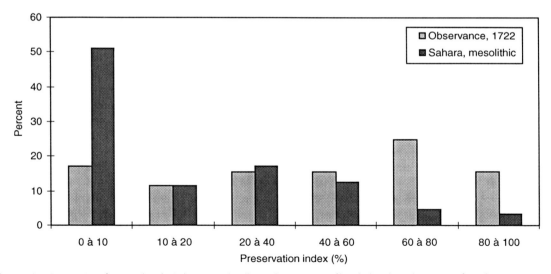

Figure 13.4 Preservation of osteoarchaeological series: each collection has its own profile, which is dependent on specific taphonomic conditions. In this example, the Saharan collections (Dutour 1989) are in poor state of preservation, in comparison with the modern series from plague (Marseille 1722) (Bello 1997).

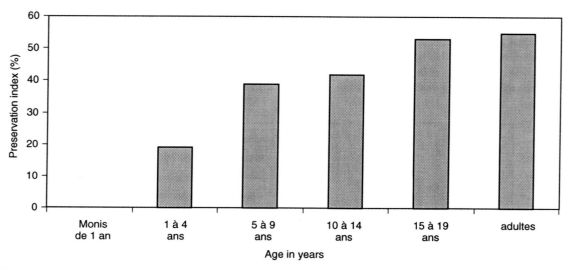

Figure 13.5 Preservation indexes as a function of age. The study on the L'Observance mass grave was performed on 174 skeletons. The preservation seems to increase with age, indicating that the state of preservation of each age category and the 'palaeodemographic' structure of the osteoarchaeological series needs to be known in palaeoepidemiology, especially to study the prevalence of the disease involving frequently or mainly children (as TB).

parameter, as some parts of the skeleton (hand and foot, ribs, spine) are more delicate and, consequently, more often missing than other parts of the skeleton (Fig. 13.6). This differential preservation must be compared with the skeletal distribution of the diseases studied. For example, as the osseous involvement of TB very frequently concerns the spine and extremities, it is of interest to know something about the preservation of these skeletal elements in the series (Fig. 13.7). Furthermore, concerning the calculation of prevalence of tuberculosis, very few methods had been developed, taking taphonomy into account. In most of the studies, the estimations concern the 'crude prevalence' (number of cases / total number of skeletons), which did not consider the preservation state of the osseous remains, and thus cannot represent the real frequency of the disease in past populations. Waldron (1994, 1999) introduced this notion and recommended calculation of the prevalence using only the preserved bones in the sample. This TB prevalence is estimated for each anatomical site, taking into account the number of affected sites and the corresponding skeletal material represented in the sample (Waldron 1999). The calculation is based on the modification of the denominators of the

'crude ratio'. One of us (Ardagna 2000) recently developed the re-evaluation of the crude ratio by taking into account the number of observable vertebrae that are essential for the diagnosis of vertebral TB. The calculation multiplies the crude prevalence by the ratio of theoretical number of vertebrae in the series and the number of effective observable vertebrae of the series. The purpose of these works is to provide a method of calculating prevalence adapted to each sample and valid for comparative studies. According to Bello (2000) reliable results can be obtained if one compares prevalence in series showing a similar preservation pattern.

At a molecular level, taphonomical processes slow down DNA extraction. If contamination should be the obsession of the aDNA specialists, DNA preservation in skeletal samples is also a major problem. Low temperatures seems to allow better preservation of aDNA, but some authors (Poinar and Stankiewicz 1999) claimed that DNA should not be preserved 'for longer than 10 000 to 100 000 years in most environments'. According to Höss (1996a,b) molecular taphonomy depends on two common processes affecting DNA in archeological material: hydrolytic damage, which leads to deamination of bases (depurination, depyrimidination), followed

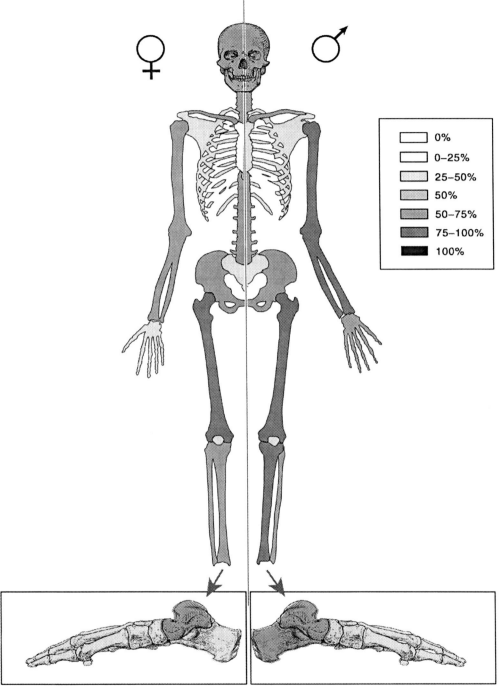

Figure 13.6 Preservation indexes in both genders (L'Observance, adults) showing differences of preservation depending on topography and on gender (drawing C. Tatilon, CNRS).

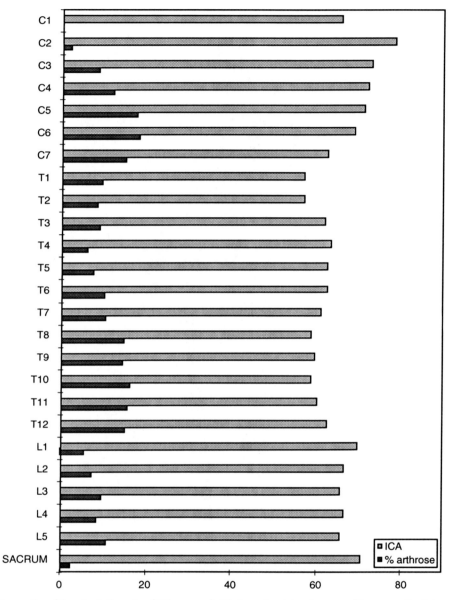

Figure 13.7 Preservation of the spine in the series of L'Observance (in %). For each vertebra, the mean of the preservation indexes (light grey) was calculated for the adult part of the sample (128 individuals) (Bello 1997). This result showed a good preservation profile of the spine (mean index about 80%) in this sample, making it appropriate for palaeoepidemiological studies of diseases involving the spine. This figure presents the analysis of frequency of vertebral osteoarthrosis (dark grey). For example, the mean preservation of the 5th lumbar vertebra is 65% and it is involved in osteoarthrosis in 15% of cases.

by oxidative damage, caused by ionizing radiation, resulting in modification of bases. According to Poinar *et al.* (1999), no direct correlation between the age of the sample and the preservation of its endogenous DNA is really observed. The taphonomy of human nucleic DNA needs a better understanding (Höss *et al.* 1996a,b). As for bacterial DNA (*M. tuberculosis, Mycobacterium leprae*), that seems to be more 'time resistant', like mtDNA, probably because of its similar structure. Anyway, for these bacteria, the question of the meaning of a negative or a positive molecular result is still broadly open from a palaeoepidemiological point of view: a negative result can signify either lack of infection or molecular taphonomical problem, a positive one, excepting contamination, should not be systematically related to disease, as it can also identify testimony of exposure to the infection.

13.3.4 Archaeology

The constitution of a skeletal series depends mainly on archaeology. Let us consider once again our example (2000 burials, 10th- to 14th-century). It is possible that only a part of the cemetery has been discovered, and excavations may have been carried out on no more than a segment of the unearthed part of the cemetery. The recovery may then concern only a part of the excavated area, or only a part of the skeletons (skull and long bones, for example, which are considered the most informative elements for anthropologists). The anthropological study may be carried out in a limited manner (gender and age distribution only), and the storage of these skeletal series can make them difficult to study in totality by the paleoepidemiologists. Hence, we see the implications of other parameters of sample selection. The ideal case would be the excavation of a site in its totality, without any selection in the recovery of the osteological samples, with appropriate storage, contributing to the constitution of skeletal libraries.

We can thus appreciate the challenge more clearly: paleoepidemiology concerns the study of diseases in different skeletal samples, the latter having been to a greater or lesser degree selected from the past populations by different factors in quite variable and totally unknown proportions. Most of the time these skeletal collections do not represent their original population because of the numerous biases in their constitution. If it is impossible to use any appropriate samples to predict some of the biases, and, moreover, if the samples are only representative of themselves rather than of past populations, is it a farewell to paleoepidemiology, like the one pronounced on paleodemography (Bocquet-Appel and Masser 1982).

13.4 The plague epidemics as paleoepidemiological models

Skeletal series from plague epidemics present very special characteristics in comparison with other osteoarchaeological samples and might be useful in the development of paleodemography and palcoepidemiology (Dutour et al. 1994b; Signoli and Dutour 1997a,b; Signoli et al. 1997b,c). These peculiarities can indeed minimize or even cancel out some of the common biases observed in skeletal collections.

13.4.1 Structure of the dead population

The structure of the dead population from plague epidemics corresponds to the 'Pompeii model' i.e. the breakdown of the number of deaths, has the same distribution as the living population and constitutes a kind of 'picture' of the living population. The lack of any selection of victims in plague epidemics can be clearly evidenced as demonstrated by the analysis of historical records. We compared the demographical data of two communities numbering 4000–6000 (using census figures for 1702–16) in Provence (Martigues and Aubagne, small towns near Marseilles, in southeastern France), with the age distribution of deaths (about 2000 victims) due to the 18th-century plague epidemic (1720–1) in the same city. The structure of the living population (which is very similar to that obtained for present-day undeveloped societies—see above) and the age distribution of the dead population (plague victims) present very similar aspects (Fig. 13.8), indicating that the plague did not result in a selection of victims within the population. On the contrary, the normal mortality profile (average mortality excluding demographic crises) is absolutely different (Fig. 13.8). This lack of selection seems to be specific to the plague, which is clearly different qualitatively from other demographic crises that occurred, such as other acute epidemics (probably measles) or famine, in which cases we noted that a segment of the population (either the very young or the old) in the same community at the same period had been affected (Fig. 13.9). The lack of selectivity in plague victims is very interesting for paleoepidemiology because it enables the dead

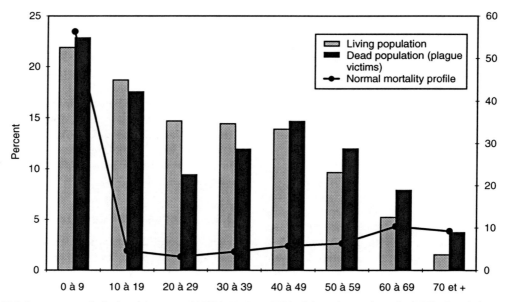

Figure 13.8 Percentage age distribution of the census of 1702 in Martigues (5888 inhabitants), normal mortality (1702–19, excluding demographic crises), and of the deaths from the plague epidemic in 1720–1 in the same city (1986 deaths). The comparison of the two curves clearly demonstrates the non-selectivity of plague epidemics.

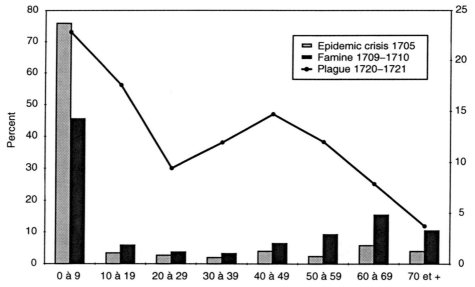

Figure 13.9 Percentage age distribution (PAD) of deaths for three different demographic crises in Martigues: increased mortality of the youngest individuals in the 1705 crisis (probably measles); increased mortality of the youngest and oldest part of the population in the 1709–10 crisis (famine). These two curves are compared with the PAD for the plague epidemic (1720–1), which is similar to the PAD for the living population.

part of the population to be treated as if it were the living one, without having to 'translate' the results obtained from a community of dead to the living population.

On the other hand, the confirmation of the diseases can be made by sequencing fragments of *Y. pestis* DNA in skeletal material. Pioneer works were carried out on our skeletal collections from

Provence dating from 1590 and from 1722 by Drancourt *et al.* (1998). New methods such as 'suicide PCR' are applied by the same team to other series dating from Black Death (Raoult *et al.* 2000). Other teams in Great Britain (Gilbert and Cooper 2001; Voong *et al.* 2001) failed in the amplification of *Y. pestis* DNA on skeletal material from the Black Death in London. One explanation could be taphonomical; differences in DNA preservation could be related to the different climatic conditions between northern and southern Europe.

13.4.2 Time effect

The distortion that might be caused to the extended period or time during which the skeletal data become constituted is, in this case, totally absent. The case under consideration provides a unique opportunity to study a skeletal population that can be dated from historical records with an accuracy corresponding to the month of death of the victims. The plague epidemics are characterized by short durations of time (*c.* 1 year), with the greatest effect being felt during the first 2–4 months. Some 75% of the mortality occurred during the first 3 months (Fig. 13.10).

13.4.3 Burial assemblage

During acute epidemics, and particularly during plague epidemics, social and other distinctions within the population tend to be set aside; the main practical problem is to bury a great number of corpses with the least possible delay, usually in mass graves. Burial practices distinguishing victims according to age, gender, or social origin are usually irrelevant during such an acute crisis, as is borne out by the many historical sources that regret that there was not enough time to bury the victims 'properly' [i.e. to organize burial assemblage (Signoli *et al.* 1998a,b, 2002)].

13.4.4 Archaeological records

Because of the pressure to bury all the victims rapidly (in mass graves, trenches, or in special cemeteries), the sites are archaeologically well defined and limited. Under such conditions, if the entire burial area is available for excavation, this can easily be an exhaustive one. In any event, because of the absence of selection, even a partial excavation would not, in theory, modify the sample structure.

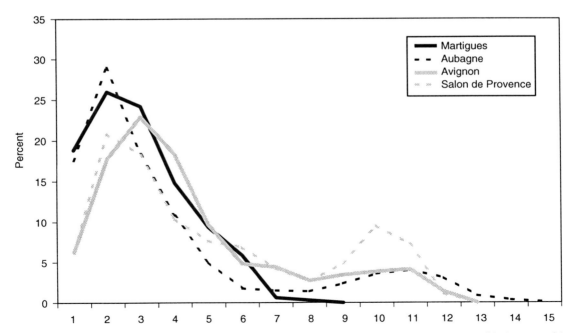

Figure 13.10 Distribution of the deaths per month (in percentage of the total number of deaths) for the plague epidemic of the beginning of the 18th century in four cities of Provence. These curves show the rapidity of the epidemic: 75% of the total deaths occurred during the first 4 months.

13.5 Epidemiological factors: practical examples

In order to study the variability in the makeup of sample skeletal populations from plague epidemics, we compared the structure (age distribution percentage) of three osteological series analysed to a reference curve of deaths by plague as derived from an analysis of historical data (Fig. 13.11). These series present different peculiarities attested to by the historical and archaeological data (Signoli *et al.* 2001). The first one (Les Fédons, Lambesc) corresponds to the cemetery of a quarantine infirmary dating from 1590 (Reynaud *et al.* 1996; Signoli 1998) and is composed of 133 individuals. The structure of this sample is quite similar to the reference curve of plague mortality, with the exception of a moderate excess of the 5–14 and 15–19 cohorts. The second one (Le Délos, Martigues) is smaller (39 individuals) and corresponds to the first months of the epidemic of 1720–1 (Signoli *et al.* 1995; Signoli 1998). Despite the small size of this sample—which could explain the underrepresentation of the 35–44 cohort—the

general distribution is very close to that of the plague mortality curve. The third one (L'Observance, Marseille, 216 individuals) corresponds to a special feature: the epidemic relapse in the spring of 1722, a year after the epidemic that killed half the population of Marseilles (50 000 victims). We believe that the differences in the structure of the sample (mainly the underrepresentation of the youngest individuals) are due to the modification of the structure of the population itself 1 year after this dramatic mortality crisis (Dutour *et al.* 1994b; Signoli 1998).

The analysis of the structure of these three series shows that, even if the plague material seems to be very appropriate for developing a paleoepidemiological study, there is an additional parameter of variation, i.e. the phases of plague epidemic (beginning, acute period, relapse). However, knowing the reference morality curve and the structure of the population itself, it is possible to correct the variation of the distribution of the skeletal population due to the epidemic phases in order to make a model of the prevalence of diseases in this material.

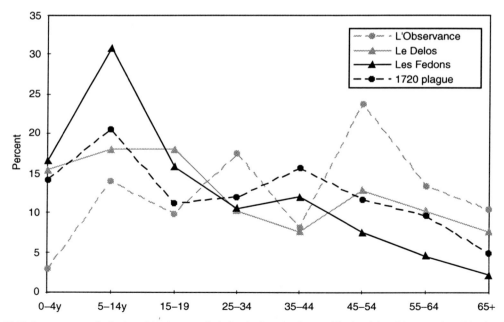

Figure 13.11 Percentage age distribution of the three osteological series from the plague epidemics of the 16th century (Les Fédons) and of 1720–2 (Le Délos and L'Observance) compared with the deaths by plague (1720, historical data). The palaeodemographic differences observed in the sample structures are due to the peculiarities of L'Observance, which corresponds to an epidemic relapse in spring 1722.

13.5.1 Tuberculosis

Tuberculosis is a good example of a re-emerging disease. Its prevalence is once more on the rise, and recent statistics place its mortality rate higher than that of AIDS. TB might become a major problem, especially if we take into account the antibiotic-resistant bacteria, which are on the increase, and its future may very well be similar to its past. What we know of its past is limited, mainly to the 19th and early 20th centuries, based on mortality records (Fig. 13.12). The classical data are that TB infection increased in the 19th century because urbanization and overcrowding facilitated its spread (Wilson 1994). In late 19th-century France, the mortality rate from phthisis (TB) was between 3.08 and 3.69 per 1000 (Signoli *et al.* 1997a; Bello *et al.* 1999); during the same period in Germany, the mortality from TB was 2.6 per 1000 (Alfer 1892, in Ortner and Putschar 1985). However, we must take care lest we are influenced by preconceived notions. Our knowledge of the situation prior to this period is very poor, limited to some rare historical records of mortality, such as the London Bills of Mortality beginning in the 17th century, which indicated that death by 'consumption' (pulmonary TB or primary lung infection) accounted for 20% of all deaths during non-plague years (Clarkson 1975). The accuracy of diagnosis in the 17th century, however, was poor. A more reliable gauge is skeletal populations.

Since there is considerable uncertainty concerning the assignation of TB as a causative agent to many macro-morphological bone changes, on which detection of TB infection has been based mainly on osteoarchaeological material, attention has been directed to the molecular level in search of more reliable diagnosis, and, therefore, more relevant disease frequency in past populations. As a consequence, molecular biological techniques developed during the past decade have largely broadened the diagnostic horizon in palaeopathology not only by confirming the macroscopical diagnosis as a result of providing direct, demonstrative proof of tuberculous infection, but also by helping to identify new criteria for differential diagnosis.

Morphological techniques often do not allow the recognition of TB lesions, and the more specific identification of the disease agents is even more difficult, since human- and bovine-hosted TB, caused by the two main human-affecting members of the *M. tuberculosis* complex [MTC; comprises *M. tuberculosis*, *Mycobacterium bovis*, *Mycobacterium microti*, *Mycobacterium africanum*, *Mycobacterium canetti* (Niemann *et al.* 2000)], produce anatomically

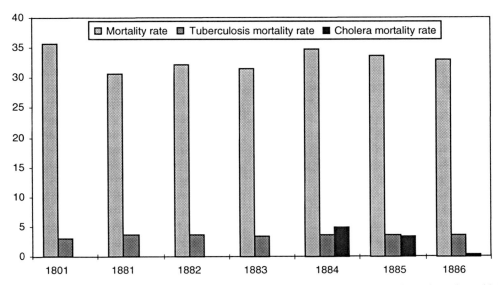

Figure 13.12 Mortality rate in the 19th century in Marseilles. TB is the most important cause of death except during the epidemics (cholera). It should be noted that the TB mortality rate is very regular and quite similar to the mortality rate from cholera (Signoli *et al.* 1997a).

similar bone changes (Ortner 1999). However, in spite of the fact that the members of the MTC share many common characteristics, they, as well as other mycobacteria, can be differentiated at the biomolecular level. The biomolecular analysis of archaeological human remains for TB proved to be efficient using two biomarkers that persist in ancient bone material and can be diagnostic of TB infection and the pathogen: DNA fragments and mycolic acids (Gernaey et al. 1999).

The most spectacular results have been gained from the search for and examination of mycobacterial DNA fragments. Such studies have been conducted in mummies (Salo et al. 1994; Nerlich et al. 1997; Crubézy et al. 1998; Pap et al. 1999; Zink et al. 2001), bone remains (Spigelman and Lemma 1993; Baron et al. 1996; Taylor et al. 1996, 1999; Dutour et al. 1999; Fearman et al. 1999; Haas et al. 2000a), and even in calcified tissues (Donoghue et al. 1998; Pálfi et al. 1999), which proved that fragments of ancient mycobacterial DNA might long survive, in all probability due to the tough cell wall, and can provide direct evidence for TB infection. These studies furnished evidence from distinct genetic loci for the presence of DNA fragments from mycobacteria (65 kDa antigen gene) and more specifically from organisms belonging to the MTC (IS6110, rpoB) (Taylor et al. 1999; Haas et al. 2000a; Mays et al. 2001).

This method is based on the extraction of aDNA, which is subjected to PCR, with a primer pair targeting the MTC-specific segment, most commonly the 123 bp segment of the repetitive sequence IS6110 (Eisenach et al. 1990). To ascertain that PCR is not inhibited and amplifiable DNA is present in the samples, segments of human β-actin gene (202 bp) or the amelogenin gene is amplified in parallel. To rule out contamination blank controls are also used during the analyses (Zink et al. 2001).

There have been further assays developed, such as PCRs for mtp40, oxyR, pncA, M. bovis-specific segments, as well as spoligotyping (spacer oligotyping) and direct sequencing, in order to distinguish M. tuberculosis and M. bovis, which at the level of bone alterations cannot be done (Taylor et al. 1999; Mays et al. 2001).

Mycolic acids, which are stable fatty acids and major components of the cell envelope of M. tuberculosis and other mycobacteria, seem to be another type of reliable biomarker for ancient tuberculosis. The chemical and chromatographic (HPLC) methods used for the sensitive detection of mycolates, providing different profiles for different mycobacteria, have been suggested as an alternative to PCR amplification (Gernaey et al. 1999).

With the help of such biomolecular analyses, a more reliable diagnosis can be set up concerning both typical and atypical morphological alterations, and, therefore, new diagnostic criteria can be determined, as in the case of the so-called early-stage TB conditions (such as vertebral hypervascularization (Ménard 1888; Baker 1999), rib periostitis (Kelley and Micozzi 1984; Roberts et al. 1994), and endocranial changes (Schultz 1999). An important source of tuberculous alterations can be found in anatomical collections, where the cause of death is recorded. In search of new diagnostic criteria studies were made of the Hamann–Todd (Kelley and Micozzi 1984) and Terry collections (Roberts et al. 1994) as well as the Coimbra Identified Skeletal collection, where M. tuberculosis infection was also confirmed by the use of biomarkers (Santos and Roberts 2001).

However, mycobacterial DNA can even be detected in bones without morphological changes (Fearman et al. 1999; Zink et al. 2001). This point leads to the question of infection or exposure to infection, which is specially relevant in molecular paleoepidemiology. However, it is the thorough biomolecular examination of archaeological human remains that can contribute to the more precise evaluation of TB infection in past populations, and might give information on the evolution and transmission of the disease pathogens, too.

Very roughly, we attempted to estimate the minimal prevalence of TB in plague material and other material studied. We encountered several methodological problems during our first attempt.

1 Reconstruction of TB prevalence from skeletal lesions depends on the frequency of TB skeletal involvement out of the total TB infections. According to the literature (Nathanson and Cohen 1941; Lafond 1958; Davies et al. 1984; Kelley and Micozzi 1984; Ortner and Putschar 1985), it varies between 3% and 9%. Hence, to avoid overestimation, we can assume a minimal prevalence from the minimal frequency of skeletal involvement.

2 TB skeletal infection affects mainly the youngest part of the population (Sorrel and Sorrel-Dejerine 1932), with more than 60% of all victims being under the age of 20. Poor preservation or absence of the youngest individuals in an osteoarchaeological series will minimize the reconstruction of the frequency of skeletal TB.

3 Bone repartition of TB involvement affects mainly the spine (25–50% of skeletal TB cases; Steinbock 1976) and the extremities. Of 24 paleopathological cases of TB from Hungary, the diagnosis was established on the spine in 19 cases (Pàlfi and Marcsik 1997). These parts of the skeleton are, unfortunately, often poorly preserved.

4 A paleopathological diagnosis is established on morphological (osteological and radiological) criteria, defined by a comparison of clinical, radiological, and pathological records (Sorrel and Sorrel Dejerine 1932). The significant TB prevalence in the past, as evidenced by documentary sources (Cronje 1984) contrasts with the paucity of paleopathological evidence (Stirland and Waldron 1990). It has been suggested that the usual paleopathological diagnostic criteria for skeletal TB are inadequate (Roberts *et al.* 1994). Biomolecular analysis of *M. tuberculosis* DNA in presumed paleopathological cases may confirm the diagnosis (Spigelman and Lemma 1993; Salo *et al.* 1994; Baron *et al.* 1996; Taylor *et al.* 1996; Dutour *et al.* 1999, Pàlfi *et al.* 1999; Nerlich *et al.* 1997; Crubezy *et al.* 1998; Haas *et al.* 2000a; Mays *et al.* 2001; Zink *et al.* 2001). If we take a rough look at some large osteoarchaeological collections (numbering a total of 5803 skeletons) from Hungary, dating from the 7th to the 17th centuries, the reconstruction from the skeletal lesions of the minimal prevalence of TB infection in the population showed variations depending on chronology (Pàlfi and Marczik 1997): between the 7th and 8th centuries (the Avar era); 23% during the 10th century Hungarian conquest,

0% (but some cases of leprosy have been described; Pàlfi 1991); between the 11th and 13th centuries, 8.6%; and for the period from the 14th to the 17th centuries, 15%. This osteoarchaeological material does not provide us with the criteria desirable for paleoepidemiology, i.e. short periods of time, absence of selection, Pompeii-like paleodemographic structure.

On our plague material, for one of our series (L'Observance) the diagnosis was established both morphologically and molecularly on three individuals: vertebral lesions evoking TB infections in two cases (burial #43 and #141) and one with costal lesions (burial #82) similar to those described as TB involvement by Sorrel and Sorrel-Dejerine (1932). These three samples gave positive results (Zink *et al.* 2001); other 'control' samples with no lesions, remained negative. The reconstruction of the minimal prevalence of TB (based on 3% of skeletal involvement) from the plague material showed different frequencies in the three series: 25% for Les Fédons (16th century) and 46% for L'Observance (1722). No cases have been identified in the material from Le Délos, but this sample is small (only 39 individuals). The prevalence seems to be very high for the 18th century material; however, we must keep in mind the frequency of tubercular infection observed in undeveloped countries 30 years ago: 37% in Phnom-Penh in 1966 (Ngyuen Tang Am 1988).

Even if those results are very preliminary, we intend to demonstrate the need for reliable material for paleoepidemiology, especially for the reconstruction of past infectious diseases. The plague material provides a unique opportunity for this type of study. In the near future the development of 'molecular paleoepidemiology' on these series will allow us to obtain a more precise knowledge of the prevalence of human infections, especially those due to mycobacteria.

CHAPTER 14

The archaeology of enteric infection: *Salmonella, Shigella, Vibrio,* and diarrheagenic *Escherichia coli*

James P. Nataro, O. Colin Stine, James B. Kaper, and Myron M. Levine

All summer the debate over Quarantine had raged. The merchants and the landowners led by Lord Londonberry argued against it—self servingly saying that a suspension of trade could only further bankrupt the poor and drive them more rapidly to their deaths. Clanny and a good number of the other doctors, backed by the London Board of Health and thus His Majesty, claimed it was their only protection. They'd all read the numbers in the newspaper: thirty thousand dead in Cairo in twenty-four hours, five hundred cases a day in Saint Petersburg, anti-doctor poison riots in Hungary, Paris and Berlin. Cholera morbus must not be allowed into the realm. This was a disease to enrage the people, bring down governments, and with Reform Riots already flaring in Nottingham and Bristol, His Majesty decided it was in the country's best interest to quarantine all incoming vessels. (The Dress Lodger, Sheri Holman, 2000)

14.1 Introduction: enteric infection and history

Enteric illness has probably afflicted *Homo sapiens* since the origin of our species. Indeed, diarrhea is nearly ubiquitous among mammalian species, the result of convergent and divergent evolution by a wide variety of viral, parasitic, and bacterial agents.

Tracing the phylogenetic history of pathogens is more than an academic exercise: diarrheal diseases have mounted devastating assaults on human populations throughout history, making the stories of these diseases inseparable from our own. In an earlier volume on the same subject, Ewald (1998) speculated that epidemic diarrhea may have been responsible for the collapse of more than one primitive civilization. Diarrheal outbreaks have also figured prominently in human migrations and have influenced countless wars. Thus knowing which pathogens were present in a population at a given time can enrich our understanding of human history.

In the pages that follow, we will

• discuss the potential contributions of molecular biology to infectious disease archaeology;
• analyze current evidence as to the history of bacterial enteric pathogens;
• highlight *Vibrio cholerae* as a specific test case for the potentially fruitful application of molecular approaches.

14.2 Molecular biology tools in the study of infectious disease archaeology

Biotechnology provides unprecedented tools for investigating the archaeology of human pathogens. The most powerful new avenues are genomics, multi-locus sequence analysis, amplification of ancient DNA (aDNA), and microarrays. The potential of these new approaches is only beginning to be realized.

14.2.1 Genomics

With current technology, the complete genome of the average enteric pathogen can be deciphered in a matter of months. With this high-throughput capacity, we can now generate sufficient genomic data to compare the sequences of multiple bacterial pathogens, both within and between species.

One common approach to infer the evolutionary history of virulence genes from genomic data is the G+C calculation. The % G & C ratio of the locus in question is compared with that of the surrounding genetic 'backbone'; if there is substantial discrepancy between the two values, the investigator infers a foreign origin for the locus. However, recent data suggest that % G & C may not be a reliable means of assessing the origin of genetic loci or the timing of acquisition (Koski *et al.* 2001). The G+C calculation is dependent upon a 'molecular clock' assumption, i.e. a constant rate of change for a given gene. But molecular clocks can only be as good as the data used for their calibration; at best, they produce a rough age estimation, since they can run fast during times of population expansion or in the face of positive selection, or slow during times of population contraction or negative selection.

Genomic data can also suggest the relative order in which genes are acquired among related strains. For example, phylogenetic trees have been useful in tracing the development of pathogenic *Escherichia coli* (Donnenberg and Whittam 2001) and *V. cholerae* (Chakraborty *et al.* 2000). In essence this exercise is similar to classical phylogenetic reconstructions. An example could be constructed involving organisms with the genotypes x,y,z (organism A); y,z (B); w,y,z (C); w,z (D). We can draw a hypothetical 'family tree' for these organisms by inferring, for example, that all four organisms have common ancestry represented by z, with a separate branch represented by acquisition of w, followed by loss of y from organism D. If, instead of whole alleles, the parameters in question were nucleotides in a single gene, a similar phylogenetic tree could be generated by scoring for loss or gain of specific mutated nucleotides (with the caveat that mutation back to the original nucleotide occurs with some frequency). The most powerful application of this approach is multi-locus sequence analysis.

14.2.2 Multi-locus sequence analysis

Multi-locus sequence analysis typically involves the complete determination of the sequence in several distinct loci for a large number of bacterial strains. Using this approach, a phylogenetic tree may be constructed in which closely related sequences are paired or clustered and more distantly related sequences are located more distantly on the phylogenetic tree. In addition, this analysis provides the opportunity to determine the timing of divergence to a limited degree, either by calculating the number of mutations or, for more distantly related homologues, by determining the ratio of synonymous to non-synonymous mutations (i.e. the ratio of mutations that do not change the amino acid codon to those that do). These data provide only an approximate estimate of the timing of divergence and the ages of various pathogens, although the combination of these data with other evidence can provide some useful clues to genetic natural history. An important caveat is that the rate of change may be affected both by the necessity of maintaining gene function (mitigating against changes) and also by the pressures exerted by the immune response (mitigating for changes). Moreover, there is selective pressure to maintain the codon bias that is most suitable for appropriate expression of the locus in question. To complicate the analysis still further, the existence of hypermutable pathogenic strains (Foster 2000), which favor rapid bacterial evolution, offers an as yet indecipherable variable.

The shortcomings of multi-locus sequence analysis may be circumvented once a larger database of multi-locus sequence data is available. Currently, the extent of variation at the sequence level within and between strains and closely related species of bacteria is largely unknown. A larger database may provide insight into what the appropriate questions are and how they can best be addressed.

14.2.3 Ancient DNA

The most direct clue to the presence of pathogens within a given human population is the detection of the pathogen or its genomic DNA in a preserved sample. This approach has been used with dramatic success to demonstrate *Yersinia pestis* genetic material in ancient dental pulp (Drancourt *et al.* 1998). Notably, in this communication the sequence of the virulence gene *pla* in ancient samples was found to be identical to that in modern strains of *Y. pestis*.

Although PCR detection of aDNA offers irresistible promise, the approach also entails several critical caveats (Cooper and Poinar 2000). Clearly, the ability of an investigator to attribute an amplified target DNA product to a particular time depends entirely on the quality of the specimen in question. The sample must be free of even remotely possible contamination with more recent target DNA. Sources of such contamination could be geologic or meteorological events, the activity of disrupting animal or plant life (e.g. tree roots, earthworms, birds, etc.) and unintentional or intentional human contamination. For enteric pathogens, these limitations render impractical or potentially erroneous the molecular analysis of the vast majority of archaeological samples. But some opportunities exist: potentially promising specimens would include the undisturbed intestines of well-preserved (i.e. mummified or frozen) human specimens.

In addition to the limitations imposed by potential contamination there is the inherent instability of DNA over time in most environments. In soil, for example, nucleic acids may degrade spontaneously, be degraded by contaminating nucleases, or be metabolized by soil microorganisms. Thus specimens subjected to analysis must be preserved under conditions that preclude such damaging influences. Ancient coprolites have been shown to undergo crosslinking between reducing sugars and amino groups, thus inhibiting standard PCR (Poinar *et al.* 1998). Poinar *et al.* (1998) have reported that *N*-phenacylthiazolium bromide can cleave these links and render the DNA susceptible to PCR. A third obstacle to the analysis of aDNA is the potential presence of substances that inhibit amplification processes. We recently attempted to amplify genes encoding enteric virulence factors from excavated medieval latrines but found the samples inhibitory to amplification of even positive control DNA (J. Nataro, C. Gonzalez, and M. M. Levine, unpublished). A fourth limitation of amplification technology is the uncertainty of identifying PCR primers that will reliably hybridize with targets

that have undergone unknown genetic mutations. Such PCR primers should be closely spaced so that only very small DNA fragments need be present in the specimen. However, this puts even greater limitations on the investigator's ability to select useful primer pairs. Naturally, sequences that are highly conserved among modern strains are most likely to be similarly conserved in ancient isolates as well. A final limitation to archaeological study of enteric pathogens is the limited point prevalence of most pathogens. Therefore high-yield specimens would be derived from a documented outbreak or case of the infection in question or should be derived from concentrated fecal sources (e.g. a latrine).

14.2.4 Microarray technology

An alternative methodology for detecting the presence of pathogens in uncontaminated ancient samples is microarray technology (Stine 2001). Microarrays consist of up to 8000 DNA probes spotted onto a surface, often a treated glass slide. Each probe can be specific for a DNA target or gene. Thus a microarray may contain DNA probes for virulence and housekeeping genes from each of a series of pathogens that might be present in the sample. The aDNA sample is subjected to random primer amplification and labeling and is then hybridized to the microarray. In addition to being able to test for multiple pathogens at once, the length of the target DNA can be shorter for microarrays (75–100 bp), and does not need to encompass the entire hybridization target. In contrast, target DNA should be on the order of 200–500 bp for PCR, and targets need to span the distance between the two PCR primers.

14.3 Natural histories of bacterial enteric pathogens

14.3.1 The evidence for bacterial enteric infections in ancient populations

Application of the above methods indicates that many of today's bacterial enteric infections have been with man for thousands of years. Below we examine the available evidence for some of the most important bacterial enteric agents.

Salmonella

Infections due to *Salmonella enterica* occur in a large number of vertebrates, both warm-blooded and cold-blooded. Molecular phylogenic studies suggest that *Salmonella* and *E. coli* shared a common ancestor *c.* 100 million years ago (Lawrence and Ochman, 1998), a date that is probably more recent than the advent of the taxon Mammalia (Wyss 2001). Presumably, the coming of the mammalian intestine provided the original niche for both of these microorganisms. This does not presume that pathogenicity comprised the original niche of *Salmonella*, but the absence of significant numbers of extant non-pathogenic salmonellae, in contrast to *E. coli*, for example, offers some evidence for this possibility. *Salmonella* serotypes are capable of causing diarrheal disease in multiple vertebrate hosts, an epidemiologic feature that accounts for a substantial burden of human food-borne disease. Thus it is possible that *Salmonella* has been a human pathogen for as long as man has lived in intimate contact with other vertebrate species. This situation has prevailed at least since the earliest domestication of mammals for human use, estimated at *c.* 10 000 years ago (Diamond 1997). Thus non-typhoid *Salmonella* may reasonably be expected in human samples of early populations who kept domesticated animals.

Salmonella Typhi represents a special case. Molecular pathogenetic studies confirm that *S. Typhi* harbors a distinctive package of virulence-related loci. Indeed, the molecular differences between *S. Typhimurium*, the cause of murine enteric fever, and *S. Typhi* are remarkable, and constitute an obstacle to the use of this animal pathogen as a model for human typhoid pathogenesis.

S. Typhi is extraordinarily host adapted to *H. sapiens*; no animal reservoir has ever been identified. Moreover, *S. Typhi* has a propensity to cause the enteric fever syndrome, with reticulo-endothelial dissemination accompanied by fever, headache, and malaise without diarrheal illness. Long-term carriage in the human gall bladder completes the epidemiological picture, which conspicuously lacks an extra-human reservoir.

Nearly all *S. Typhi* strains can be traced to a single molecular clone, implying common origin with worldwide dissemination (Moshitch *et al.* 1992). Moshitch *et al.* (1992) have suggested that the predominant clone originated in Indonesia, although common origin in Africa has also been suggested (Selander *et al.* 1990). Whichever is correct, the data strongly support an ancient origin for this organism. Moreover, it can be inferred that strict adaptation of the typhoid clone to man necessitates the existence of a critical human population density. At the same time, the phenomenon of gall bladder colonization would allow *S. Typhi* to persist within a pre-civilized population until new susceptibles are encountered in the course of travel or warfare. Thus we infer that early human civilizations may have tolerated a low-level endemic burden of this still-feared human scourge.

Shigella

All *Shigella* serotypes are pathogenic for man. Each is human host adapted and causes watery diarrhea, often followed by dysentery (blood and mucus in stools), frequently with evidence of invasion and inflammation of the colonic mucosa. All shigellae share plasmid-borne and chromosomal virulence genes. The most virulent shigella serotype, *Shigella dysenteriae* 1, is a hypervirulent clone that has been lysogenized with Shiga toxin-encoding phage (McDonough and Butterton 1999). This Shiga toxin induces epithelial and endothelial cell death and thereby exacerbates local and systemic complications.

Various lines of evidence suggest that *Shigella* originated from *E. coli* and probably on several independent occasions (Pupo *et al.* 1997, 2000). Unlike *E. coli*, however, *Shigella* spp. are well adapted to the pathogenic state; a long-term carrier state is only rarely observed (Levine *et al.* 1973) and there is no extra-human reservoir. Accordingly, it might be expected that *Shigella* spp. have not only acquired genes that meet pathogenic requirements (invasion, host immune system avoidance, etc.) but also that the organism could dispense with genes specifically required for *E. coli*'s commensal existence. Indeed, Maurelli *et al.* (1998) have shown that *Shigella* spp. have undergone a conserved chromosomal deletion event (the 'black hole' of pathogenicity), which may actually promote diarrheagenicity.

In contrast to non-typhoid *Salmonella* with its animal reservoir, the requirement for continuous human transmission necessitates a critical population size for perpetuation of *Shigella* spp. within human communities. This suggests that the pathogen could only arise when confluence of tribal societies was prevalent. However, phylogenetic studies estimate that the origins of *Shigella* serotypes is of the order of 35 000–270 000 years ago, a period that overlaps or predates the early spread of *H. sapiens* (or related hominids) through much of the Old World (Diamond 1997). Therefore the data predict that early humans carried their *Shigella* strains with them during their migrations. Whether or not modern *Shigella* strains could have survived under such circumstances is not clear, but it is likely that the organism has undergone further molecular evolution in response to changes in the human condition. The molecular evidence provided by *E. coli* phylogenetic studies (see below) suggests that horizontal acquisition or loss of genetic loci may occur over very short time frames. In any case, *Shigella* species are likely to be encountered in remains of early *H. sapiens*, and genomic analysis of early *Shigella* strains could test some of these conjectures.

E. coli

E. coli is ubiquitous in the mammalian intestine and thus this bacterium has probably co-evolved with the earliest mammalian species. However, several pathogenic varieties of *E. coli* have also arisen. Currently six pathotypes of diarrheagenic *E. coli* have been characterized (Nataro and Kaper 1998); each of these pathotypes has distinct clinical, epidemiological, and pathogenetic features.

The origins of pathogenic *E. coli* have been the subject of intensive investigation. The best studied is the enterohemorrhagic or Shiga-toxin-producing *E. coli* (EHEC). The most distinctive pathogen of this group, *E. coli* serotype 0157:H7, causes the hemolytic uremic syndrome by virtue of the elaboration of Shiga toxin (see *S. dysenteriae* above). Because of the highly characteristic nature of this clinical syndrome, it has been suggested that fully virulent *E. coli* 0157:H7 was first introduced to the USA around 1955 (Nataro and Kaper 1998). However, the recently determined genomic sequence of 0157:H7 suggests the presence of a remarkable 1 million BP of DNA over what is present in the ancestral *E. coli* genome (Perna *et al.* 2001). Much of this DNA has been found to encode likely virulence factors. These observations belie the possibility of an abrupt

adaptation to the pathogenic state. Whittam and his group (Reid *et al.* 2000b) have suggested that *E. coli* 0157:H7 has undergone a stepwise acquisition of chromosomal and plasmid-borne virulence factors, although the timing of this sequence cannot be estimated with confidence. Moreover, EHEC may share a common ancestor with a highly prevalent diarrheal pathogen of infants, enteropathogenic *E. coli* (EPEC), whose presence in widely dispersed human populations and great genetic diversity suggest longevity as a human pathogen (Donnenberg and Whittam 2001). Indeed, the specificity of EPEC as a pathogen for infants less than 6 months old would serve as an impediment to rapid worldwide dissemination of an emerging clone: infants in early populations move with their families, and the success of such movement is heavily dependent on the availability of sustaining resources. In contrast, adults may move rapidly in pursuit of commerce or the spoils of war. Thus although no precise estimations have been offered as to the age of diarrheagenic *E. coli*, evidence suggests that they have co-evolved with man for many generations.

Many strains of enterotoxigenic *E. coli* (ETEC) elaborate the heat-labile (LT) enterotoxin, which closely resembles the *V. cholerae* cholera toxin (78% identical at the amino acid level). It is impossible to date accurately the timing of this divergence, but it is not likely to be a very recent event by human standards. Interestingly, some LT-producing ETEC are animal pathogens, whereas *Vibrio cholerae* is not. Since the cholera toxin is found chromosomally in a lysogenized phage, it is possible that the toxin originated in an *E. coli* ancestor and was transferred via a phage to *V. cholerae*.

14.4 Cholera: a case study

Notably among the enteric pathogens, *V. cholerae* has undergone remarkable epidemiologic and microbiologic variations during recent history. This fact produces an unusual opportunity for the application of archaeological techniques.

14.4.1 Background

Cholera is a diarrheal disease that holds a special place in the annals of public health (reviewed in

Kaper *et al.* 1995). Patients stricken with its severe form, cholera gravis, purge voluminous, electrolyte-rich stools with the appearance of rice water; dehydration, hypovolemic shock, renal shutdown, and death ensue if proper and aggressive rehydration therapy is not initiated. From the time of onset of the first diarrheal stool, cholera gravis can lead to death in 6–8 hours; few infectious diseases can be fatal with such a short clinical course. One characteristic mark of the epidemiological behavior of cholera is its propensity to occur in explosive outbreaks; a second is its capacity to spread in true pandemics involving several continents over many years in time.

In the 1990s cholera once again taught the modern world the devastating public health problem that it can pose. Cholera returned to South America in 1991, after an absence of a century, and over a 3-year period resulted in approximately 1 million clinical cases, as it spread in all directions following its introduction in Peru (Sanchez and Taylor 1997). In Goma, Zaire, cholera appeared among Rwandan refugees in 1994, causing an explosive outbreak that within 3 weeks led to *c.* 68 000 cases and 12 000 deaths among the 700 000 refugees (Goma Epidemiology Group 1995).

Most authorities interested in the history of cholera recognize that seven distinct pandemics have taken place (Pollitzer 1959). While sporadic cases or small outbreaks of cholera probably occurred for many centuries, particularly in the Ganges river delta of Bengal, it is only in the first quarter of the 19th century that cholera emerged as a disease capable of pandemic spread, which periodically reached populations in Europe, North and South America, and Africa (Pollitzer 1959; Barua 1992). The first six pandemics began in Bengal, the seventh broke out on the island of Sulawesi in Indonesia.

Isolation and characterization of the etiologic agent responsible for cholera occurred during the fifth pandemic with Robert Koch's description of the comma bacillus (1884) subsequently designated *V. cholerae*. Later bacteriologic studies showed that cases of endemic Asiatic cholera in the Ganges delta and the agent of the sixth pandemic of cholera, which spread in the first decades of the 20th century were caused by *V. cholerae* O1, which was further categorized into two serotypes, Inaba and

Ogawa. It later became recognized that there are many distinct serogroups of *V. cholerae* in addition to O1, with these other serogroups inhabiting brackish water environments. Lastly, it was observed that there exist variants of serogroup O1 vibrios such as the El Tor biotype; however, prior to the early 1960s, these variants were not considered capable of causing epidemic disease.

The spread of the seventh pandemic of cholera from Indonesia in the early 1960s constituted an epidemiological and microbiological surprise, because the vibrio responsible, while bearing the serogroup O1 antigen, exhibited characteristics that constituted a distinct biotype, so-called El Tor. In contrast, *V. cholerae* O1 strains responsible for the sixth pandemic and strains isolated from cases of endemic cholera in Bengal were of the 'classical' biotype. While the existence of El Tor vibrios had been known for many years, as had their ability to cause sporadic cases of diarrhea, they were not considered 'true' cholera vibrios until the early 1960s. Ironically, the seventh pandemic, the so-called El Tor pandemic, has become the most extensive cholera pandemic of all in geographic spread and in duration.

It had always been assumed that the first five pandemics, like the sixth, were caused by *V. cholerae* O1 of the classical biotype and that the seventh pandemic, the only one to originate outside the Ganges delta, was an aberration since it was caused by biotype El Tor. However, certain events of the early 1990s shook cholera dogma to its core and raised fundamental questions about the microbiological nature of the organisms responsible for the early cholera pandemics.

In 1992, simultaneously in India and in Bangladesh, epidemic cholera appeared that was typical in its clinical form (causing dehydrating diarrhea) and epidemiological behavior (resulting in explosive outbreaks) (Cholera Working Group Ic 1993; Nair *et al.* 1994). However, the *V. cholerae* organism responsible for these epidemics was not serogroup O1; rather, it was serogroup O139 (Cholera Working Group Ic 1993; Nair *et al.* 1994). Over the next 2 years, O139 cholera spread to many neighboring countries in Asia, including Burma, Thailand, Nepal, Pakistan, and China, and cases among travelers were reported in the USA and the UK. An eighth pandemic, so it seemed, had begun

(Swerdlow and Ries 1993). For reasons that are not well understood, since 1995 the incidence of O139 cholera has diminished in India and Bangladesh and it has essentially disappeared from other countries in Asia. Continued epidemiological and microbiological surveillance worldwide will reveal whether O139 cholera will reappear in Asia or in other venues.

Based on the molecular characterization of O139 strains isolated from patients with cholera in several different geographic sites since 1992, it seems that epidemic O139 originated by genetic recombination of a *V. cholerae* O1 strain. The phenotypic and genetic characterization of virulence genes of O139 strains shows that they are identical to El Tor strains (Berche *et al.* 1994; Hall *et al.* 1994; Johnson *et al.* 1994; Rhine and Taylor 1994; Waldor and Mekalanos 1994). Apparently, the virulent O139 epidemic strain initially arose by deletion of *c.* 22 kb of DNA from the *rfb* region of the chromosome of an El Tor strain, followed by introduction of *c.* 35 kb of DNA from the *rfb* region of an O139 strain (Comstock *et al.* 1995, 1996).

A large epidemic of O139 cholera, extending over several years and spreading to several countries in South Asia, was a sobering event for epidemiologists and microbiologists interested in cholera, as it challenged the dogma that preached that only serogroup O1 vibrios were capable of large outbreaks and pandemic behavior. Moreover, it raised questions about the assumptions that had been made with respect to the early pandemics of cholera, when that disease had emerged in the 19th century as a significant threat to public health. It had been presumed that all the early pandemics originating from the Ganges delta were caused by *V. cholerae* O1 of classical biotype. However, the O139 outbreak of 1992–4 assaults that dogma and raises the distinct possibility that some of the early pandemics might have been caused by other serogroups, such as O139, or by other biotypes of O1, such as El Tor.

14.4.2 Use of molecular approaches to address hypotheses concerning *V. cholerae* natural history

The hypotheses

We propose to utilize molecular techniques in an attempt to characterize the *V. cholerae* strains that were responsible for the early pandemics of cholera, particularly the second pandemic. Polymerase chain reaction (PCR) primers are available to amplify genes indicating the presence of cholera toxin-producing *V. cholerae* (Shirai *et al.* 1991; Fields *et al.* 1992). Amplification of additional genes can determine whether the *V. cholerae* was serogroup O1 and, if so, whether they were of classical or El Tor biotype. Primers are also available to amplify genes specific for serogroup O139 (Comstock *et al.* 1996). We hypothesize that if these primers are used to amplify genes from appropriate archaeological specimens, such as material from the graves of individuals who died of cholera or from appropriately dated latrines, it may be possible to characterize the *V. cholerae* responsible for cholera in the early or mid-19th century.

Why focus on the second pandemic?

The second pandemic of cholera, which most authorities suggest extended from 1831 to 1850, was the first to achieve wide dissemination throughout Europe, the Middle East, and North America, and its progression is remarkably well documented (Koch 1884; Pollitzer 1959; Rosenberg 1987). In addition, this pandemic is especially interesting for molecular archaeological studies for several reasons: precise descriptions are available that characterize the disease clinically as cholera gravis (O'Shaughnessy 1832), reports of clinical chemistry studies documented rice water stools containing high concentrations of sodium and base (O'Shaughnessy 1832), and early reports even demonstrated the efficacy of intravenous administration of saline (Latta 1832). For these reasons, we consider the second pandemic to be of particular interest for pursuing molecular epidemiological studies.

Sources of material for PCR studies

In the large cities of Europe (e.g. London, Paris, Dublin) and North America (e.g. Montreal, New York, Philadelphia, Baltimore), the death toll during the first wave of the second cholera pandemic was high (Rosenberg 1987). Consequently cemeteries from that era in Europe and North America hold marked graves of persons who died of cholera. In some cultures and religions it is considered unethical to deliberately disturb graves for any reason; other cultures condone the exhumation of graves if there are compelling indications. Furthermore, old graves are sometimes disturbed in connection with construction of houses, buildings, and roads, or are inadvertently discovered in the course of such construction. Material from the era of the second pandemic would be particularly interesting to submit to PCR analysis. We propose to contact civil authorities in several jurisdictions in Europe and the USA to explore the feasibility of obtaining permission for the exhumation of the graves of several individuals who died of cholera during the second pandemic. In addition, we are beginning an international search for pathology specimens from the intestines of patients who died of cholera during the 19th century. These specimens will have to be in a preservative that does not modify DNA; thus, for example, formalin-preserved specimens would not be useful. It is likely that if such specimens are available, they are from the late, rather than the early, part of the century. Should they become available, such material would be worthy of testing.

Extraction procedure

We have been able to circumvent the effects of PCR inhibitors in ancient fecal specimens by performing a DNA cleaning step using a silica-based resin (MERmaid spin kit, Qbiogene, Carlsbad, CA, USA), according to the manufacturer's instructions. This resin is specially designed to capture small DNA fragments (10–200 bases) and works well to isolate the small DNA templates potentially present in ancient specimens.

Genes to be amplified

V. cholerae strains, whether they are O1 or O139, must express cholera toxin in order to cause the clinical syndrome of cholera gravis. Therefore the amplification of cholera toxin genes provides evidence for the presence of DNA from virulent *V. cholerae*. Primers for sequences specific for the *rfb* regions of O1 and O139 would then be used to determine whether the *V. cholerae* was of these O serogroups. If O1 DNA is amplified, then primers specific for genes of El Tor versus classical biotype of O1 would be used. For example, the gene that encodes the structural subunit of toxin co-regulated pili (TCP), *tcpA*, differs between the biotypes (Rhine and Taylor 1994). By using carefully selected primers, one can amplify biotype-specific *tcpA* sequences. Similarly, diagnostic primers can allow the amplification of *hlyA*, which is specific for the two biotypes of *V. cholerae* O1. El Tor strains have a complete *hlyA*, whereas classical biotype strains harbor *hly* that carries an internal deletion (Alm *et al.* 1988; Rader and Murphy 1988).

Positive modern controls

We have been able to amplify spiked genes for cholera toxin and *E. coli* LT in ancient latrine samples at a sensitivity of <100 genome equivalents; however, no authentic templates have yet been found. Targets sought thus far include LT and cholera toxin, *E. coli* heat stable toxin, Shiga toxin, and mycobacterial IS*6110* targets.

14.5 Conclusion

Modern molecular techniques provide promise in our ability to illuminate pathogen archaeology, which in turn can assist us in understanding the complex roles of infection in human history. However, the stringent limitations attendant on current techniques have been obstacles to our understanding of the history of enteric infection. It is hoped that continued development of and attention toward these issues, as presented in this book, will lead to important breakthroughs in this area.

CHAPTER 15

Palaeobacteriology with special reference to pathogenic mycobacteria

Mark Spigelman and Helen D. Donoghue

Today I am sick; I may die of this disease, having always had a presentiment that I would die young. My mother was consumptive, and I have lived so as to contract this disease or aggravate my only inheritance. (Alexander Dumas, Camille, 1848)

Skeletal deformities and the occasional lesion in a preserved corpse or mummy are often highly suggestive of ancient microbial disease. However, until recently it appeared that archaeologists were doomed never to fully identify the origins, extent, and types of these ancient diseases, thus making it impossible to understand the aetiology of epidemics that were so important in determining the course of history. A new science of palaeomicrobiology can now give insight into the aetiology and nature of many microbial diseases. This chapter deals in the main with bacterial infections, while the chapter that follows—a study in paleo- or archaeovirology—examines a major viral infection, influenza (Taubenberger *et al.* 2000). In the next sections we shall examine the skeletal evidence for ancient bacterial activity.

15.1 Structural observations and palaeopathology

A number of conditions leave skeletally diagnosable changes on the infected bones. Some are so specific that they are considered diagnostic on purely morphological grounds. For example, syphilis (Fig. 15.1), yaws, tuberculosis (Fig. 15.2), brucellosis, and leprosy will all leave the bone so altered that most palaeopathologists have attributed a bone deformity to a specific disease. The skull of Rhodesian man (40 000 BP) contains a clearly identifiable mastoid abscess (Ryan 1992). In

others, a non-specific osteitis or osteomyelitis is the only diagnosis that can be made with any degree of certainty (Fig. 15.3). Yet the fact that many diseases leave no evidence of their presence on bones, does not mean that all trace of the microorganisms has been lost. If organisms are in the bloodstream (septicaemia) at the time of death, then some of these organisms may be present in the bones of the subject *post mortem*. Hence it is possible that some residual DNA from these organisms will remain in ancient bones and be available for analysis. Important diseases that may be diagnosed in this manner include malaria (Taylor *et al.* 1997), bubonic plague (Drancourt *et al.* 1998), syphilis (Kolman *et al.* 1999), and Chagas' disease (Madden *et al.* 2001).

15.2 Long-term bacterial survival

Before using molecular technologies in confirming palaeopathological diagnoses on the basis of surviving DNA or bacterial products, it is worth considering the theoretical possibility that ancient microorganisms themselves may survive. Kennedy *et al.* (1994) gathered a database of 62 separate entries which involved the work of 37 independent research teams and found reports in the literature of 5000 different but viable microorganisms. The authors noted that whilst claims can be individually disputed, the sheer mass of the results indicated to them that long-term microbial survival is

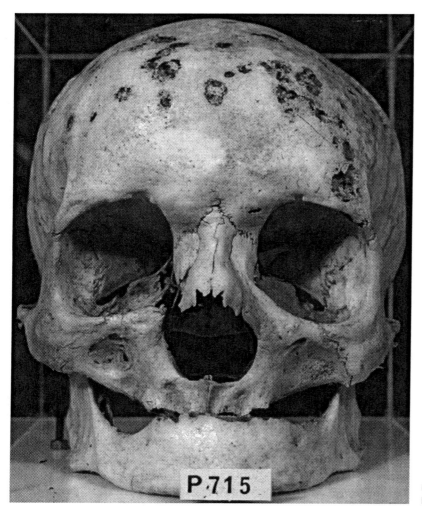

Figure 15.1 Syphilitic skull, mediaeval, English.

possible. The oldest known bacteria thus far identified, and designated as micrococci on morphological grounds, were found in pre-Cambrian rocks of Montana (Moodie 1923). In the present volume (Chapter 4), Cano presents data on the successful germination and culture of endospores of *Bacillus* spp. entombed in amber for millions of years. Amongst the reports of bacterial survival over thousands of years, Goldstein has described living intestinal bacteria in the stomach contents of a 11 500-year-old mastodon, found in a peat bog in Ohio (Park 1991).

It has been suggested that the great majority of microorganisms in the natural environment are in a physiological state described as 'viable non-cultivable' or VNC (Oliver 1993; Barer *et al.* 1998) and which enables bacteria to persist under unfavourable conditions. This condition can be induced in the laboratory as a result of exposing organisms to stress and is discussed further below. It is conceivable that some long-term persistent viable organisms may have survived as a result of entering this state.

It is feasible that some of the frozen bodies such as the Tyrolean Ice Man or those constantly being found in Russia could yield an enormous amount of information if researchers could reach them before thawing, and were able to take bacterial samples immediately before any significant contamination

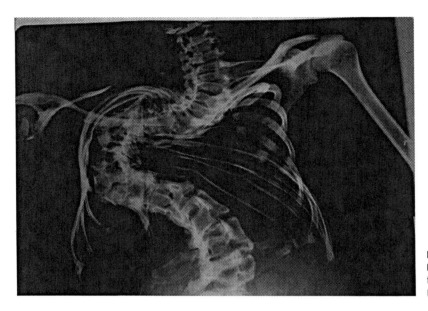

Figure 15.2 A female from Vác, Hungary who died aged 37. Material from her right chest was found to be MTB-positive.

had occurred. It is possible that in this type of body the bacterial gut flora may still be viable. Indeed, Cano (1997), who studied the DNA of the gut contents of the Ice Man, detected intestinal microflora but did not try to grow the bacteria as the body had thawed before he had access to it. Similarly, Ubaldi *et al.* (1998) had the same experience with a Peruvian mummy.

15.3 Persistence of bacterial products

Microorganisms contain a broad range of macromolecules that may be preserved in the geological and archaeological record (Bada *et al.* 1999; Briggs 1999). These include lipids, amino acids, and DNA (Pääbo 1989), although these are susceptible to taphonometric changes (Höss *et al.* 1996b). There is evidence that DNA survives better in bone than in comparable soft tissue samples (Hagelberg and Clegg 1991; Lassen *et al.* 1994). Therefore, bones, teeth (Montiel *et al.* 1991), and calcified pleura (Donoghue *et al.* 1998) are all potentially good sources of ancient microbial DNA. Bone serves as a useful matrix for any remaining bacterial DNA as it protects the DNA from the effects of leaching and bacterial attack. Other matrices protect ancient biomolecules from complete degradation. One is palaeofaeces, which can be found in a variety of

archaeological contexts. If found in bog bodies or mummies, it can be attributed with certainty; others are more difficult to interpret. If there is difficulty in establishing human origin of a particular faecal specimen, a number of tests are available, including searching for human DNA. Animal and human stools have distinct morphological differences (Holden 1990) and the presence of coprosterols and high concentrations of eggs of *Tricuris* and/or *Ascaris* are regarded as suggestive of human paleofaeces (Jones 1983) although further confirmation is required for certainty of diagnosis. The eggs themselves may yield valuable information about the parasite DNA and changes that have occurred over time (Loreille *et al.* 2001).

All faeces contain the bacterium *Escherichia coli*. In addition, the human gut sheds its mucosal cells every 5 days. Therefore, DNA from both sources may be detected by polymerase chain reaction (PCR) analysis. The presence of DNA from various sources in faeces has been shown by experiments on the excrement of brown bears where Höss *et al.* (1992) identified DNA both from the animal and from dietary plants. In Lindow man, an English bog body from the Iron Age, *E. coli* DNA was detected in the gut contents (Fricker *et al.* 1997).

A recent study of faecal samples more than 2000 years old showed that palaeofaeces from dry cave

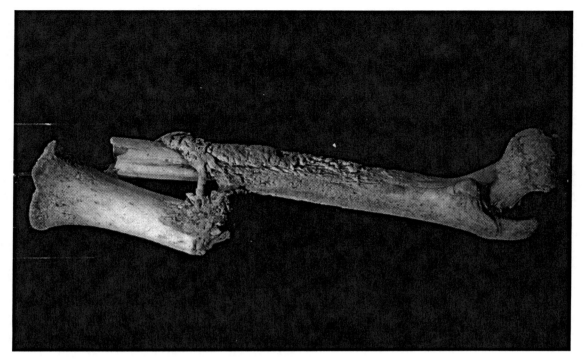

Figure 15.3 Non-specific osteomyelitis causing non-union in a fractured left femur.

and rock shelter sites represent a source of ancient DNA (aDNA) that is relatively abundant, and from which the DNA is more reliably retrievable than is the case for human skeletal remains. In addition, palaeofaecal DNA offers valuable information on the meat and plant components of ancient diets, which complements and extends information obtainable by morphological and biochemical analyses (Poinar *et al.* 2001b).

15.4 Microbial DNA persistence

15.4.1 Differences between microbes and host cells

Why does microbial DNA survive any length of time and why is it not overwhelmed and hidden by the mass of mitochondrial DNA that is present in vastly larger numbers of copies? We believe the answer lies in two differences between prokaryotic and eukaryotic DNA.

First, there is a significant difference in chemical composition and permeability between prokaryotes such as the mycobacteria and actinomycetes, compared with eukaryotic cells. For example, mycobacteria, which include the causative organisms of tuberculosis and leprosy, produce cell walls of unusually low permeability, which contribute to their resistance to therapeutic agents. Their cell walls contain large amounts of C_{60}–C_{90} fatty acids, mycolic acids, that are covalently linked to arabinogalactan. Recent studies have clarified the cell wall structure, which includes many extractable lipids. Most of the hydrocarbon chains of these lipids assemble to produce an asymmetric bilayer of exceptional thickness. Structural considerations suggest that the fluidity is exceptionally low in the innermost part of the bilayer, gradually increasing toward the outer surface (Brennan and Nikaido 1995).

Another factor to be borne in mind is that prokaryote DNA is not susceptible to host degradative enzymes. As far back as 1985 Ginsburg noted that no enzyme is present in a mammalian host that can destroy a bacterial cell wall, unless it is first deacetylated. Secondly, prokayotic and eukaryotic DNA do not appear to be degraded in an identical manner (Ginsburg 1988). Some bacterial DNA seems to survive better than the host mammalian DNA. Our work on ancient tuberculosis indicates that the quantity of *Mycobacterium tuberculosis* (MTB) DNA is in most cases more abundant than the host's mitochondrial DNA when analysed under identical conditions, suggesting that the prokaryotic DNA has preferentially persisted (Matheson *et al.* 2000). Furthermore, experimental evidence on various bacteria has identified mechanisms that may support this differential preservation. It is in differences between survival of the bacteria and their stress response mechanisms (see below) that the answer to the preferential survival of prokaryotic DNA may lie.

15.4.2 Survival during degradation of the host

The preferential survival of prokaryote DNA is due to the mechanisms of DNA protection that exist in prokaryotes that do not occur in eukaryotes. It is these mechanisms that enable the prokaryote DNA to persist after the death of the host and that protect the prokaryote DNA long after death of the microorganism.

Growing bacterial cells have the capacity to respond rapidly to changing environments to enable them to cope with stresses that generate cellular damage. They have evolved mechanisms to protect their DNA in life but in stationary phase the constraints placed on the cell require other mechanisms to deal with the damage (Grant *et al.* 1998). Once in this stationary phase a bacterium can withstand extremes of temperature, oxygen content, water content, pH, osmolarity, radiation, and chemistry. In our study of the Hungarian Vác mummies (see below), the host condition of dehydration and cold preservation is believed to have contributed greatly to the conservation of the bacterial DNA. Some of the proteins that aid bacteria to preserve their DNA have been identified, e.g. phage-shock protein, PspA (Kobayashi *et al.* 1998), and Dps (DNA binding protein from starved cells) in *E. coli* (Almirón *et al.* 1992) .

Morphological changes aid in the protection of prokaryotes. The protective mechanisms are induced when a prokaryote is placed under stress such as limited nutrients, change in pH, temperature, osmotic pressure, etc. In the case of a pathogen,

this occurs with the death of the host. Starved rod-shaped bacteria usually change shape into small spheres to reduce the cytoplasm, surface area, and water content (Reeve *et al.* 1984). This reduces the rate of protein synthesis and metabolism (Lange and Hengge-Aronis 1991). The morphological change results in the development of a more resistant state, which in many organisms leads to the persistent VNC condition, and though not so dramatic a change as in endospore formers, is still quite substantial. Whilst this has yet to be reported in mycobacteria, an analogous state of metabolically inactive, non-acid-fast granular forms has long been recognized and is believed may play a role in pathogenicity (Chandrasekhar and Ratnam 1992; Grange 1992). However, we have demonstrated in one of the Hungarian mummies from Vác, lung macrophages within whose parenchyma rounded bacilli were visible which were still acid fast (Pap *et al.* 1999). The lung tissue from this individual was strongly positive for MTB DNA and the acid-fast stain indicates that the cell wall retained its integrity to some extent.

The stress responses are conserved amongst a broad range of organisms, and gene sequence homologies for products such as heat shock proteins are found between prokaryotes and eukaryotes. However, a response apparently restricted to prokaryotes is based upon protection of their DNA. As stated above, in their natural environment, bacteria are rarely exposed to conditions that enable them to have continuous growth. Thus, bacterial life cycles are characterized by long periods of nutritional deprivation followed by short periods of plenty that allow for fast growth (Kolter *et al.* 1993). Mechanisms have developed that cope with this phenomenon, such as endospore formation, microbial aggregation, and the VNC state. One of the most studied responses has been noted in *E. coli* and other Gram-negative species when confronted by various threatening stresses. These have been observed morphologically and studied microbiologically (Martinez and Kolte 1997; Kobayashi *et al.* 1998; Frenkiel-Krispin *et al.* 2000). The outcome is the development of a crystal structure, which protects the DNA by sequestering it in a chemically stable conformation within the cell (Grant *et al.* 1998). Some of the proteins produced in this response appear to have preferential binding qualities to different types of DNA (Yamada *et al.* 1990).

The process has been termed biocrystallization (Wolf *et al.* 1999). The DNA binding proteins have been found to have homologues in many other prokaryotes (Almirón *et al.* 1992). They are induced under stress and their production increases and continues through the stationary phase of the prokaryote when all other gene expression is reduced (Almirón *et al.* 1992). Kolter *et al.* (1993) showed the reduction of protein synthesis in cells at the onset of stationary phase and its significant decline after 10–20 hours. Although not found to occur specifically in *Mycobacterium*, we are certain that similar mechanisms contribute to the preservation of MTB DNA during the early period of host autolysis.

15.5 Precautions necessary in the detection of microbial aDNA

The earliest work on aDNA was based on eukaryotic organisms, principally mammals (Pääbo 1989) but also plants and arthropods (Bada *et al.* 1999). The work on prehistoric human remains gave rise to particular concerns about cross-contamination from the investigators themselves, and this resulted in a series of recommendations for good practice (O'Rourke *et al.* 2000). One of the criteria recommended is that in order to verify the finding of microbial DNA from a specimen, it should be possible to obtain evidence of the corresponding host DNA. The comments outlined above (Section 15.3) indicate why we do not believe that this gives a reliable indication of the state of preservation of microbial DNA. The other precautions are: physical separation of pre- and post-PCR activities; strict protocols to prevent and monitor the introduction of modern DNA; the use of negative controls; replication of samples to confirm results, preferably in different laboratories; assessment of sequence data to confirm that they make phylogenetic sense; and observation of an inverse relationship between fragment size and PCR efficiency.

Difficulties can be encountered if there is too little sample available to permit repetitive sampling. In addition, because microbial DNA is unlikely to be distributed evenly throughout a specimen, it is important to ensure that the correct site is sampled, for example, an obvious lesion rather than intact surrounding bone. Even the physical distribution of material to different laboratories may pose unexpected hazards such as exposure to ionizing radiation, which is known to reduce the amount of amplifiable DNA (Götherström *et al.* 1995). Not only are specimens frequently exposed to radiation in the laboratory, but they may also be exposed if sent by airmail or on air flights, and the use of photographic pouches may be worth considering.

15.6 Strategies for the detection of ancient microbial DNA

Initial work on ancient microbial DNA relied upon species-specific PCR that had been developed in clinical diagnostic laboratories. As more organisms are sequenced, the opportunities for highly targeted PCRs arise, which enable particular characteristics such as virulence markers to be explored. The great majority of work done in palaeomicrobiology has adopted this approach.

Other strategies include *in situ* DNA hybridization; the use of magnetic beads to capture specific microbial DNA of interest; and separation of PCR products obtained with general primers by techniques including denaturing gradient gel electrophoresis (DGGE) for applications such as the visualization of DNA bands representing dominant bacterial species within communities, or the detection of mutations. One technique that has recently been applied to aDNA, in an attempt to overcome the problems due to the damage that has occurred over time, is that of pre-amplification extension PCR. The aim of this method is to carry out general DNA repair before targeted PCR is performed. Initial work looks promising (Pusch *et al.* 2000).

Where the contents of the specimen are unsuspected, one technique that has been applied is to use universal PCR primers and to subsequently clone and sequence the resulting PCR products to obtain an identification. The 16S rRNA gene is often used as the target site (Cano *et al.* 2000; Zink *et al.* 2001). This is potentially of great interest, although it must be appreciated that the existing databases contain only limited information on much of the environmental and commensal gut flora. In addition, there is the persistent problem of the production of chimaeric DNA molecules as a result of PCR amplification (Wang and Wang 1997).

15.7 Evidence for the detection of mycobacteria and their DNA

The following are reasons why mycobacteria, in particular *M. tuberculosis* and *Mycobacterium leprae* are good organisms for study.

- They leave readily identifiable bony changes (Møller-Christensen 1961; Buikstra 1981).
- *M. tuberculosis* and *M. leprae* have no environmental reservoir because they are obligate pathogens;
- as *M. tuberculosis* is located within a caseating mass of tissue in an active case of pulmonary tuberculosis, it is physically protected from the bacteraemia that accompanies death of the host;
- *M. tuberculosis* is known to survive long after the death of the host (Weed and Baggenstoss 1951), possibly due to its thick wall rich in mycolic acids (Daffe and Draper 1998), which also facilitate the formation of a pseudo-capsule protecting it from harmful attack;
- the high fatty acid content in mycobacteria also makes the cell membrane less fluid and more durable;
- *M. tuberculosis* is extremely slow-growing and *M. leprae* is uncultivable, so clinical microbiologists have done considerable work to identify and assess PCR-based identification methods for these organisms and to set up international databases;
- the *M. tuberculosis* and the *M. leprae* genomes have been fully mapped, so many different target sites can be explored, with single or multiple copies in the genome.

Therefore, the survival of *M. tuberculosis* and *M. leprae* DNA in archaeological samples is likely to be due both to its physical inaccessibility and the chemical composition of the surrounding cell wall and cytoplasm. These afford protection from the barrage of lytic enzymes produced from autolysis of the host cells, the other microflora and fauna of the host on death, and the first-stage decomposers that invade *post mortem*. This has allowed survival of viable *M. tuberculosis* after fixing in formalin and

staining (Gerston *et al.* 1998) and even through the death of the host and its transmission from the corpse to a new host, 1 year after burial (Sterling *et al.* 2000). Indeed, there are at least two published reports of the presence of MTB DNA and MTB-specific mycolic acids in archaeological specimens. The finding of both types of biomolecule, the DNA being amplified and the mycolic acids detected directly by mass spectroscopy helps to confirm the diagnosis of tuberculosis (Donoghue *et al.* 1998; Gernaey *et al.* 2001).

The first published report on the finding of bacterial DNA from ancient bones was in June 1993, with the isolation and amplification of a sequence specific for the *M. tuberculosis* complex (Eisenach *et al.* 1990) from the insertion sequence IS6110, in bones showing morphological evidence of clinical tuberculosis (Spigelman and Lemma 1993). In this initial report, four bones contained this sequence out of 11 tested, which included four bones believed to be TB-free and which were used as controls. The ages of the specimens ranged from 300 to 1400 years BP. The bones came from Europe, Turkey, and pre-European-contact Borneo. There was criticism of the first paper subsequently as not all the now-current criteria for DNA identification were performed in this first experiment (Blondiaux and Stanford 1995). However, it has since been confirmed in experiments carried out both in the London laboratory with different scientists and at the Kuvin Center in Jerusalem (Spigelman *et al.* 2003).

Subsequently, Salo *et al.* (1994) repeated the experiment using identical primers, with DNA from the lung tissue of a Peruvian mummy about 1000 years old. In several other cases that were published subsequently from various laboratories, the DNA isolated was from the IS6110 target site (Arriaza *et al.* 1995; Baron *et al.* 1996; Taylor *et al.* 1996; Nerlich *et al.* 1997; Braun *et al.* 1998; Donoghue *et al.* 1998; Nuorala 1999; Haas *et al.* 2000a; Ubelaker *et al.* 2000). Work is now being published with reports of other MTB-specific sequences being detected (Faerman *et al.* 1997; Taylor *et al.* 1999; Mays *et al.* 2001; Rothschild *et al.* 2001). These first experiments in this field have answered questions that have long puzzled palaeopathologists, such as proving that human tuberculosis pre-existed European contact in both Asia and the New World.

It is well known that the mycobacterial diseases of tuberculosis and leprosy have existed since ancient times (Haas and Haas 1996) and tuberculosis has remained at epidemic levels since the 17th century. Indeed, the Bills of Mortality for London in 1665, when there was a major outbreak of the bubonic plague or 'Black Death', show that consumption (TB) was endemic and the greatest cause of death during plague-free weeks. During the plague epidemic, deaths from consumption actually increased. During the 19th century, Alexandre Dumas in his memoirs noted that 'it was all the fashion to suffer from chest complaint; everybody was consumptive, poets especially'. Tuberculosis has ever-increasing importance in modern medicine and the potential to become the scourge of the early part of this millennium. Reichman and Tanne (2001) noted that one-third of the world's population is infected with MTB, and that tuberculosis due to drug-resistant MTB could well become an untreatable epidemic, 'and there are no new diagnostic tests no new drugs and no magic cures'.

Is it possible that we may actually learn from the remains of past sufferers something of significance to modern medicine?

We detail below some of our work which encourages us to believe that this may indeed be the case. Initially, detection of *M. tuberculosis* DNA in archaeological specimens focused on confirming the osteological diagnosis of disease. However, the complete sequencing of the MTB genome and comparative genomic studies of members of the *M. tuberculosis* complex, have revealed additional targets for study. We aim to evaluate these markers in the analysis of archaeological specimens and assess their contribution towards increasing knowledge of the evolution and epidemiology of modern day tuberculosis.

15.8 The Hungarian mummy project and tuberculosis

In 1994 the remarkable discovery was made of 263 human remains, many of which were naturally mummified, buried in a sealed and forgotten crypt of the Dominican church in Vác, Hungary between 1731 and 1838 AD (Fig. 15.4). The mummification

was in the main excellent. We believe the low crypt temperature and a constant air current may well be responsible for the mummification, though it is noted that the coffins were of pine and that the bodies appear to have been laid on a bed of pine shavings. It is possible that the turpinoids in pine, long known to have bacteriostatic properties, could have contributed to the mummification process. Tuberculosis was then widespread in Hungary and the rest of Europe, accounting for one in four deaths from the 16th to 18th centuries (Hutás 1999) and examination by computerized tomography (CT) full body scans of a number of bodies from Vác indicated three individuals who showed evidence of tuberculosis (Fig. 15.5). This was confirmed by sampling one of these individuals—a 56-year-old man who died in 1783. Histological examination of one sample showed acid-fast intracellular bacteria and PCR amplification detected MTB complex-specific DNA (Pap *et al.* 1999).

There is a contemporaneous archive, which occasionally indicates occupation and a brief description of cause of death. Therefore, in many cases it is

Figure 15.4 Arrangement of coffins found in the crypt of the Dominican Church of Vác, Hungary.

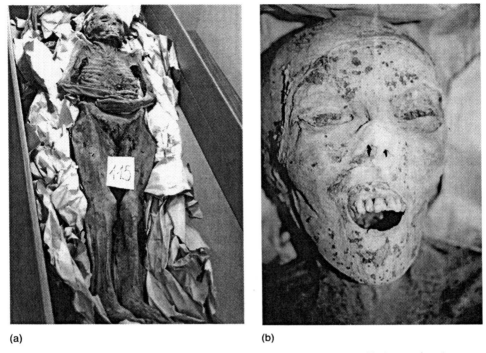

Figure 15.5 (a) Cachectic body of a woman from Vác, Hungary. Samples were strongly MTB-positive. (b) Close up of another cachectic subject.

possible to place individuals into provisional family groups, and to examine the MTB incidence in relation to age, gender, and the other data from the archive. The church archival records, combined with the appearance of some of the bodies, suggested that several were suffering from tuberculosis at the time of death. For example, we know that Antal Simon, Abbot and Director of the 'institute for deaf mute persons' (as translated from the Hungarian) died aged 36 after 'daily spitting of blood'. He was found strongly positive for MTB DNA in his lungs and abdomen.

Examination of the Vác samples has become a major project of our laboratory in collaboration with a number of associated laboratories, where according to the current standard protocols in aDNA (O'Rourke *et al.* 2000) we have obtained independent confirmation of our results, although budgetary constraints make it impossible to confirm each and every result.

Over 400 samples from 168 of these 18th-century individuals have been obtained, most from naturally mummified soft tissues, and a lower rib from skeletalized remains (approximately one-third of the individuals). The samples from mummified bodies with intact body cavities were taken under strict aseptic conditions with modern (Stortz) endoscopes. Initial work has shown by PCR that over 50% of these individuals were infected with MTB (Fletcher *et al.* 2002).

The long-term objectives of this study are to determine:

• the location of MTB DNA at different sites in the body, and its characteristics including genotype (Sreevatsan *et al.* 1997; Frothingham *et al.* 1999) and virulence markers;
• the molecular epidemiology of MTB strains in this population in relation to family group, using spoligotyping and other PCR-based methods;
• the molecular characteristics of a group of modern MTB strains obtained from Hungary, including the area around Vác, to enable comparisons between 18th- and 20th-century strains.

This study is intended to shed light on the changes that have occurred in MTB over the last 300 years, from the pre-antibiotic era and more than 200 years before its first cultivation in the laboratory.

To validate the findings a two-laboratory, three-workstation strategy is used for DNA extraction and PCR amplification. Stringent precautions are taken against cross-contamination: equipment and surfaces are cleaned thoroughly, and pre-aliquoted reagents and filter tips are used at all stages of the procedure. DNA extraction is performed using a modified silica-based method (Boom et al. 1990). No work with modern MTB DNA is ever done in the laboratories where DNA extraction and PCR set-up is done.

DNA preservation was excellent in some of the remains and several different DNA target sequences have been successfully amplified, including one target sequence of >350 bp. The IS6110 nested PCR was the most sensitive. With single-stage PCR, there is an inverse relationship between amplicon size and MTB DNA positivity. To date, 52% of the individuals have been found to be MTB positive and 79% of individuals aged between 51 and 60 years. Infection with MTB in endemic areas is believed to occur early in life, so results have been analysed in relation to year of birth. Although numbers are low, this shows an overall incidence of 44% in individuals born between 1685 and 1740 but every person in this cohort born between 1760 and 1780, 13 individuals in all, was positive for MTB. According to local records, this coincided with settlement of rural agricultural workers, thereby leading to an increase in population, and a major civic building programme.

M. tuberculosis and M. bovis DNA can be distinguished by a raft of tests (Sales et al. 2001): a polymorphism in the pseudogene oxyR codon 285; the presence of the mtp40 element in the phospholipase A gene plcA; the detection of a deletion in region RD7; and examination of a region in the plcD cutinase region which is a hot-spot site for IS6110 insertion and is subject to deletion events (Ho et al. 2000). Spoligotyping, which is a form of molecular fingerprinting based on PCR of spacer sequences in the direct repeat region (Kamerbeek et al. 1997) was performed in two different laboratories, using methods described by van der Zanden et al. (1998). In 12 different individuals, where the DNA was best-preserved, almost complete molecular fingerprints (spoligotypes) were obtained. This work was carried out in St Mary's Hospital Medical School, London and at Bilthoven, the Netherlands, and demonstrates the reproducibility of the method.

It was clear that not all individuals infected by MTB died from tuberculosis. For example, the individual of greatest age at death was a 94 year old who was fairly obese. Samples from her abdomen, and right and left lungs were negative for MTB DNA. However, a radiograph revealed that there was some pleural calcification in the right upper chest. This was sampled separately and was the only TB-positive site. We believe this was a Ghon lesion healed and calcified in childhood, where although intact and probably viable tubercle bacilli were retained for up to 90 years, they never caused her active disease. The question is 'why?'

15.8.1 A family study

There were certain Vác families in which long life and lack of TB was prominent and others in which TB seemed rampant. Future studies of known host susceptibility/resistance genes may tell us if there is a genetic explanation. In several cases the DNA preservation was sufficiently good to permit in-depth study of the M. tuberculosis strains present in the bodies. Of particular interest was a family group consisting of a mother and two daughters, each of whom had died within the period 1793–7 AD.

The mother of the two other individuals reported here was 55 years old and died on 16 December 1793. Her height was noted as 1.45 m. Soft tissue was taken from the tracheal region and abdomen. The older of the two daughters of the above died on 25 December 1797 aged 28 years. No abnormalities were detected in a chest radiograph but she was noticeably small for her age and of cachectic appearance. A tissue sample from the chest was examined. The younger daughter, who died on 2 March 1795 was described in the archive as being aged 14 years, although when examined we initially believed her to be aged only 8–9 years due to her very small size and cachectic appearance. No abnormalities were seen in a chest radiograph.

Samples were taken from the abdomen and from possible calcified pleura in the chest.

The small size of the two daughters in comparison with many of the bodies found in the crypt is unlikely to be due to malnutrition, as it is known that only wealthy, middle-class families used the crypt. The cachectic appearance of these bodies is consistent with active tuberculosis infection and this is supported by the molecular findings. All three individuals had tuberculosis, and the strains of *M. tuberculosis* were distinct in each case, based on genotyping, the spoligotype, and the *plcD* cutinase gene region, where it appears a deletion had occurred in the MTB DNA from the younger daughter (Table 15.1).

Although tuberculosis occurred in antiquity, it is thought that the modern TB epidemic began in Europe in the 1700s then spread to the New World and Africa (Stead *et al.* 1995). Consistent with these theories, we have demonstrated that spoligotypes common today were present in Europe in the 18th and 19th centuries. The detection of small-scale genomic deletions is a useful technique for exploring the molecular epidemiology, microbial evolution, and pathogenesis of tuberculosis (Zumárraga *et al.* 1999; Ho *et al.* 2000). Here, spoligotyping demonstrated a difference between the strain infecting the mother and those infecting her two daughters, with the loss of spacer region 31 (Sola *et al.* 1999). In addition, the *plcD* PCR revealed an apparent deletion in the *M. tuberculosis* strain from the younger daughter. About one-third of modern clinical isolates of *M. tuberculosis* studied appear to have undergone a deletion in this region.

We conclude that the three members of this family group were infected with three distinct strains of *M. tuberculosis*. Compared with the strain infecting the mother, we detected a deletion event

(or possibly a rearrangement) in the elder daughter and a further deletion in the younger, supporting the theory that *M. tuberculosis* undergoes more deletion events as it evolves. This demonstration of spoligotyping combined with screening of small-scale deletions for determining the molecular epidemiology of *M. tuberculosis* in antiquity clearly demonstrates the value of molecular genetic analysis of historical cases of tuberculosis.

15.8.2 Microbial ecology in 18th- to 19th-century Vác

There is archival information about other diseases that were prevalent in Vác during the 18th century, which include bubonic plague, cholera, smallpox, and typhus amongst the human population. Although the causative organisms are less likely than MTB to have surviving DNA, as the material is available and the relevant dates of outbreaks are known, it is hoped to include some preliminary work, initially on *Yersinia pestis* (Drancourt *et al.* 1998). Another area for preliminary investigation is that of the commensal gut flora and the presence of any antibiotic resistance markers.

15.9 Leprosy (Hansen's disease)

Leprosy is primarily a disease of peripheral nerves and skin but affects bones as well. Its clinical effects vary from the slowly developing pauci-bacillary or tuberculoid leprosy, to the multi-bacillary lepromatous leprosy. In this lepromatous state there is direct invasion of soft tissues around the face and mouth by *M. leprae* (Andersen and Manchester 1992; Roberts and Manchester 1995). The type of infection appears dependent on the immune response of the host and only in the lepromatous state does the

Table 15.1 Summary of results obtained for a mother and two daughters

Samples analysed	MTB complex IS6110	Differentiate between MTB complex strains		M. bovis-specific RD7	RvD2 deletion plcD
		Genotypic group	Spoligotype		
Mother: abdomen	+	2	53	–	–
Elder daughter: chest	+	3	50	–	–
Younger daughter: chest	+	3	50	–	+

organism ever appear in any numbers in the affected limbs. The study of *M. leprae* DNA in archaeological material is also of interest to ourselves and others (Haas *et al.* 2000b), for the reasons outlined above. A number of bones, which appeared from their morphology to be from individuals afflicted by leprosy, have been examined in an attempt to extract species-specific DNA. This has proved successful in samples dating from the 1st, 7th, 10th–11th, and 14th–15th centuries, with one of the largest pieces of aDNA ever recovered from ancient bone, isolated and sequenced. The detection of unexpectedly large DNA fragments supports the suggestion that the *M. leprae* cell wall may play a role in the survival of the mycobacterial DNA.

15.9.1 A summary of the samples examined in our laboratory

1 A specimen of a metatarsal bone from a skeleton buried at the site of the Monastery of St John the Baptist from ancient Palestine, dating from 624 AD.
2 A case of leprosy from a mediaeval burial ground in Suraz, Poland. Samples of this specimen were examined from two areas—the metatarsals and around the nasal region. This was the body of a 40- to 50-year-old male with characteristic changes in the nose on X-ray and in the leg and toe bones (Gladykowska-Rzeczycka 1976).
3 A metatarsus ('pencil' form) and a fibula (tibia with periostitis) were examined from grave 923 in a 14–15th century site from Ópusztaszer-Monostor, Hungary.
4 Two bodies from graves 222 and 503 in a 10th–11th century site at Püspökladány, Hungary, were examined. Both samples were of *cavum nasale* with periostitis.
5 A sample from a specimen previously labelled as Madura foot and where there was some controversy regarding the diagnosis (Hershkovitz *et al.* 1992, 1993; Manchester 1993).
6 Four samples from the Roman period cemetery at the Daklah Oasis, supplied by J. El Molto and C. Matheson (Lakeside University, Canada) for confirmation of their results.

We successfully demonstrated *M. leprae* DNA in the first specimen (Rafi *et al.* 1994) using primers for

the 36 kDa antigen gene (Hartskeerl *et al.* 1989). All three *cavum nasale* samples, from graves 222 and 503, at Püspökladány, Hungary; and from Suraz, Poland were positive for *M. leprae* DNA using the primers for both the multicopy *RLEP* site and the single-site 18 kDa antigen (Donoghue *et al.* 2001). *M. leprae* is known to localize to Schwann cells and nasal epithelial cells (Andersen and Manchester 1992), and this can result in an overwhelming rhinitis associated with very high numbers of the organism. The three *cavum nasale* specimens demonstrated the typical palaeopathology associated with leprosy, so the presence of *M. leprae* DNA in these samples was anticipated. In addition, a 531 bp sequence of *M. leprae* DNA from the 32 kDa antigen gene was obtained from the Püspökladány nasal samples, which suggests extremely good DNA preservation (Donoghue *et al.* 2002).

In the case of the possible Madura foot, we were able to shed light on the primary diagnosis, as *M. leprae* DNA was detected (Spigelman and Donoghue, 2001), but showed that it was not possible to confirm the diagnosis of Madura foot as the organisms causing this conditions are common soil commensals so could be present in the lesion due to disease or to contamination *post mortem*.

The absence of detectable *M. leprae* DNA in metatarsal specimens and the fibula sample does not necessarily imply that the remains were from individuals who were not suffering from the disease. Although many skeletons have been found with changes associated with lepromatous leprosy, the peripheral bones are likely to contain relatively little *M. leprae* DNA compared with the nasal region. In addition, most observers consider the changes in the limbs to be due to secondary infection, which is the result of loss of sensation due to nerve damage by the organism (Møller-Christensen, 1961). The presence of paleopathological changes due to secondary infections ensures that there will always be some uncertainty about whether or not *M. leprae* infection was the primary cause of the pathological changes observed, if no DNA can be identified.

Two of the Daklah Oasis specimens were positive for *M. leprae* DNA. Interestingly, one of these positive samples and the remaining sample were found to be positive for MTB DNA. In addition, one of the positive *cavum nasale* samples from Püspökladány was also positive for MTB, indicating a co-infection. This should not be surprising, as many authors have reported upon the frequency with which leprosy and tuberculosis occur in the same individuals and Hansen himself, reporting in 1895, found tuberculosis to be the most common cause of death among leprosy patients in Norway (Fine 1984).

15.10 Conclusions

This exciting field of science is developing rapidly. Initially, the questions related to whether or not it was possible to detect the microbial DNA from putative pathogens indicated by the paleopathology of the specimens. Even this has led to findings of archaeological interest, such as the demonstration of MTB DNA in the Americas and Far East prior to known European contact. In the case of the pathogenic mycobacteria, this work has gone considerably further. It is now possible to distinguish between the individual species of the MTB complex and somewhat to everyone's surprise, *M. tuberculosis* rather than *M. bovis* has been identified (Taylor *et al.* 2001). Where the preservation of the material is as good as in the mummified remains from 18th-century Vác, it is proving possible to examine the molecular characteristics of the infecting MTB strains and even to compare these with modern-day strains. A recent article (Brosch *et al.* 2002) shows how this type of investigation can support modern research. The article presents a new evolutionary scenario for the *M. tuberculosis* complex. Cited among the evidence that supports the hypothesis are the findings of the Hungarian mummy project, the 17 000-year-old bison bone (Rothschild *et al.* 2001), as well as the work reported by Salo *et al.* (1994), Nerlich *et al.* (1997), and Mays *et al.* (2001). We believe that as more information becomes available we will not only learn when evolutionary changes occurred but also the prevailing conditions that may have been responsible. Without the historical groundwork this would not be capable of even a speculative argument. Even with the information currently available, the data support the hypothesis that *M. tuberculosis* is of greater evolutionary age than *M. bovis*. The proposed evolutionary bottleneck is also consistent with current data and is believed to have occurred 15 000–20 000 years BP.

Work is now to our knowledge in progress in various laboratories worldwide on the pathogens that cause plague, syphilis, brucellosis, Chagas' disease, malaria, schistosomiasis, and ascariasis, whilst in addition to work on influenza and human papilloma viruses, smallpox and measles viruses are certain to be targeted in the future. We have also attempted, without success, to isolate *Helicobacter pylori* from alcohol-preserved specimens from the 18th-century Hunterian collection at the Royal College of Surgeons of London (Barnes *et al.* 2000).

The outcome of these efforts should not only throw light on the history of infectious diseases and medical anthropology, but may prove of direct benefit in understanding the evolution of microbial pathogenicity and host susceptibility to disease.

Acknowledgements

The Hungarian mummy project was part-funded by the UK Wellcome Trust with initial work funded by the Wenkart Foundation (Australia). Much of the work reported here on the Vác samples was carried out by Dr Helen Fletcher, in the UCL Department of Medical Microbiology, with valuable contributions from Dr G. Michael Taylor (St Mary's Hospital Medical School, Imperial College London).

Many other researchers and laboratories have collaborated in various aspects of our work. These include The Kuvin Center for the Study of Tropical Diseases at the Hebrew University Medical School (Chuck Greenblatt) thanks to grants from The Centre for Emerging Diseases in Israel and the Australian Ancient Ills Modern Cures Fund; the Department of Medical Microbiology and Infectious Diseases, Gelre Hospitals, Apeldoorn, The Netherlands (Adri G.M. van der Zanden): Paleo-DNA Laboratory, Lakehead University, Ontario, Canada (Carney Matheson); and the Department of Biochemistry and the Center for Molecular and Cellular Biology, University of Queensland, Australia.

None of this work would have been possible without the provision of archaeological specimens and samples for examination. Therefore, in addition to those mentioned above, we wish to thank the following individuals: Ildicó Pap (Anthropology Department, Hungarian Natural History Museum, Budapest, Hungary); Judyta Gladykowska-Ryyeczycka (Department of Anatomy and Anthropology, Academy of Physical Education, Gdansk, Poland); Israel Hershkovitz (Department of Anatomy and Anthropology, Tel Aviv University, Tel Aviv, Israel); Antónia Marscik (Department of Anthropology, Natural Sciences Faculty of the University of Szeged, Hungary); and Joe Zias (Hebrew University of Jerusalem Science and Antiquity Group, Israel).

CHAPTER 16

Archaevirology: characterization of the 1918 'Spanish' influenza pandemic virus

Jeffery K. Taubenberger and Ann H. Reid

One nurse found a husband dead in the same room where his wife lay with newly born twins. It had been twenty-four hours since the death and the births, and the wife had no food but an apple which happened to be within reach. (Survey, October 19, 1918)

In Chicago a nurse found a delirious eight-year-old in his nightdress on the sidewalk in the rain. She took him into his home and found his father sick, exhausted, and frantic with worry. Four children, including the boy in the street, and their mother had temperatures over 104°. The father's temperature was 101.6°. He had given his wife a spoonful of camphorated oil instead of the castor oil he intended. (Westphal, ME, Public Health Nurse, February 1919, quoted from: America's Forgotten Pandemic, A. W. Crosby, 1989)

16.1 Introduction

16.1.1 Archaeology of infectious diseases using pathology archives

The advent of molecular genetic pathology has opened a new venue for the study of infectious diseases, especially the ability to study 'fossil' organisms not currently circulating. Formalin-fixed, paraffin-embedded (FFPE) tissue samples, the most common form of pathologic tissue collections, can be used not only for histological and immunohistochemical analyses, but also for molecular genetic analyses of DNA and RNA (Krafft *et al.* 1997). Indeed, the advent of the polymerase chain reaction (PCR) has made it possible to perform molecular genetic analyses of DNA and RNA from the vast majority of alcohol- and formalin-fixed tissue samples. For infectious diseases, such analyses have found routine use in clinical diagnosis (e.g. whether a liver biopsy is positive for hepatitis C virus) and in research. Tissue archives have also made it possible to identify previously unknown infectious agents [e.g. the identification of human herpes virus 8 in Kaposi's sarcoma samples using the subtractive DNA hybridization method 'representation difference analysis' (Chang *et al.* 1994)]. Since influenza viruses that circulate in humans are constantly evolving, archival tissue specimens can be used to characterize 'fossil' influenza strains that no longer exist. A prime example of this type of analysis is the characterization of the influenza virus that caused the 'Spanish' influenza pandemic of 1918 using autopsy lung tissues of 1918 flu victims (Taubenberger *et al.* 1997). Such analyses may to help answer questions about the origin and virulence of this pandemic strain.

16.1.2 Influenza virus as a continually emerging disease

Influenza A viruses, single negative-stranded RNA viruses of the Orthomyxoviridae, circulate in humans in yearly epidemics and antigenically novel strains emerge sporadically as pandemic viruses (Cox and Subbarao 2000). In the United States, influenza kills 20000 people in an average year (Simonsen *et al.* 2000). Occasionally, and unpredictably, influenza sweeps the world, infecting 20–40% of the population in a single year. In these pandemic years, which have occurred every 10–40 years for at least several centuries, the numbers of deaths can be dramatically above average. In 1957–8, a pandemic caused 66 000 excess deaths in the United States (Simonsen *et al.* 1998). In 1918, the worst pandemic in recorded history caused *c.* 546 000 excess deaths in the United States, and killed up to 40 million people worldwide (Patterson and Pyle 1991).

Influenza A viruses evolve constantly by the mechanisms of antigenic drift and shift (Webster *et al.* 1992). Mutation in the antigenic sites of the prevailing strain (**antigenic drift**) confers selective advantage allowing viruses with the mutation quickly to become the new predominant strain. Evolution of the prevailing subtype by antigenic drift provides a long-term strategy for survival of a particular subtype in humans; the H3N2 subtype has thrived in humans for over 30 years by antigenic drift. Nevertheless, history suggests that long periods of drift are punctuated by sudden changes in antigenic subtype by **antigenic shift** or the production of novel influenza viruses by genetic reassortment. Since the genome of influenza A viruses consists of eight separate gene segments (Fig. 16.1), co-infection of one host with two strains can result in novel, recombinant strains. Replacement of one or both major surface protein-encoding genes results in a strain able to spread in humans without facing the challenge of overcoming pre-existing immunity. While this pattern of long periods of drift, interrupted by sudden shifts, can be traced back over hundreds of years, the details of exactly when and how new strains will emerge is poorly understood.

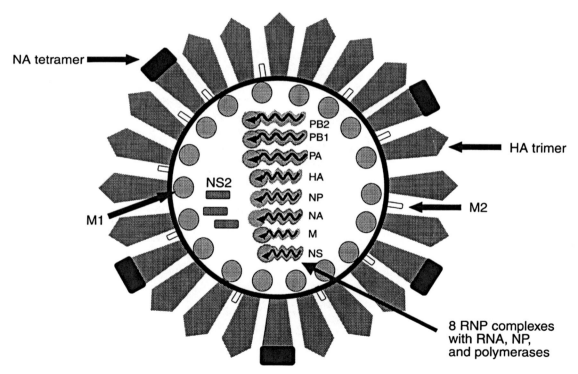

Figure 16.1 Diagrammatic representation of an influenza A virus. The two major surface glycoproteins, HA and NA, are embedded in a lipid bilayer. Small numbers of the matrix 2 (M2) ion channel protein are also inserted into the lipid envelope. The matrix 1 (M1) protein underlies the envelope and interacts with the helical ribonucleoproteins (RNPs). Both M1 and M2 are encoded by the matrix gene segment (M). The RNPs consist of the eight negative-stranded RNA segments, nucleoprotein (NP), and small amounts of the polymerase complex (PB2, PB1, and PA). Small amounts of the non-structural 2 protein (NS2) are also packaged with the virus (Murphy and Webster 1996).

Pandemic influenza strains have emerged three times in the last century: 1918 ('Spanish' influenza), 1957 ('Asian' influenza), and 1968 ('Hong Kong' influenza) (Webster *et al.* 1992; Cox and Subbarao 2000). Studying the extent to which the 1918 influenza was like the subsequent pandemic strains may help us to understand how pandemic influenzas emerge in general. On the other hand, until and unless we can determine what made the 1918 influenza different from other pandemics, we cannot use the lessons of 1918 to predict the magnitude of public health risk that a new pandemic might pose. The unique features of the 1918 pandemic include unprecedented morbidity and mortality (especially significant in young adults), unusual prevalence of viral pneumonia, and a heightened potential to develop secondary bacterial pneumonias (LeCount 1919; Linder and Grove 1943).

The last two pandemics, in 1957 and 1968, resulted when gene segments coding for viral surface protein(s) from avian-adapted influenza viruses reassorted with the prevailing human-adapted virus (Scholtissek *et al.* 1978). Interestingly, genetic analyses have also shown that the PB1 polymerase gene segment was also replaced by reassortment with an avian influenza-derived segment in both the 1957 and 1968 pandemic strains (Kawaoka *et al.* 1989). It appears that the combination of antigenically novel surface proteins with human-adapted internal viral proteins (possibly plus a novel PB1 segment) produces a virus that is able to replicate in human cells and spread from human to human—the prerequisites for the emergence of a pandemic strain.

The pandemic of 1918–19 occurred before influenza viruses could be isolated. Therefore its causative strain was thought lost for subsequent study. It was not known whether the 1918 strain was also the result of a reassortment between avian- and human-adapted viruses or whether it may have arisen through another mechanism. Because the 1918 pandemic was substantially more severe than any other known pandemic and had a uniquely devastating effect on young adults, the possibility that it had unusual origins seems heightened. Recent advances in molecular biology have made it possible to isolate and characterize the genetic material of the 1918 virus from archival and frozen tissue samples. Nearly half of its genome has been sequenced and an effort is underway to understand the origin of this strain. As the most deadly influenza virus ever experienced, the 1918 strain offers a unique potential to understand the connection between genotype and virulence.

16.2 Overview of the 1918 influenza pandemic

The influenza pandemic of 1918 was exceptional in both breadth and depth. Outbreaks of the disease swept not only North America and Europe but also spread as far as the Alaskan wilderness and the most remote islands of the Pacific. It has been estimated that 28% of the world's population (500 million people) may have been clinically infected during the pandemic (Frost 1920). The disease was also exceptionally severe, with mortality rates among the infected of over 2.5%, compared with <0.1% in other influenza epidemics. Total mortality attributable to the 1918 pandemic ranges from 20 to 40 million, with the higher number likely being more accurate (Patterson and Pyle 1991).

Unlike most subsequent influenza strains, which have developed in Asia, the 'first wave' or 'spring wave' of the 1918 pandemic seemingly arose in the United States in March 1918. However, the near simultaneous appearance of influenza in March-April 1918 in North America, Europe, and Asia makes definitive assignment of a geographic point of origin impossible (Jordan 1927). It is possible that a mutation or reassortment occurred in the late summer of 1918, resulting in significantly enhanced virulence. The main wave of the global pandemic, the 'fall wave' or 'second wave', occurred in September–November 1918. It has been estimated that the influenza epidemic of 1918 killed 675 000 Americans, including 43 000 servicemen mobilized for World War I. The impact was so profound as to depress average life expectancy in the USA by over 10 years, and may have played a significant role in ending the World War I conflict.

The majority of individuals who died during the pandemic succumbed to secondary bacterial pneumonia, since no antibiotics were available in 1918. However, a subset died rapidly after the onset of symptoms often with either massive acute pulmonary hemorrhage or pulmonary edema, often in less than 5 days (LeCount 1919). In the hundreds of autopsies performed in 1918, the primary pathologic findings were confined to the respiratory tree and death was due to pneumonia and respiratory failure (LeCount 1919). These findings are consistent with infection by a well-adapted influenza virus capable of rapid replication throughout the entire respiratory tree (Reid and Taubenberger 1999; Taubenberger *et al*. 2000). There was no clinical or pathological evidence for systemic circulation of the virus (LeCount 1919).

Another distinctive feature of the 1918 pandemic, which might be related to the interaction of the virus with the human immune system, is the unique age incidence of mortality from the 1918 virus. The 1918 influenza produced an unusually high proportion of cases that developed pneumonia, especially in young people. Normally, influenza causes only mild illness in young adults. The graph of death rate by age is shaped like a U, with high rates for the very young and the very

old (Simonsen *et al*. 1998). The graph for the 1918 flu, in contrast, is shaped like a W, with a steep peak for 15–45 year olds (Collins 1931; Linder and Grove 1943) (Fig. 16.2). The death rate from pneumonia cases was somewhat higher than in other pandemics suggesting that the pneumonic complications were more serious (Collins 1931), but not dramatically so. These two facts—more pneumonia in general, and more pneumonia among young adults in particular—need to be examined separately.

The seriousness of an influenza infection is determined by how many (and which) cells the virus infects before being stopped by the body's immune system. A virus that can infect more cells, either because it replicates exceptionally well or because it infects cells not normally targeted by influenza will cause a more severe infection. However, specific antibodies can stop even an exceptionally virulent virus. The severity of a pandemic, then, will be determined by the inherent virulence of the virus and by the immune status of the population. For example, the severity of the 1968 pandemic was muted by widespread immunity to its

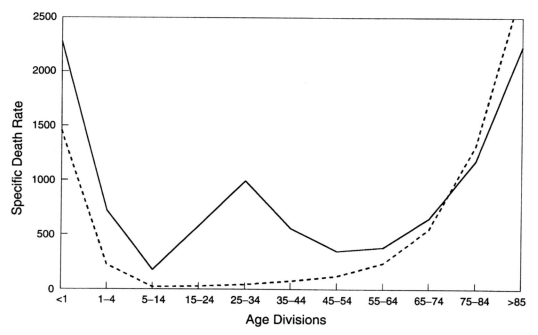

Figure 16.2 Influenza and pneumonia specific mortality by age, United States, including the pandemic year 1918 (solid line), and the average of the inter-pandemic years 1911–15 (dashed line) is shown. Specific death rate is per 100 000 of the population for each age division (Linder and Grove 1943).

neuraminidase (NA) protein, which it shared with its predecessor.

If we assume that the 1918 virus was more virulent than other pandemic viruses, we might expect that the normal U-shaped curve would be shifted upward with death rates uniformly higher across all age groups. In fact, this is true only for parts of the curve; for those under 15 years of age and for those between 41 and 60. For 15–40 year olds, where we would expect the rate to be very low, the curve instead forms a distinct peak. In the elderly, the death rate falls below that experienced in the 1892 pandemic, even lower than their average influenza death rate in the non-pandemic years of 1911–17 (Fig. 16.2). The unusual shape of the 1918 death rate curve suggests that while its agent may have been exceptionally virulent, that fact alone can not adequately explain its impact.

Our understanding of the 1918 flu would be incomplete if it did not also explain the unique age distribution of those deaths. Simonsen *et al.* (1998) noted that younger age groups account for a higher percentage of excess deaths in all pandemics, with older age groups accounting for increasingly more excess mortality in subsequent years. Whereas all age groups are similarly vulnerable to a new virus, young people appear to acquire more complete immunity from their initial encounter with the pandemic virus. In subsequent years the young are thus better protected from the virus's descendants. In 1957 and 1968, 36% and 48%, respectively, of the excess deaths were in people under 65 years of age. Most of these excess deaths were in 45–64 year olds. In 1918, fully 99% of excess deaths were among people under age 65. The majority of these were between 15 and 35 years of age; the death rate of 45–64 year olds being only somewhat higher than in other pandemics. The pattern of greater pandemic vulnerability in the young was maintained, but the degree of the shift in vulnerable age groups was more marked. In the United States, if the 15- to 45-year-old age group had experienced its usual low death rate, the number of deaths would have been reduced to 285 000, about double the average number of deaths per year from influenza in those pre-antibiotic years.

What was it about the 15–35 year olds of 1918 that made them particularly susceptible to the pandemic virus? That many of them were soldiers living in miserable conditions is not sufficient explanation; the same death rates were seen in young people unaffected by the war. Did the robust immune systems of young adults overreact to the novel virus? One might expect such a phenomenon to contribute to higher death rates in this age range in all pandemics. The sharpness of the mortality peak in the 15–35 year range also argues against such an explanation since the decline of the immune system is a gradual process. Furthermore, the pathological evidence does not point to a massive immune response; in many cases death seemingly came too quickly for a naïve immune system to have responded. However, differences in immune status could have been a factor in the odd death rate curve of 1918. If the virus replicated extremely efficiently, the difference between a mild case and full-blown pneumonia could be made by an only slightly slower immune response.

Hemagglutinin (HA) subtypes can recycle in the human population after enough time has passed that the majority of the population has no immunity (Masurel and Marine 1973). The lower death rates among the elderly in 1918 might indicate that an H1-subtype virus was circulating before 1870. Since the elderly experienced a lower than expected death rate (unlike 1957 and 1968) perhaps the 1918 virus was more similar to the previously circulating strain than was the case in the later pandemics. Those people between 15 and 45 years of age in 1918 would have been exposed to a different set of influenza viruses than people of other ages, perhaps resulting in an anti-influenza immune status particularly ill-suited to the virus of 1918.

16.3 Molecular analyses of the 1918 influenza virus

Even though contemporary observers were unable to isolate the causative agent of the pandemic, many detailed studies of the epidemiology and pathology of the disease were performed. Subsequently, as more was learned about influenza viruses and the origins of pandemics, some characteristics of the 1918 virus could be deduced. As the sequence of the virus is determined, it does not emerge into a vacuum. There is a body of historical and biological knowledge with which analysis of the sequence of the 1918 virus must be consistent. Answering questions about the origin and virulence of the 1918 virus requires integration of what is known about the history of the pandemic, influenza virus biology, and the actual characteristics of the virus as they are revealed through sequencing the 1918 virus (Taubenberger *et al.* 1997; Reid *et al.* 1999, 2000a; Basler *et al.* 2001). For the first time, it has become possible to test hypotheses about where the 1918 influenza virus came from, and what made it so deadly.

Frozen and fixed lung tissue from three fall-wave 1918 influenza victims has been used to examine directly the genetic structure of the 1918 influenza virus. Two of the cases analyzed were US Army soldiers who died in September 1918, one in Camp Upton, New York and the other in Fort Jackson, South Carolina. The material available consists of FFPE autopsy tissue, hematoxylin and eosin-stained microscopic sections, and clinical histories. These samples reflected two 1918 influenza autopsy cases out of approximately 100 such cases stored since 1918 in the National Tissue Repository of the Armed Forces Institute of Pathology (AFIP). This tissue collection, the word's largest, grew out of the US Army Medical Museum, founded in 1862. From that time to the present the collection has accumulated millions of cases representing all aspects of human disease. The samples consist of tens of millions of glass slides, FFPE tissue blocks, and clinical records. A third sample was obtained from an Alaskan Inuit woman who had been interred in permafrost in Brevig Mission, Alaska, since her death from influenza in November 1918.

The first case identified was a 21-year-old male stationed at Fort Jackson, South Carolina. He was admitted to camp hospital on 20 September 1918 with influenza and pneumonia. He had a progressive course with cyanosis, and died on 26 September 1918. At autopsy it was noted that he had a fatal secondary lobar bacterial pneumonia in his left lung while the right lung showed only focal acute bronchiolitis and alveolitis, indicative of primary influenza pneumonia. FFPE right lung tissue was positive for influenza RNA [A/South Carolina/1/18 (H1N1)] (Taubenberger *et al.* 1997; Reid *et al.* 1999).

The second case was a 30-year-old male stationed at Camp Upton, New York. He was admitted to camp hospital with influenza on 23 September 1918 and had a very rapid clinical course with death from acute respiratory failure on 26 September 1918. Autopsy showed massive bilateral pulmonary edema and focal acute bronchopneumonia. FFPE lung tissue was positive for influenza RNA [A/New York/1/18 (H1N1)]. RNA templates larger than 150 nucleotides were not amplifiable in these two cases (Reid *et al.* 1999, 2000a).

Contempory FFPE tissues may yield larger amplifiable RNA templates, at least to 250 bp and amplifiable DNA templates up to about 600 bp (Krafft *et al.* 1997). Thus, these 80-year-old tissue blocks had more degraded RNA than is typically found in modern fixed tissue. Nevertheless, sensitive reverse transcription (RT)–PCR methodology using incorporation of ^{32}P-labeled nucleotides into the amplified product allowed amplification and detection of the small influenza RNA fragments present in the 1918 tissue samples (Taubenberger *et al.* 1997).

An additional 1918 influenza case was found by examining lung tissue from four 1918 influenza victims exhumed from a mass grave in Brevig Mission, on the Seward Peninsula of Alaska. These cases were contributed to the laboratory by Dr Johan V. Hultin, who performed the exhumation in 1997 after obtaining permission from the village council. Brevig Mission (called Teller Mission in 1918) suffered extremely high mortality during the influenza pandemic in November 1918. While individual case records were not available, historical records show that influenza spread through the village in about 5 days killing 72 people, representing about 85% of the adult population (Fosso 1989). Victims were buried in a mass grave in permafrost. Four of these victims were exhumed in August 1997. Frozen lung tissues were biopsied *in situ* from each and tissues were placed in formalin, alcohol fixatives, and in guanidine HCl. While histologic analysis was hampered by freeze artifact, these cases showed evidence of acute massive pulmonary hemorrhage and edema. One of the cases, an Inuit female, age unknown, was influenza RNA positive [A/Brevig Mission/1/18 (H1N1)]. In this case, RNA templates greater than 120 nucleotides were not amplifiable (Reid *et al.* 1999, 2000a).

16.3.1 Hemagglutinin gene

The HA protein is expressed on the surface of the virus as a homotrimer. It must be cleaved by proteases to become active, whereupon it binds to receptors on host cells and initiates infection. Antibodies against the HA protein prevent receptor binding and are very effective at preventing re-infection with the same strain. In order for a pandemic influenza strain to emerge, the virus must have a HA protein distinct from the one currently prevailing. This HA subtype cannot have circulated in humans for 60–70 years, and the virus must be transmissible from person to person (Kilbourne 1977). While it is generally accepted that drift is responsible for inter-pandemic influenza outbreaks and shift for pandemics, there are exceptions to this rule. In 1947, drift in the prevailing H1 strain resulted in vaccine failure and outbreaks of influenza on a pandemic scale (Kilbourne 1997). In 1977, an H1N1 virus re-emerged but failed either to cause a pandemic or to replace the prevailing H3N2 subtype.

There is indirect evidence that the HA of the 1918 pandemic was of the H1 subtype. In the 1930s, it was shown that survivors of the 1918 flu had antibodies that neutralized classic swine flu (Shope 1936). These antibodies were rarely found in people born after 1920 (Dowdle 1999). Furthermore, contemporary accounts of the fall wave of the 1918 flu described simultaneous outbreaks of flu in pigs and humans (Koen 1919). It seemed likely that the 1918 flu had spread from humans to pigs, where it survived as the classical swine flu lineage. While the serological data suggested that the 1918 hemagglutinin would resemble the H1-subtype swine flu hemagglutinin isolated in 1930, the avian origin of the 1957 and 1968 HAs (Scholtissek et al. 1978), made it possible that the 1918 H1 would more closely resemble an avian H1.

Using the three 1918 influenza RNA-positive tissue samples, small fragments, 80–150 bp, of the HA gene were amplified. Since no 1918 sequence was previously known, it was initially unclear whether the 1918 HA would more closely resemble avian, swine, or human H1 HA sequences. To circumvent this problem, conserved areas of the gene were used to design degenerate PCR primers that should amplify all H1 HAs. The fragments that were amplified were cloned and sequenced. Since the primers were degenerate, the sequence obtained from the region of the primers was not considered, only the inter-primer sequence. Once 1918-specific sequence was obtained, specific primers could be designed to produce overlapping PCR products that allowed us ultimately to obtain the complete sequence of the 1918 HA gene segment. In order to confirm the sequence, the entire process was performed twice starting with RNA isolated from separate tissue sections of each case. Further confirmation was obtained by generating this sequence from each of the three cases. Ultimately the full-length sequence of the 1918 HA was obtained and it shows that it is most closely related to the human and swine influenza strains of the 1930s (Reid et al. 1999).

Although these cases were widely separated geographically, there is very little heterogeneity amongst them at the sequence level, suggesting the virus was optimally adapted to infect a majority of the human population in 1918. Only two nucleotide differences in the HA1 domain were noted among the strains (Reid et al. 1999).

Influenza is a zoonotic disease, affecting many species of birds and mammals. In order for influenza to spread in a new host, the HA protein must acquire the ability to bind to the new host's cells (Murphy and Webster 1996). Pandemic influenza results when an influenza strain emerges with a HA protein to which few people have prior immunity (Kilbourne 1977). It is thought that the source of HA genes new to humans is the extensive pool of influenza viruses infecting wild birds. Of the 15 HA subtypes found in birds, only three (H1, H2, and H3) are known to have caused pandemics in man (Kilbourne 1997). Recently avian H5N1 and H9N2 viruses caused illness in a limited number of people in Hong Kong (Lin et al. 2000). How new HA genes emerge, and where they adapt from their characteristic avian form to a form that spreads successfully in humans is not understood. In both 1957 and 1968, the pandemic strains had new HA genes which were very similar to known avian strains (Schafer et al. 1993). We sought to determine whether the HA gene of the 1918 pandemic strain also had avian characteristics.

The sequence of the 1918 HA is most closely related to the oldest available 'classical' swine flu strain, A/Swine/Iowa/15/30 (H1N1). However, despite this similarity the sequence has many avian features. Of the 41 amino acids that have been shown to be targets of the immune system and subject to antigenic drift pressure in humans, 37 match the avian sequence consensus, suggesting that there was little immunologic pressure on the HA protein before the fall of 1918. Another mechanism by which influenza viruses evade the human immune system is the acquisition of glycosylation sites to mask antigenic epitopes. Modern human H1N1s have up to five glycosylation sites in addition to the four found in all avian strains. The 1918 virus has only the four conserved avian sites (Reid *et al.* 1999).

In addition to acquiring an antigenically novel HA, a pandemic influenza virus must also acquire specific adaptations to spread and replicate efficiently in a new host, including functional HA receptor binding and interaction between viral and host proteins. Human-adapted influenza viruses preferentially bind sialic acid receptors with $\alpha(2,6)$ linkages. Those strains adapted to birds preferentially bind $\alpha(2,3)$ linked sugars (Matrosovich *et al.* 1997). The HA receptor binding site consists of a subset of amino acids that are invariant in all avian HAs but vary in mammalian-adapted HAs. In H1-subtype viruses like the 1918 pandemic virus, to shift from the avian receptor-binding pattern to that of swine H1s requires only one amino acid change, E190D (Reid *et al.* 1999). All three 1918 cases have the E190D change. In fact, the receptor binding site of one of the 1918 cases (A/New York/1/18) is identical to that of A/Swine/Iowa/15/30. The other two 1918 cases have an additional change from the avian consensus, G225D. Since swine viruses with the same receptor site as Swine/Iowa/15/30 bind both avian and mammalian-type receptors, A/New York/1/18 probably also had the capacity to bind both. The change at residue 190 may represent the minimal change necessary to allow an avian H1-subtype HA to bind mammalian-type receptors, a critical step in host adaptation.

Certain avian influenza subtypes (H5 and H7) have insertional mutations of extra basic amino acids at the H1–H2 cleavage site which makes them extremely virulent by expanding their tissue tropism (Webster and Rott 1987). The 1918 HA gene did not have a mutation of this type (Taubenberger *et al.* 1997; Reid *et al.* 1999).

16.3.2 Neuraminidase gene

The principal biological role of NA is the cleavage of the terminal sialic acid residues that are receptors for the virus's HA protein (Palese and Compans 1976). The active site of the enzyme consists of 15 invariant amino acids, which are conserved in the 1918 NA. The functional NA protein is configured as a homotetramer in which the active sites are found on a terminal knob carried on a thin stalk (Colman *et al.* 1983).

The complete sequence of the 1918 NA gene has been determined (Reid *et al.* 2000a). In many ways analyses of the 1918 NA sequences give results similar to that of the 1918 HA. As with HA, the functional and antigenic sites of A/Brevig Mission/1/18 (Brevig/18) NA closely resemble those of avian isolates. The 1918 virus had an N1-subtype NA. The 15 conserved amino acids making up the active site of the molecule are retained, as are the seven glycosylation sites found in all avian strains. Twenty-two amino acids have been identified as antigenic in the N2 subtype (Colman *et al.* 1983). Of the homologous amino acids in N1, 15 have shown variation in human strains. Brevig/18 matches the avian consensus at 14 of the 15 residues, suggesting little or no antigenic pressure on the protein before 1918. Human strains from the 1930s show extensive drift at these sites.

Several early human strains have deletions of 11–16 amino acids in the stalk region of the NA, which may affect the activity of the protein (Castrucci *et al.* 1993). The 1918 strain does not have a stalk deletion, suggesting that the various deletions found in early human strains are likely to be artifacts of their extensive culture in various hosts.

Certain mouse-adapted H1N1 influenza viruses have a mutation leading to the loss of a glycosylation site at residue 146, the absence of which contributes to the extended tissue tropism and neurotropism (Li *et al.* 1993; Goto and Kawaoka 1998; Taubenberger 1998). The 1918 viral NA gene does not have this mutation (Reid *et al.* 2000a).

Therefore, neither surface protein-encoding gene has known mutations that would allow the virus to become pantropic. Since clinical and pathological findings in 1918 showed no evidence of replication outside the respiratory system (LeCount 1919), mutations allowing the 1918 virus to replicate systemically would not be expected. However, the relationship of other structural features of these proteins (aside from their presumed antigenic novelty) to virulence remains unknown. In their overall structural and functional characteristics, the 1918 HA and NA are avian-like but they also have mammalian-adapted characteristics.

While evidence suggests that the 1918 HA and NA gene segments were new to humans, the source of the 1918 influenza surface proteins and how they became part of a human-adapted influenza virus remain unclear. The absence of information about pre-1918 human influenza strains, the lack of influenza strains from birds and pigs around 1918, the gap between the 1918 strain and human and swine strains from the 1930s, and the lack of pre-1918 human serum samples all conspire to make solution of the question of the 1918 influenza's origin exceedingly difficult. However, despite these problems, sequence data from the 1918 influenza virus itself is finally providing a basis for answering these questions.

16.3.3 Non-structural genes

Recently the complete coding sequence of the 1918 non-structural (NS) segment was completed (Basler *et al.* 2001). The entire 1918 NS coding sequence (838 nucleotides) was determined from the frozen sample obtained from Brevig Mission, Alaska [A/Brevig Mission/1/18 (H1N1)].

It has been shown that the influenza A virus NS1 protein functions as a type I interferon (IFN) antagonist (Garcia-Sastre A *et al.* 1998) and is required for influenza A virus virulence. A mutant influenza virus, influenza delNS1 virus, which lacks the NS1 open reading frame and therefore produces no NS1 protein, was found to grow poorly on substrates in which type I IFN-induced antiviral pathways are intact (Garcia-Sastre A *et al.* 1998). However, this virus grows similarly to wild-type virus [influenza A/PR/8/34 (H1N1) virus] on substrates which do not mount an effective type I IFN response (Garcia-Sastre A *et al.* 1998). Thus, influenza delNS1 virus is highly attenuated in wild-type mice but can cause lethal infection in mice, such as STAT1$^{-/-}$ mice or PKR$^{-/-}$ mice, which are unable to mount an effective type I IFN response (Garcia-Sastre A *et al.* 1998).

One of the distinctive clinical characteristics of the 1918 influenza was its ability to produce rapid and extensive damage to the respiratory epithelium (LeCount 1919). Such a clinical course suggests a virus that replicated to a high titer and spread quickly from cell to cell. Lack of an antibody-based immune response because of novel HA and NA proteins provides only a partial explanation of this phenomenon, since other influenza strains with novel surface proteins, such as the 1957 pandemic, were not as virulent as the 1918 strain. An NS1 protein that was especially effective at blocking the type I IFN system might have contributed to the exceptional virulence of the 1918 strain. The genetic characteristics of NS1 that contribute to its ability to block IFN have not yet been mapped and the likelihood that the 1918 NS1 would be effective at blocking IFN could not be assessed simply by examining its sequence. Therefore, in order to begin to understand the role that the NS gene products, NS1 and NS2, may have played in virulence, the 1918 virus NS1 gene and the entire 1918 virus NS segment were reconstructed, and transfectant influenza viruses bearing these genes were generated (Basler *et al.* 2001). In the first set of viruses, the open reading frame encoding the 1918 NS1 or a control NS1 protein was introduced into virus. In the second set, the entire 1918 NS segment was introduced into virus. The virulence of these viruses in mice was then determined.

In both cases, viruses containing 1918 NS genes were attenuated in mice compared with wild-type A/WSN/33 controls (Basler *et al.* 2001). The attenuation demonstrates that NS1 is critical for the virulence of A/WSN/33 in mice. The 1918 NS1 varies from that of WSN at 10 amino acids. The amino acid differences between the 1918 and A/WSN/33 NS segments may be important in the adaptation of the latter strains to mice and likely account for the observed differences in virulence in this set of experiments. Thus, further experiments using different viral and animal backgrounds may be necessary to test the hypothesis that the 1918 NS1 specifically contributed to virulence.

16.4 Phylogenetic analyses of the 1918 influenza virus genes

Since virulence can not yet be adequately explained by sequence analysis of the 1918 HA and NA genes, what can these sequences tell us about the origin of the 1918 virus? The best approach to analyzing the relationships among influenza viruses is phylogenetics, whereby hypothetical family trees are constructed that take available sequence data and use them to make assumptions about the ancestral relationships between current and historical flu strains (Gammelin *et al.* 1990; Fitch *et al.* 1991; Scholtissek

et al. 1993). Since influenza genes are encoded by eight discrete RNA segments that can move independently between strains by the process of reassortment, these evolutionary studies must be performed independently for each gene segment.

Phylogenetic analyses of the 1918 HA and NA genes allow us to place them within the context of a wide range of H1- and N1-subtype genes. Phylogenetic analyses based upon HA nucleotide changes (either total, synonymous, or non-synonymous) or HA amino acid changes always place the 1918 HA with the mammalian viruses, not with the avian viruses (Fig. 16.3) (Reid *et al.* 1999). Its placement is compatible with its being the ancestor of all subsequent human and swine H1-subtype strains. Some analyses place the 1918 HA in the human

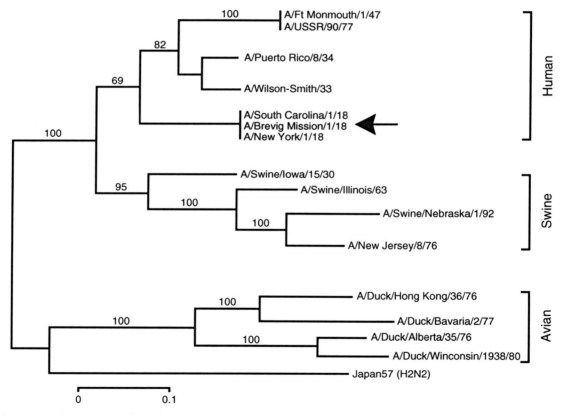

Figure 16.3 Phylogenetic tree of the influenza virus HA gene segment, HA1. Sequences were analyzed for phylogenetic relationships by the neighbor-joining method with proportion of sequence differences as the distance measure (0.1 p-distance ~ 22.35 synonymous differences). Synonymous substitutions were analyzed and bootstrap values (100 replications) are given for selected nodes. Human, swine, and avian clades are identified with brackets. The arrow identifies the position of the 1918 sequences. Distance bar is shown below the tree. Influenza strains used in the analyses are described in Reid *et al.* (1999).

clade and some in the swine clade, suggesting that it shares characteristics of both. The HA proteins of human influenza viruses are subject to substantial immune pressure; HAs that have acquired mutations changing or masking antigenic sites have a selective advantage for spread in humans. Therefore, by the time influenza strains were isolated from humans in the 1930s, many antigenic sites had drifted from the 1918 sequence. In swine the substitution rate is lower. As a result, the earliest swine influenza strain, isolated in 1930, resembles the 1918 strain more closely than the 1930s human strains, especially at the antigenic sites.

The placement of the 1918 HA at the root of the mammalian clade is compatible with the historical record and with what is known about influenza evolution. However, it does not pinpoint where the HA gene of the 1918 virus came from or when it began circulating in humans. Both the 1957 and 1968 pandemic strains had HA proteins that were very similar to those found in wild birds. The rapid spread of the 1918 flu suggests that it also had acquired a novel HA.

Of all mammalian H1s, those of the 1918 viruses are most similar to their avian counterparts. In spite of its many avian characteristics, the 1918 HA is nevertheless phylogenetically distinct from current avian H1s. However, the 1918 HA gene has accumulated enough non-selected, mammal-associated changes to place it consistently in the mammalian clade phylogenetically (Fig. 16.3). It is possible that the HA involved in the pandemic did not pass directly from an avian source to its pandemic form but rather spent some unknown amount of time adapting in a mammalian host. The 1918 HA1 sequence differs from its closest avian relative by 26 amino acids, whereas the 1957 H2, the 1968 H3, and 1997 H5 HAs had 16, 10, and three differences, respectively, from their closest avian relatives (Schafer *et al.* 1993; Claas *et al.* 1998; Subbarao *et al.* 1998). Phylogenetic analyses of the 1957, 1968, and 1997 HAs place them in avian clades and suggest that all three strains derived from the Eurasian group of avian strains. By contrast, phylogenetic analyses invariably place the 1918 sequence outside the avian clade and suggest that the 1918 sequence is phylogenetically equidistant from the Eurasian and North American avian strains. It is possible

that current avian strains may have drifted substantially from their 1918 form and no longer closely resemble the HA that made the jump to a human virus. Without samples of avian viruses from 1918, it is difficult to choose between these possibilities.

Phylogenetic analyses of the NA gene also show that of all mammalian isolates, the 1918 sequence is the most closely related to avian isolates, but also suggest that the 1918 sequences share enough characteristics with mammalian isolates to distinguish them from the avian clade (Reid *et al.* 2000a). The placement of the 1918 NA nucleotide sequence in the phylogenetic trees is usually within and near the root of the mammalian clade, suggesting that it is very similar to the ancestor of all subsequent swine and human isolates. At the same time, and in contrast to the results with HA, phylogenetic analyses of the NA protein sequence place it within the avian clade. In these cases branch lengths are very short and bootstrap values are low, suggesting that there are not enough differences among the sequences to place them unambiguously.

Phylogenetic results and structural characteristics suggest that the 1918 NA sequence is intermediate between avian and mammalian sequences, and are consistent with the idea that the 1918 pandemic virus acquired its NA gene directly (with little modification) from avian viruses. Nevertheless, the 1918 NA differs at 26 amino acids from its nearest known avian relative [A/Duck/Alberta/35/76 (H1N1)] (Reid *et al.* 2000a). In contrast the 1957 pandemic N2 and the N1 from the 1997 Hong Kong H5N1 outbreak differ by only 18 and two, respectively, from their nearest avian relatives. Again, either avian sequences have drifted away from their ancestral sequences over the past 80 years, or the 1918 genes acquired mammalian-specific changes in a mammalian host in the years preceding the 1918 pandemic. That the ultimate source of the 1918 NA was avian is supported by the phylogenetic analyses, but the precise path of the gene from its avian source to its pandemic form cannot be determined by sequence alone.

Identifying the minimal changes necessary to allow a virus with avian surface proteins to replicate and be transmitted efficiently in mammalian hosts is extremely important for our understanding of the emergence of pandemic influenza viruses. Besides placing viral sequences in their evolutionary context, phylogenetic analyses can be used to identify features of viral proteins involved in host adaptation. For example, when the 1918 HA is compared with strains derived from the independent introduction of an avian H1N1 into European swine in the late 1970s, the E190D change is found in both (Reid *et al.* 1999). In the NA, five amino acids were found to change in both. Two of these positions have replacements that are either chemically similar or identical, suggesting that these two sites (residues 285 and 344) may be particularly important in host adaptation of NA (Reid *et al.* 2000a).

In mammalian-adapted influenza proteins, the accumulation of amino acid changes from a hypothetical ancestral sequence occurs in a linear fashion, with the slope representing the number of amino acid changes per year (Buonagurio *et al.* 1986, Kanegae *et al.* 1994). If the substitution rates of human and swine influenza are projected back in time, the 1918 HA falls near the intersection of the human and swine lines (Taubenberger *et al.* 2000). This, in turn, is consistent with the historical record wherein concurrent outbreaks of influenza in swine and humans were reported in the USA, Europe, and Asia (Koen 1919). In the USA, the disease became established in swine and has recurred yearly since 1918 (Shope and Lewis 1931; Zhou *et al.* 1999).

Such a regression analysis is shown for human and swine H1 and N1 proteins using available strains from 1930 to the present and excluding the 1918 sequence (Fig. 16.4). When the 1918 HA and NA data points are then placed on the graph they are located close to the hypothetical common ancestor of human and swine H1N1 strains. The x-intercept of the lines suggests that the surface protein-encoding genes of the virus may have entered mammals just prior to the 1918 pandemic. Together, the phylogenetic analyses support the conclusion that the sequences derived from the 1918 cases are very similar to those of the hypothetical common ancestor of both human and swine H1N1 strains.

Similarly, phylogenetic analysis of 63 NS1 nucleotide sequences produced a tree with nine

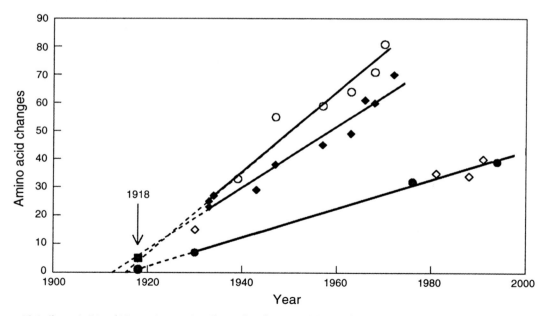

Figure 16.4 Change in HA and NA proteins over time. The number of amino acid changes from a hypothetical ancestor was plotted versus the date of viral isolation for viruses isolated from 1930 to 1993. Open circles, human HA; closed diamonds, human NA; closed circles, swine HA; open diamonds, swine NA. Regression lines were drawn, extrapolated to the x-intercept and then the 1918 data points, closed square 1918 HA, closed circle 1918 NA, were added to the graph (arrow) (Taubenberger *et al.* 2000).

clades: avian clades 1–5, equine clades 1 (H7N7) and 2 (H3N8), human, and swine. Phylogenetic analyses of 61 NS2 (NEP) nucleotide sequences produced a tree with the same nine clades seen in the NS1 tree. The A/Brevig Mission/1/18 NS gene was within and near the root of the swine clade for NS1, and within and near the root of the human clade for NS2 (NEP) (Basler *et al.* 2001).

This placement of the genes of the 1918 NS segment suggests that the 1918 NS is the ancestor of all subsequent classic swine and human NS genes. However, in the case of NS, the bootstrap values are low and branch lengths are short, suggesting that there are not enough differences among the sequences to place them unambiguously. Because the NS1 and NEP sequences are much more conserved than HA and NA (Reid *et al.* 1999, 2000a), their phylogenetic trees are less informative. The known functional domains of NS1 and NEP are highly conserved across species-adapted strains; no amino acid changes necessary for host adaptation have yet been identified even though both proteins have been shown to interact with cellular proteins during viral replication (O'Neill *et al.* 1998; Chen *et al.* 1999; Basler *et al.* 2001). Consequently, it is not possible to determine whether the 1918 NS segment derived from a novel (avian) source at about the same time as the 1918 HA and NA gene segments, or was already part of a previously circulating human-adapted strain. In the absence of pre-1918 human and avian influenza samples this question will be difficult to address.

16.5 Origin of the 1918 pandemic influenza virus

If one of the requirements for a pandemic influenza virus is that it have, at least, a novel HA protein, another is that is must be readily transmissible from person to person. A pandemic virus faces the twin challenges of being 'new' to its host, while being supremely well adapted to it. This condition has been fulfilled in recent pandemics by reassortment: combining surface proteins novel to humans with human-adapted internal proteins. A trade-off is implied—the more avian genes, the less recognition by the human immune system, but probably the less well adapted to growth in human cells (Reid and

Taubenberger 1999). The 1997 outbreak of H5 influenza in Hong Kong may be a case in point. It was an entirely avian virus and, while it caused severe illness in several people, it apparently spread extremely poorly, if at all, from human to human (Subbarao *et al.* 1998, Claas *et al.* 1998; Katz *et al.* 1999). We have obtained sequence from five RNA segments of the 1918 virus: HA (Reid *et al.* 1999), NA (Reid *et al.* 2000a), NS (Basler *et al.* 2001), NP and MA (Taubenberger *et al.* 1997; and unpublished data). The sequences of all the genes examined appear to be more closely related to old human and swine strains than to avian strains.

The difficulty in determining whether any or all of these genes shifted in 1918 is that the virus it replaced is not available for analysis. Our data confirm an H1N1 shift around 1918, but cannot determine with precision when these genes entered the human population. If all human influenza viruses ultimately derive from avian sources, the oldest human strain will necessarily be more closely related to avian strains than all subsequent strains. In the case of HA and NA, the 1918 genes do not seem to be as closely related to avian strains as were the surface proteins of the 1957 and 1968 pandemics. For the other genes, even with sequence in hand it will be difficult to determine whether a gene was new to humans in 1918 or had entered the human population earlier, although phylogenetic analyses may provide clues. Future work will concentrate on finding archival influenza RNA- positive cases from before 1918. Such cases would provide a means to ascertain which gene segments were novel in the 1918 strain.

16.6 Conclusions

In many respects, the 1918 influenza pandemic was similar to other influenza pandemics. In its epidemiology, disease course, and pathology, the pandemic generally was different in degree but not in kind from previous and subsequent pandemics. However, there are some characteristics of the pandemic that appear to be unique. Mortality was exceptionally high, ranging from five to 20 times higher than normal. Clinically and pathologically, the high mortality appears to be the result of a higher proportion of severe and complicated infections of the respiratory tract, not with systemic infection or involvement of organ systems outside the influenza virus's normal targets. The mortality was concentrated in an unusually young age group. Finally, the waves of influenza activity followed on each other unusually rapidly, resulting in three major outbreaks within a year. Each of these unique characteristics may find their explanation in genetic features of the 1918 virus. The challenge will be in determining the links between the biological capabilities of the virus and the known history of the pandemic.

Acknowledgements

This work was supported by grants from the Department of Veteran's Affairs and the American Registry of Pathology, and by the intramural funds of the Armed Forces Institute of Pathology. The opinions or assertions contained herein are the private views of the authors and are not to be construed as official or as reflecting the views of the Department of the Army or the Department of Defense. This is US government work; there are no restrictions on its use.

Part IV

CHAPTER 17

Lessons from the past

Many molecular biologists are becoming historians. Little by little they start to approach genomes not so much as polymers, but as old manuscripts which document the history of life on our planet. As they learn more about old genes they open a door to unexpected historical events that might later reflect on our ideas of molecular biology itself.
(Richard Pollack, Signs of Life. The language and meanings of DNA, 1994)

Why knowing the causative agents of ancient infectious diseases can help us understand modern emerging diseases?

Greenblatt: Its my prerogative as editor to see how we all feel about the field of ancient pathogens. Are we making headway? Let's divide the discussion into several parts:

17.1 Are we endangering humankind by digging up the dangerous pathogens of the past?
17.2 Are we asking the right questions?
17.3 Are our techniques advancing?
17.4 Lessons learned.

17.1 Are we endangering humankind by digging up the dangerous pathogens of the past?

In two universities I attempted to obtain biosafety guidelines for the culture of amber. The answers, not committed to writing were, 'These are ordinary environmental microbes, and women have been rubbing their amber beads on their chests for years.' But still the fears persist. In Italy, when I attempted to culture bones from Pompeii, one student simply freaked out.

A question to Jeff Taubenberger: I know that the 'American expedition' that exhumed the Eskimo remains did this with little fanfare. The Canadian/British group, on the other hand, were in biosafety tents with a high level of precaution.

Jeff answers: Johan Hultin and I agreed that there was virtually no chance of recovering infectious

influenza virus from the permafrost graves. Flu viruses, as single stranded RNA viruses, are very labile. Despite its name, permafrost actually hovers between about −10 and +10°C, the worst possible case of freeze/thawing for maintaining viral viability. This would hold true for all viruses, even much more stable ones like the dsDNA orthopox viruses. Thus I think the chance of digging up infectious viruses is remote.

To Mark Spigelman: I know you were in Spitzbergen with the Canadian/English group. Did they speak about the imminent danger of their undertaking?

Mark Spigelman: I recall when John Oxford first came to visit us at the University College London, Department of Microbiology. One of his major concerns was that the virus might be inadvertently released by the digging up of the bodies. He detailed their preparations. In the actual Spitzbergen expedition they not only had a double entry tent, they wore space suits for excavating, and sprayed people in the middle chamber before they could leave. They were quite concerned that they might release the virus into the community and designed specialized hollow metal tubes to biopsy the bodies. They were thus able to take a core from the frozen cadavers. These were designed to go into special containers for transport to the Defence, Science, and Technology Laboratory, Porton Down, where work on the most dangerous pathogens is carried out in the UK.

I recall advising John regarding the anatomical features of where to take the core; for example, a transverse core; was to be taken which would biopsy the lungs and bronchi. We also decided where to biopsy for stomach and colonic contents, as we hoped to find live bacteria; but we were particularly interested in seeing if the lungs contained the living virus. The best negative controls in ancient pathogen DNA work is to take samples from areas where the pathogen is not likely to be present.

The whole aspect of pathogens and their investigations from ancient tissues has had a lot of criticism. Some people are even claiming that we should not be investigating any ancient diseases that have counterparts today, for fear of contamination, not only may we bring back ancient pathogens, but because our results will be confused by the presence of today's organisms. For example, we are in a department that does study modern-day TB. This is in a separate biosafety laboratory well away from our ancient DNA facility. Yet we have been told it is inappropriate for us to study ancient TB DNA. I see the point, but it is wrong. We look at a number of DNA sequences from the TB genome, so we can distinguish between the organisms. Spoligotyping has provided us with patterns that can show distinctions between strains, should we ever believe there is a contamination problem. If these 'fingerprints' are different then we know there is no contamination. In fact, due to this level of discrimination, I believe we are on safer ground in the area of contamination with pathogens than those studying mitochondrial DNA.

Bernd Herrmann: The materials we study are so degraded that there is no chance of recovering them intact. Thus we can just go for the proof of presence—if we know what we are going for. The fear of innocently awakening a megavirulent pathogen from the past is science fiction, Patho-Park.

BUT: who knows, time, money, and staff given, you can make a movie.

We are not creating pathogens, they are already there. And we are not reanimating them. For me there are no reasons, visible and imaginary, to support such fears. People may behave hysterically because they are uneducated. This is one of the reasons to write books like this.

Chuck Greenblatt mentioned his concern about persistence and dispersal of pathogens. This also came up in the chapters of Görtz and Michel, and

Cano. As vectors of diseases, one should think about aerosols flying around the earth. Ducks were recently reported to fly at 12 km. If their droppings contain viruses able to infect pigs or humans, how big is the area that would be infected by such flying bombs? Likewise, consider the wind as a global vector, and mineral dust. The Amazon ecosystem is 'excreting' minerals, biomasses, etc. via the river systems into the oceans. It is not fed by the slopes of the mountains in the hinterland but it is maintained by the 'dust' that is brought in via air flows from the Sahara and bordering areas. So: quite an area of one continent is depending on another, without any direct linkage. And the amounts of materials must be really huge. Imagine, bacteria or 'pathogens' are likely also transported … you would end up in a scenario that is different from Diamond's assumption, as America could have been not that independent in terms of microorganisms. (However, this is imagination, or, as Rothschild would say, mythology, as it has not been tested, etc).

Chuck Greenblatt: I remember that W. Hamilton speculated on dispersal through cloud formation. The bacteria that exploited this atmospheric system were successful in causing water condensation.

Spigelman: You may remember the story of Howard Carter, the tomb of Tutenkamun and the 'Curse of the Mummy'. Most archaeologists I have worked with will tell you of mysterious boils and skin infections noted when digging up human remains. Each infection on an individual basis can be explained, but there are a lot of these stories. Nobody has ever tried growing these bacteria! Maybe we should try.

Hans-Dieter Görtz and Rolf Michel: Modified, this question is relevant even for pathogens coming from protozoa. However, it is not the study of such infections in the laboratory that bears a risk, since infected protozoa are present in natural environments anyhow. Rather, humans are confronted with potential pathogens from protozoa as a result of civilization. New technologies such as air conditioning, etc. bring about the risks. Protozoa as hosts may serve as ecological niches within a poor or hostile environment.

Despite the fact that most bacterial species serve as main food sources of free-living amoebae some bacteria have the ability to prevent intracellular digestion by their host, survive, and replicate intra-cellularly as endosymbionts. It was shown first for *Legionella pneumophila* and later for other pathogens or their relatives such as members of the genera *Listeria*, *Pseudomonas*, *Mycobacterium*, *Chlamydia*, *Burkholderia*, and *Vibrio*. Some of the intracellular bacteria, for example, members of the Chlamydiales are obligate endosymbionts; while others, e.g. *Legionella* or *Burkholderia*, behave as fac-ultative endoparasites of *Acanthamoeba* species. A considerable number of hitherto unidentified iso-lates of endosymbionts are waiting for species determination. In all probability various—perhaps unexpected—pathogens will be found among them.

Mark Spigelman: I can certainly support this, I am aware that there is research in the UK at a vet-erinary institute where they found *Mycobacterium bovis* living happily in fields that had been free of cattle for a year. Where were they? Intracellularly in the amoebae in the soil!! So no surprise that they have all the equipment to transfer to our own macrophages—stealth technology is transferable?

17.2 Are we asking the right questions?

A little philosophy first. To my surprise I discov-ered that I was a 'historical scientist'. Some hint of this came to Mark Spigelman and I when a confer-ence committee invited us to a meeting of the American Society of Historical Archaeology. Since all of my previous work had been experimental it was only then that I realized that in history some-one else had done the experiments. It also pene-trated through my usual haze how important the archival material can be in building medical archaeology. Mark knew this from his Hungarian mummy work.

Jared Diamond has argued that:

'The discipline of history is generally not consid-ered to be a science, but something closer to the humanities, … Most historians do not think of themselves as scientists and receive little training in acknowledged sciences and their methodologies … One cannot deny that it is more difficult to extract general principles from studying history than from studying planetary orbits. However, the difficulties seem to me not fatal … But recall that the word 'science' means 'knowledge' … to be obtained by whatever methods are most appropriate to the par-ticular field … Historical sciences in the broad sense (including astronomy and the like) share many features that set them apart from non-historical sci-ences such as physics, chemistry, and molecular biology. I would single out four: methodology, cau-sation, prediction, and complexity.'

Now listen to what Martin wrote in this volume, 'Truth is not only elusive in science; in a very real sense, it cannot be demonstrated. What we accept as true is the convergence on a common answer of two or more independent lines of argumentation. Diagnosis of fossil disease needs to be tested by more than one line of evidence and, whenever possible, confirmed in many individuals where epidemiology may itself become an additional and independent test. This is especially true for a his-torical science like palaeopathology … In 'his-torical' sciences, the experiments have already occurred and we are confronted with answers. It is our job to find the right question for a particular result, and our manipulation of how we frame that question provides controls on the experiment. In many ways, the differences between experimental and historical sciences are not profound. This is rec-ognized in physics where astronomy is embraced.'

So apparently we don't have to convince our-selves that we can arrive at the 'truth', but a conver-gence of mutually confirming approaches. Martin in a way turns things upside down—the answers are provided, go out and ask the appropriate questions.

Mark Cohen and Gillian Crane-Kramer wrote that paleoepidemiology is based on three types of evidence. The three are 'the assumption that the laws of biology do not change in a serendipitous manner, ethnographic or contemporary observa-tions of the presence and distribution of diseases among both human and non-human primate popu-lations', and 'skeletal evidence'. They score all three of these as 'often incomplete or simply wrong', and this would include the past evolutionary history of the causative agent. Especially important to this volume are their statements: 'The representative nature of skeletal samples (whether they accurately reflect real health in the once living populations) has been challenged (Ortner 1992; Wood *et al.* 1992)', and 'estimates of incidence or prevalence in a once-living population based on cemetery samples, the ultimate goal of all paleodemographic research' doubted. The 'observation that diseases are slow to develop in the skeleton so that an apparent increase in skeletal pathology might actually represent greater life expectancy with more people living long enough for the disease to register in the skeleton rather than an increase in actual disease'.

Bruce Rothschild comments:

1 Cohen and Crane-Kramer have made several good comments, but have a number of unfounded comments. The uniformitarian assumptions that they refer to represent a mythology. For those diseases which have been studied to date, there has been no evidence of change in skeletal impact. Ewald has also abandoned that misconception. While an interesting theory, there is no documentation, so it falls in the area of mythology.

2 Contemporary distribution of diseases should not be the arbiter for past distribution, and not just because of human activities. The bones speak for themselves and we should listen! Ortner's osteologic paradox is smoke and mirrors. It has nothing to do with the science. We can only discuss entities that we can identify/measure. Bone health can be discussed. Osseous health is the measure of osseous health. Attempting to say that it is meaningless or inversely related to organism health is not valid. They are different phenomena. As there has been no documentation of relationship, the issue is in the realm of mythology!

3 The assumption that diseases are slow to develop has no bearing on the diseases discussed. As long as the individuals in the cemetery lived long enough to have contracted and manifest the disease, the sample is appropriate for study of that disease. While osteoarthritis and spondylosis deformans require older populations, they actually cannot be validly studied by current anthropologic methods—as those methods do not seem valid beyond age 50! Thus the questions are not directly answerable where anthropologists have traditionally applied them—and are valid for treponemal disease and forms of arthritis other than osteoarthritis.

Furthermore, the assumption that cemeteries are not valid for study of palaeopathology is patently false. If it were true that cemeteries did not reflect populations, then any study of anthropology would be invalidated!!! What is true is that biases do not generally affect palaeopathology considerations. If the disease in question was not a determinate of burial, then its presence in the cemetery should represent the population. Exceptions are of course hospital cemeteries, but even they have problems. Hospital cemeteries reflect what was diagnosed at the time. Given the various diseases that were lumped with leprosy, it is perhaps not surprising that proven leprosy does not match what they have reported.

Oliver Dutour: I feel I must comment on Bruce Rothschild's remarks.

1 I do believe in the osteological paradox (or in a part of it), and if it is right that bones are speaking to us, we must also listen to the sound of their silence. In infectious diseases, bone involvement is often the result of a slow process and/or good immunization of the populations. 'Better health makes the worst skeletons' is not a wrong idea.

2 Cemeteries are valid for palaeopathology, that's pretty sure! But it is also clear that skeletal samples cannot reflect the past populations in their living status. Biases concerning the structure of the skeletal samples do affect paleopathological considerations at the level of the frequency of a given disease observed in skeletal collections. This is the case if we want to consider it from an epidemiological perspective in order to know something about the past prevalence. In a skeletal collection, absence of proof is not proof of absence, but what is the real meaning of its frequency observed in skeletal collections (differential taphonomy, burial practices, quality of excavations, etc.). How **much** was the prevalence in the living population (reverse structure from the dead one in cemeteries), and **when** was it (time distortion). In this approach, considering the same disease, period of time, location, population, the reconstructed prevalence obtained from skeletal samples from 'normal' cemeteries and from plague mass graves are **very** different. The 'Pompei effect' is not mythic.

Jake Baum on Larry Martin: I like the scenario that Martin has painted very much. However, the detailed introduction, which I feel lacks relevance in places to the general thesis he is putting forward for changes in primate evolutionary history and the emergence of disease, does not fully justify the way that he has applied theoretical approaches to reconstructing the distant evolutionary history of human and primate diseases. How much of Martin's wonderfully rich imagery of evolutionary change is based on speculation and how much has scientific backing with real data?

Baum continues: The general drive of the paper also suffers from being rather Homo-centric, with the primate side of things being left somewhat neglected, perhaps the title might have more accurately been 'Human evolution and the emergence of disease'? This is not to take away from the very thorough and enjoyable evolutionary picture for the emergence of human disease that Martin has presented.'

{Editor's note—Martin wrote, 'The lack of a theoretical framework for many interesting problems with living pathogens has probably impeded palaeopathology more than any inadequacy of the fossil record. In contrast, recent attempts towards a theoretical framework for emerging diseases provided new opportunities for palaeopathology (Ewald, 1998). [He then cited the conference this volume was based on.] One of the most interesting aspects of this conference was the application of population theory to the evolution of pathogen virulence by Paul Ewald (1998).}

Larry Martin: Baum's remarks about speculation and scientific backing seem a little redundant. Of course I'm speculating about how disease may have responded to past environments, but the past was very different from the present and constantly changing. It invites speculation, and from that speculation we hope to derive hypotheses capable of testing and 'real science'.

Bernd Herrmann: Jared Diamond's is not a strong argument, as he is pretty weak in historical tools (see his America and China chapters). His suggestions are a revival of the old geographic determinism, which has been overcome at least to a certain extent in social anthropology.

Martin gives a well-known variation of the old hermeneutic principle: 'the proof does not create doubts as the proof is not doubted'. Truth is contextual and not absolute. The driving force is error.

But we are engaged in a basically economic framework that suggests that truth is slowly improved in a truth optimizing system. This is crude Darwinistic intellectualism. Truth depends on contexts. Martin is right: 'history' might be considered as long-term experiments under 'natural' conditions. The answers are there, they have to be systematized. (This is the job of history and historians.) But: there are no absolutely right answers ('Anything goes').

I note Dutour's comments: The skeletal evidence is the weakest argument, as skeletons expose only those afflictions that do not affect human reproduction (more or less). I would doubt any prediction from skeletons, as palaeopathology—even after almost 100 years—has not even tried to develop a model to predict the basic burden of morbidity in a historic population from skeletal materials. You can predict from the ratios of males and females and children the whole demography of such a population, as you rely on socio-economic prerequisites that can be found empirically. But there have been no models offered or discussed about the morbidity burden of a given ancient population from skeletal evidence. (And if there was, it could not be successful.)

Diseases in any culture and at any time have a strong 'social' implication. The more money, the easier it is to cope with a disease, to put it simplistically. I miss those thousands of papers in anthropology, epidemiology, and palaeopathology where 'social' aspects (real features and not those superficial general statements) are truly linked with reports on diseases and historical societies. What we have are thousands of case studies, but no general theory or a framework on that … We have some papers by historians, who do not know anything about the biosciences.

So: I am very much with Bruce Rothschild's comments! This issue is a strong argument to use molecular probes for ancient pathogens.

The comment on the cemeteries is poor. Not the existence of studies makes them plausible, but the plausibility makes them. At least to a certain extent cemeteries are within the number of myths physical anthropologists have successfully created. If you insistently ask what is the 'biology' of past people that can be reconstructed from graveyards: you will finally end up with just a very few and very basic self evidences. (Shakespeare was an early critic of anthropology in *Much Ado About Nothing*.)

I think Baum is right.

17.3 Are our techniques advancing?

Chuck Greenblatt: While there are many articles dealing with improving aDNA extraction methods, a little on repair, quite a bit on scavengers for PCR inhibitors, a few are trying to repair aDNA, and the use of *N*-phenyl thiazolium bromide (PTB) on breaking sugar–protein bonds is being reported. However, I'm puzzled by the fact that aDNA scientists—especially those doing ancient pathogen research—aren't yet using real time PCR, microarray analysis, multiplex PCR, etc.

Carney Matheson: The analysis of ancient or degraded DNA has always been limited by the inherent damage in the material and inhibitors that come with it. However, this limitation is beginning to be realized as a source of the methods and the lack of the ability these methods have in overcoming the inhibition and recovering the highly degraded and fragmented DNA. The field of ancient DNA in the mid 1990s had then reached a hiatus where there was little improvement of methods and more or less an expansion of applications. However, this was simply the application of the standard methods to any and every different type of tissue. It is these errors of routinely applying the same method, that without an in-depth understanding of the differences of each tissue in the different forms of inhibition that may be present or the different forms of taphonomy that each tissue has undergone, that has led to difficulties in recovery. The improvement of methodologies began with Poinar and the introduction of PTB to methodologies in an attempt to overcome the different forms of inhibition present in coprolites. Investigation has now begun seriously in these different protocols. Biochemists and chemists are now providing input in improving the extraction methods themselves to increase the sensitivity and recovery of the highly fragmented DNA and accommodate these differences. Research teams are also looking into the possibility of repairing the ancient or degraded DNA in the hope of cloning extinct species, like the Tasmanian tiger or the woolly mammoths. However, with all the inherent problems of correct ligation and replication these projects have yet to produce great results. But with further improvements in the artificial construction of chromosomes there may not be the need to repair, just the need to sequence the genetic material on the genome level, which for fragmented DNA would require a lot of time and a lot of sample. Improvement in this field as an outcome of the human genome project is already well on its way and the incorporation of microarrays and increased throughput and speed would only bring that possibility closer.

Bern Herrmann: Susanne Hummel has almost finished a book on improved handling and processing of aDNA (Hummel, 2002). What you are puzzled about is the strange phenomenon that most of the aDNA fellows are from microbiology and are not very much interested in thinking about experimental strategies, which is very important in the aDNA business. Look, for example, at the two laboratory strategies or the authentication by amino acids, both nonsense strategies, but people tend to follow names instead of using their own brains. There is also a lot of nonsense about PCR itself in the literature, including rubbish from the inventor himself ('PCR is magic' and he means it).

You are absolutely right to complain about the dismissing of multiplex PCRs, which is the only way to authenticate AND to investigate at the same time. About the others I do not know, and I have doubts about the benefits of real time PCR.

BUT: why go for advanced techniques if the 'old fashioned' ones are not even being tested to their limits? For example, we learned that many extracts contain specific aDNA and we did not realize it, because we had too much DNA in our Eppendorfs. So by simply lowering the amount of target DNA one often gets improved results. I wonder if all such knowledge has been systematically applied instead of going for instant success by using new techniques.

Helen Donoghue: Real time PCR is starting to be used by aDNA workers, at least those looking at mycobacteria. I'm aware that Mike Taylor at Imperial College London has been using real time PCR, both to optimize the amplification conditions and to examine samples. We have also just made a start with real time PCR. It is useful for optimizing individual PCRs and for comparing the amplification efficiency of different reactions, thus enabling meaningful comparisons to be made between results from archaeological material. The data produced can be converted to copy numbers per unit volume, probably more accurately than by the use of gels for densitometric analysis. A real bonus is that the data can be read directly from the sealed plate or tube, thus avoiding the possibility of cross-contamination at this stage.

The idea of using microarrays is very attractive. However, there are some very practical difficulties to be overcome. In order to compare an aDNA sequence with one from the present day, ideally the total genome sequence of the comparator, e.g. human or microbial, must be known. Working with pathogenic mycobacteria, we are fortunate that this has been completed for *Mycobacterium tuberculosis* (two strains), *M. bovis* (wild-type and BCG), and *Mycobacterium leprae*.

However, the present state of the technology requires a minimum of 1–2 μl of DNA per array, so it is difficult to envisage its use without amplification, thereby creating inevitable bias. The application of nanotechnology (Park *et al.* 2002) may make this approach feasible in the future.

Multiplex PCR has certainly been used to examine some of the excellently preserved material from Vác. For example, Mike Taylor has examined samples for the presence of molecular markers that can differentiate between strains. There may be methodological problems as there would be a need to find common conditions optimal for different reactions. In addition, I can see the dangers of doing this where the total genomic sequences of the organism under question are not known, as it could prove impossible to distinguish between chimeric PCR products and sequences obtained from environmental microorganisms.

Carney Matheson: To add to Helen's comments on multiplexing. I have done autosomal multiplexing on six Mayan Copan samples and on Y chromosomal samples and it has been working. Hans Netter from Monash University Melbourne is working on using microarrays to examine viruses in archaeological material but I don't know whether he has any results yet from archaeological material. Also, I realized that for MTB total genome microarrays you need 2–3 μg DNA/array!

H.-D. Gortz and Rolf Michel: With the availability of fluorescence *in situ* hybridization (FISH) and PCR we now have the tools for easy detection of potential pathogens. Appropriately designed probes enable detection of bacteria being more or less closely related to known pathogens.

Chuck Greenblatt: I vaguely remember that Bernd Herrmann's group hybridized aDNA PCR products to chromosome preparations. Spoligotyping also uses fluorescent PCR primers and then hybridizes the products to membrane bound oligonucleotides. Although not exactly FISH there is an analogy.

Mark Spigelman: Just one more point. Before one criticises pathogen research from antiquities one should know more than just the basics of bacteriology (and perhaps of biology as Bernd Hermann has noted above). Many of the critics suggest we should do studies such as amino acid racermization and other surrogate measures of DNA breakdown, not realizing that there are vast differences between DNA preservation between prokaryotes and eukarotes. The same rules that apply to one may not apply to the other. Bacteria have been around for some billions of years longer then nucleated cells so they have learned better how to preserve and protect their DNA.

17.4 Lessons learned

Chuck Greenblatt: Starting with genomics—I wish to tell a few stories. More than 30 years ago, Ed Korn and I discovered a peculiarity in the fat metabolism of *Leishmania*. The protozoa were desaturating fatty acids as if they possessed a chloroplast. Others established in the ensuing years that indeed a number of these early eukaryotes, possessed remnants of their heritage from the photosynthetic bacteria and these were still around and functioning. It has become clear in the last few years that protists other than the *Leishmania*, including a number of pathogens—trypanosomes, malaria, and toxoplasma were 'experimenting' with their subcellular organelles. Fred Opperdoes will shortly be in press summarizing the data from the trypanosomes and *Leishmania*. In the 'Overview' I expended considerable space telling of how endosymbionts were acquired, and contributed to the evolution of the eukaryotic cell. Only now is the potential of treating protozoal diseases with 'herbicides' being realized. In the *Leishmania* and apicomplexa (malaria and toxoplasmosis) treatments are being devised that hit at the 'plant lineages' of the causative agents. This is what I meant when I wrote that although our genome constitutes a single entity, if one pulls apart the 'gene trees' one finds that we and our pathogens have in some cases combined separate lineages and these can be exploited for our benefit in chemotherapy.

A similar, but more ecological approach can be taken by looking more closely at the vectors that carry disease. If pathogens carry evolutionary baggage so do vectors. Perhaps these behavioral and metabolic traits can be used to block disease transmission. William Black told us of mouthparts of many arthropods that were pre-adapted to biting people or animals through evolution as plant feeders. With that in mind, Yosef Schlein in our laboratory has taken advantage of this behavior, by checking out the plant preferences of sandflies, the vectors for leishmaniasis, and finding among those preferred plants natural substances, such as lectins, that are toxic to *Leishmania*. A cordon around villages of these plants, from which the sandfly vectors take sugar meals, might possibly interrupt transmission of the parasite. There are so many other vector-transmission breaking ideas, I'm afraid to open the door too wide. Many have noted that certain natural infections in vectors prevent superinfection; there has been speculation in the malaria field about genetically engineering resistance to plasmodium into mosquitoes; energy has gone into vaccinating hosts against vector mouthparts, etc., etc.

Bill Black: I've read what Ewald says about transmission blocking. Lew Miller was probably talking about another thing in regard to transmission blocking via transgenic vectors. This is an extremely complicated topic. The bottom line at this point in time is that no one has an autonomously replicating transposable element that could be used in driving genes into natural populations of vector species. Even if we did, few agree on the types of genes that should be used to render a vector species incompetent.

Chuck Greenblatt: Paul Ewald, who has so influenced this volume through his insights into pathogen evolution has gone a step further in interpreting what would happen to the poor pathogen trapped into a dead end. This approach says that if you can block transmission of virulent parasites there is a feedback on the transmission cycle so that less virulent parasites take over. Looking at malaria in situations where the human host is screened off from the vector, studies have shown that the parasite becomes less virulent—or perhaps better, less virulent parasites fill the niche.

Ewald described two other related concepts which I found exciting. He cited work in Guatemala, Peru, and Chile. Following outbreaks of cholera in these countries, the toxicogenic forms of the bacteria quickly disappeared in Chile, but in Guatemala they continued to flourish. The answer, he argues, lies in the differing sanitary conditions (toilets in Chile). Organisms that go from a very sick patient's dirty clothes and bedpans directly into the drinking water remain virulent. A second notion deals with antibiotic treatment of zoonotic infections. Vancomycin is being used by the veterinarians and the physicians, so resistant organisms able to infect both animal and human host are selected. Ewald says it would be wise to separate the animal reservoir from the human host. For example, people might receive antibiotic X, which prevents bacterial wall synthesis, and animals would be treated with antibiotic Y, which messes up ribosomal function. Thus the organisms head off in different directions.

Ewald says in the same way vaccines should target virulence factors rather than the microbe itself. An elegant example is the recent work of Naomi Balaban who has argued that it may be possible to intervene at the level where organisms signal to each other. Staphylococci have a system of RAP protein and RIP peptides, which control toxin formation. Balaban has made vaccines that target the RAP or therapy using RIP, and the organisms fail to produce toxins and kill their host. This seems like a very clever approach in which one targets the 'turning on and off' of the virulence system, but not the bacterium, which rather than becoming a killer is happy to be non-pathogenic normal flora (Balaban *et al.* 1998).

Mark Spigelman: From my days when I was a full-time surgeon I often wondered why surgeons, instead of scrubbing their hands with bactericidal soaps, which after all kill a lot of harmless commensals on their skin, do not dip their hand into yogurt or some other cocktail of harmless bacteria and make sure that the skin is covered with bacteria that are not likely to be harmful to the recipient or patient. Also to make sure that there is little or no room for the more harmful bacteria to settle on the surgeon's skin.

Similarly pre-operative surgical skin preparation may well be a 'yogurt wash', and certainly it is not far fetched to say that the way to attack the pathogenic bacterium causing wound infection could be perhaps to introduce a less toxic or aggressive strain to restrict the pathogenic bacterium from growing. If you try to kill the pathogens you are merely selecting the bacterium with the most resistant genes.

Here is another example of intervention in the host–parasite relationship 'without killing but disarming the enemy'.

From Raul Cano's chapter: 'the arthropod–microbe interaction is an essential one for the survival and development of the pathogen within its host. In most instances, one or more microbial genes are expressed only as a result of the interaction between the microbe and the arthropod. From these reports it can be extrapolated that most if not all human-pathogenic microbes with a life cycle that includes an arthropod host, possess essential genes that allow them to survive or interact within their arthropod host. [see Chapter 5] If detected, these genes could represent ideal targets for the development of antimicrobial agents through rational drug design or through natural product mechanism-based screens. Through the use of ancient DNA or microorganisms entombed in amber inclusions it could be possible not only to detect these targets, but to assess the amount of molecular change through time and to identify conserved regions within the gene that might be suitable targets for vaccines or chemotherapeutic agents.'

Chuck Greenblatt to Paul Ewald: Paul, you make much of 'transmission blocking' in your new book *Plague Time* (Ewald 2000) screening up malaria patients, safe water and toilets, safer sex, and staying home—do you really think the powers-that-be can be convinced of the cost-effectiveness of these measures?

Paul Ewald: Intervening to control the evolution of virulence is definitely not yet a mainstream approach, but the lack of attention to this possibility seems to be more attributable to oversight than to evaluation of the evidence. The evidence suggests not only that evolutionary control of virulence is feasible, but that it has already happened when interventions disfavored virulent variants. One example is the evolutionary response to the blocking of water-borne transmission. Another is the evolutionary response to diphtheria vaccination. The main hurdle will be to get people to invest money in intervention studies. If the money is made available, the research will get done. If the research demonstrates decisive amelioration of disease through evolutionary control of virulence, such measures will probably be enacted. The history of medicine shows that the transition from radical proposition to mainstream approach requires a major success story. This was true for the acceptance of antibiotics, hygienic improvements, and vaccination, and I expect that it will be true for evolutionary management of pathogen virulence. Unfortunately the evidence from interventions such as the diphtheria vaccination program don't seem to count; the evolutionary interpretations of these successes are inappropriately dismissed as 'just-so stories'. This dismissive mindset seems to owe more to human psychology than to scientific analysis. Perhaps this state of affairs is not surprising given that our minds evolved for social competition, hunting, and gathering, rather than for scientific inquiry.

References

Abbott, A. (2001). Earliest malaria DNA found in Roman baby graveyard. *Nature* **412**: 847.

Abebe, M., Cupp, M. S., Champagne, D., and Cupp, E. W. (1995). Simulidin: a black fly (*Simulium vittatum*) salivary gland protein with anti-thrombin activity. *Journal of Insect Physiology* **41**: 1001–6.

Acha, P. N. and Szyfres, B. (ed.) (1987). *Zoonoses and communicable diseases common to man and animals* (2nd edn). Publication 503, Pan American Health Organization, Washington, DC.

Acha, P. N. and Szyfres, B. (ed.) (2001). *Zoonoses and communicable diseases common to man and animals* (4th edn). Pan American Health Organization, Washington, DC.

Achtman, M., Zurth, K., Morelli, G., Torrea, G., Guiyoule, A., and Carniel, E. (1999). *Yersinia pestis*, the cause of plague, is a recently emerged clone of *Yersinia pseudotuberculosis*. *Proceedings of the National Academy of Sciences, USA* **96**: 14043–8.

Agarwal, A., Guindo, A., Cissoko, Y., Taylor J. G., Coulibaly, D., Kone, A., *et al.* (2000). Hemoglobin C associated with protection from severe malaria in the Dogon of Mali, a West African population with a low prevalence of hemoglobin S. *Blood* **1**: 2358–63.

Aikawa, M., Suzuki, M., and Gutierrez, Y. (1980). Pathology of malaria. In: *Malaria* (ed. J. P. Kreier). Academic Press, New York: 47–102.

Albuquerque, C. M. and Ham, P. J. (1995). Concomitant malaria (*Plasmodium gallinaceum*) and filaria (*Brugia pahangi*) infections in *Aedes aegypti*: effect on parasite development. *Parasitology* **110**: 1–6.

Allen, P. L. (2000). *The wages of sin: sex and disease, past and present*. University of Chicago Press, Chicago, and London.

Allen, T. M., O'Connor, D. H., Jing, P., Dzuris, J. L., Mothe, B. R., Vogel, T. U., *et al.* (2000). Tat-specific cytotoxic T lymphocytes select for SIV escape variants during resolution of primary viraemia. *Nature* **407**: 386–90.

Allison, A. C. (1954). Protection afforded by sickle-cell trait against subtertian malarial infection. *British Medical Journal* **1**: 290–4.

Allison, M., Hossanni, A., Munizaga, J., and Fung, R. (1978). ABO blood groups in Chilean and Peruvian mummies. II. Results of agglutination–inhibition technique. *American Journal of Physical Anthropology* **49**: 139.

Alm, R. A., Stroeher, U. H., and Manning, P. A. (1988). Extracellular proteins of *Vibrio cholerae*: nucleotide sequence of the structural gene (hlyA) for the haemolysin of the haemolytic El Tor strain O17 and characterization of the hlyA mutation in the non-haemolytic classical strain 569B. *Molecular Microbiology* **2**: 481–8.

Almirón, H., Link, A., Furlong, D., and Kolter, R. (1992). A novel DNA binding protein with regulatory and protective roles in starved *Escherichia coli*. *Genes and Development* **6**: 2646–54.

Amann, R., Springer, N., Ludwig, W., Görtz, H.-D., and Schleifer, K.-H. (1991). Identification *in situ* and phylogeny of uncultured bacterial endosymbionts. *Nature* **351**: 161–4.

Amann, R., Springer, N., Schönhuber, W., Ludwig, W., Schmid, E. N., Müller, K.-D., *et al.* (1997). Obligate intracellular bacterial parasites of acanthamoebae related to *Chlamydia* spp. *Applied and Environmental Microbiology* **63**: 115–21.

Ancient Biomolecules Initiative (ABI) *Newsletter* (1996). November, Issue No. 3.

Andersen, J. G. and Manchester, K. (1992). The rhinomaxillary syndrome in leprosy: a clinical radiological and paleopathological study. *International Journal of Osteoarchaeology* **2**: 121–9.

Angel, J. L. (1966). Porotic hyperostosis, anemias, malarias and the marshes in prehistoric eastern Mediterranean. *Science* **153**: 760–2.

Ardagna, Y. (2000). Method for re-evaluating the prevalence of spinal diseases in human paleopathology. Poster session: 13th European meeting of the Paleopathology Association, Chieti, Italy, 18–23 September.

Argentine, J. A. and James, A. A. (1995). Characterization of a salivary gland-specific esterase in the vector mosquito, *Aedes aegypti*. *Insect Biochemistry and Molecular Biology* **25**: 621–30.

Arriaza. B. T. (1993). Seronegative spondyloarthropathies and diffuse idiopathic skeletal hyperostosis in ancient northern Chile. *American Journal of Physical Anthropology* 91: 273–8.

Arriaza, B., Salo, W., Aufderheide, A., and Holcomb, T. (1995). Pre-Columbian tuberculosis in Northern Chile: molecular and skeletal evidence. *American Journal of Physical Anthropology* 98: 37–45.

Asquith, R. S. and Leon, N. H. (1977). Chemical reactions of keratin fibres. In: *Chemistry of natural protein fibres* (ed. R. S. Asquith). Plenum Press, New York.

Aufderheide, A. C. and Rodriguez-Martin, C. (1998). *The Cambridge encyclopedia of human paleopathology*. Cambridge University Press, Cambridge.

Austin, J. J., Ross, A. J., Smith, A. B., Fortey, R. A., and Thomas, R. H. (1997). Problems of reproducibility—does geologically ancient DNA survive in amber-preserved insects? *Proceedings of the Royal Society of London B, Biological Sciences* 264: 467–74.

Bada, J. (1985). Amino acid racimization dating of fossil bones. *Annual Review of Earth and Planetary Science* 13: 241–68.

Bada, J. L., Wang, X. S., and Hamilton, H. (1999). Preservation of key biomolecules in the fossil record: current knowledge and future challenges. *Philosophical Transactions of the Royal Society of London B, Biological Sciences* 354: 77–87.

Baker, B. J. (1999). Early manifestations of tuberculosis in the skeleton. In: *Tuberculosis past and present* (ed. G. Pálfi, O. Dutour, J. Deák, and I. Hutás). Golden Book Publishers and the Tuberculosis Foundation, Budapest–Szeged: 301–10.

Baker, B. J. and Armelagos, G. J. (1988). The origins and antiquity of syphilis. *Current Anthropology* 29: 703–20, 732–7.

Balaban, N., Goldkorn, T., Nhan, R. T., Dang, L. B., Scott, S., Rasooly, A., *et al.* (1998). Autionducer of virulence as a target for vaccine and therapy against *Staphylococcus aureus*. *Science* 280: 438–40.

Ball, G. H. (1943). Parasitism and evolution. *American Naturalist* 77: 345–64.

Bannister, B. A., Begg, N. T., and Gillespie, S. H. (2000). *Infectious disease* (2nd edn). Blackwell, Oxford.

Barbaree, J. M., Fields, B. S., Feeley, J. C., Gormann, G. W., and Martin, W. T. (1986). Isolation of protozoa from water associated with a legionellosis outbreak and demonstration of intracellular multiplication of *Legionella pneumophila*. *Applied and Environmental Microbiology* 51: 422–4.

Barer, M. R., Gribbon, L. T., Harwood, C. R., and Nwoguh, C. E. (1998). The viable but non-culturable hypothesis and medical bacteriology. *Reviews in Medical Microbiology* 4: 183–91.

Barillas-Mury, C., Charlesworth, A., Gross, I., Richman, A., Hoffmann, J. A., and Kafatos, F. C. (1996). Immune factor Gambif1, a new rel family member from the human malaria vector, *Anopheles gambiae*. *EMBO Journal* 15: 4691–701.

Barillas-Mury, C., Han, Y. S., Seeley, D., and Kafatos, F. C. (1999). *Anopheles gambiae* Ag-STAT, a new insect member of the STAT family, is activated in response to bacterial infection. *EMBO Journal* 18: 959–67

Barker, J. and Brown, M. R. W. (1994). Trojan horses of the microbial world: protozoa and the survival of bacterial pathogens in the environment. *Microbiology* 140: 1253–9.

Barnes, I., Holton, J., Vaira, D., Spigelman, M., and Thomas, M. (2000). An assessment of the long-term preservation of the DNA of a bacterial pathogen in ethanol-preserved archival material. *Journal of Pathology* 192: 554–9.

Baron, H., Hummel, S., and Herrmann, B. (1996). *Mycobacterium tuberculosis* complex DNA in ancient human bones. *Journal of Archaeological Sciences* 23: 667–71.

Barua, D. (1992). History of cholera. In: *Cholera* (ed. D. Barua and W. B. Greenough III). Plenum Medical, New York: 1–36.

Basler, C. F., Reid, A. H., Dybing, J. K., Janczewski, T. A., Fanning, T. G., Zheng, H., *et al.* (2001). Sequence of the 1918 pandemic influenza virus nonstructural gene (NS) segment and characterization of recombinant viruses bearing the 1918 NS genes. *Proceedings of the National Academy of Sciences, USA* 98: 2746–51.

Bass, W. M., Gregg, J. B., and Provost, P. E. (1974). Ankylosing spondylitis (Marie Strumpel disease) in historic and prehistoric Northern Plains Indians. *Plains Anthropologist* 19: 303–5.

Behrensmeyer, A. K. (1978). Taphonomic and ecological information from bone weathering. *Paleobiology* 4: 150–62.

Bello, S. (2000). Taphonomie des restes osseux humains. Effet des processus de conservation du squelette sur les paramètres anthropologiques. PhD thesis (biological anthropology), Université de la Méditerranée, Faculté de Médecine de Marseille et Universita, degli studi di Firenze.

Bello, S., Signoli, M., Maczel, M., and Dutour, P. (1999). Evolution of mortality due to tuberculosis in France. In: *Tuberculosis past and present* (ed. G. Pálfi, O. Dutour, J. Deák, and I. Hutás). Golden Book Publishers and the Tuberculosis Foundation, Budapest–Szeged: 93–104.

Beran, G. and Steele, J. H. (1994). *CRC handbook of zoonoses (bacterial and viral)*. CRC Press, Boca Raton, FL.

Bérato, J., Dutour, O., Williams, J., Zakarian, H., and Acquaviva, P. C. (1990). Epidémiologie des affections rhumatismales dans une population antique. Etude de la nécropole du Haut-Empire de Saint-Lambert (Fréjus, var). *Revue du Rhumatisme* **57**: 397–400.

Berche, P., Poyart, C. Abachin, E., Lelievre, H., Vandepitte, J., and Dodin, A. (1994). The novel epidemic strain O139 is closely related to the pandemic strain O1 of *Vibrio cholerae. Journal of Infectious Disease* **170**: 701–4.

Berry, W. J., Rowley, W. A., and Christensen, B. M. (1986). Influence of developing *Brugia pahangi* on spontaneous flight activity of *Aedes aegypti* (Diptera: Culicidae). *Journal of Medical Entomology* **23**: 441–5.

Berry, W. J., Rowley, W. A., and Christensen, B. M. (1987a). Influence of developing *Dirofilaria immitis* on the spontaneous flight activity of *Aedes aegypti* (Diptera: Culicidae). *Journal of Medical Entomology* **24**: 699–701.

Berry, W. J., Rowley, W. A., Clarke, J. L., III, Swack, N. S., and Hausler, W. J., Jr (1987b). Spontaneous flight activity of *Aedes trivittatus* (Diptera: Culicidae) infected with trivittatus virus (Bunyaviridae: California serogroup). *Journal of Medical Entomology* **24**: 286–9.

Berry, W. J., Rowley, W. A., and Christensen, B. M. (1988). Spontaneous flight activity of *Aedes trivittatus* infected with *Dirofilaria immitis. Journal of Parasitology* **74**: 970–4.

Billingsley, P. P. (1990). The midgut ultrastructure of hematophagous insects. *Annual Review of Entomology* **35**: 219–48.

Binks, R. H., Baum, J., Oduola, A. M., Arnot, D. E., Babike, H. A., Kremsner, P. G., *et al.* (2001). Population genetic analysis of the *Plasmodium falciparum* erythrocyte binding antigen-175 (*eba-175*) gene. *Molecular and Biochemical Parasitology*, **114**: 63–70.

Biraben, J. N. (1968). Certain demographic characteristics of the plague epidemic in France 1720–22. *Dedalus* **97**: 536–45.

Birtles, R. J., Rowbotham, T. J., Michel, R., Pitcher, D. G., Lascola, B., Alexiou-Daniel, S., *et al.* (2000). 'Candidatus *Odysella thessalonicensis*' gen.nov., sp.nov., an obligate intracellular parasite of *Acanthamoeba* species. *International Journal of Systematic and Evolutionary Microbiology* **50**: 63–72.

Black, F. L., Hierholzer, F., Pinheiro, A. S., Evans, J. P., Woodall, E. M., Opton, J. E., *et al.* (1974). Evidence for persistence of infectious agents in isolated human populations. *American Journal of Epidemiology* **100**: 230–50.

Black, F. L., Pinheiro, F. D. P., Hierholzer, V., and Lee, V. (1977). Epidemiology of infectious diseases: the example of measles. In: *Health and disease in tribal societies*. CIBA Foundation Symposium 49. Elsevier, New York: 115–35.

Blancou, P., Vartanian, J. P., Christopherson, C., Chenciner, N., Basilico, C., Kwok, S., *et al.* (2001). Polio vaccine samples not linked to AIDS. *Nature* **410**: 1045–6.

Bleker, J. (1983). Der gefaehrdete Koerper und die Gesellschaft. Ansaetze zu einer sozialen Medizin zur Zeit der buergerlichen Revolution in Deutschland. In: *Der Mensch und sein Koerper* (ed. A. E. Imhof). C. H. Beck, Munich: 226–42.

Blondiaux, J. and Stanford, J. (1995). Correspondence. *International Journal of Osteoarchaeology* **5**: 299–301.

Bocquet-Appel, J. P. and Masser, C. (1982). Farewell to paleodemography. *Journal of Human Evolution* **11**: 321–33.

Boesch, C. and Ackerman, H. B. (2000). *The chimpanzees of the Tai Forest*. Oxford University Press, New York.

Bogdan, G. and Weaver, D. S. (1992). Pre-Columbian treponematosis in coastal North Carolina. In: *Disease and demography in the Americas* (ed. J. W. Verano and D. H. Ubelaker). Smithsonian Institution Press, Washington, DC: 155–63.

Boldsen, J. L. (2001). Epidemiological approach to the paleopathological diagnosis of leprosy. *American Journal of Physical Anthropology* **115**: 380–7.

Boom, R., Sol, A., Salimans, M., Jansen, C., Wertheim-van Dillen, P., and van der Noordaa, J. (1990). Rapid and simple method for purification of nucleic acid. *Journal of Clinical Microbiology* **28**: 495–503.

Borsky, I., Hermanek, J., Uhlir, J., and Dusbabek, F. (1994). Humoral and cellular immune response of BALB/c mice to repeated infestations with *Ixodes ricinus* nymphs. *International Journal of Parasitology* **24**: 127–32.

Bowman, A. S., Dillwith, J. W., and Sauer, J. R. (1996). Tick salivary prostaglandins: presence, origin and significance. *Parasitology Today* **12**: 388–96.

Bramblett, C. (1994). *Patterns of primate behavior*. Waveland Press Inc., Prospect Heights, IL.

Braun, M., Cook, D. C., and Pfeiffer, S. (1998). DNA from *Mycobacterium tuberculosis* complex identified in North American, pre-Columbian skeletal remains. *Journal of Archaeological Sciences* **25**: 271–7.

Brennan, P. J. and Nikaido, H. (1995). The envelope of mycobacteria. *Annual Review of Biochemistry* **64**: 29–63.

Briggs, D. E. G. (1999). Molecular taphonomy of animal and plant cuticles: selective preservation and diagenesis. *Philosophical Transactions of the Royal Society of London B, Biological Sciences* **354**: 7–17.

Brooks, D. R. and McLennan, D. A. (1991). *Phylogeny, ecology, and behavior: a research program in comparative biology*. University of Chicago Press, Chicago, and London.

Brooks, D. R. and McLennan, D. A. (1993). *Parascript: parasites and the language of evolution*. Smithsonian Institution Press, Washington, DC.

Brosch, R., Gordon, S.V., Marmiesse, M., Brodin, P., Buchrieser, C., Eiglmeier, K., Garnier, T., *et al.* (2002). A new evolutionary scenario for the *Mycobacterium tuberculosis* complex. *Proceedings of the National Academy of Sciences, USA* **99**: 3684–9.

Brossard, M., Monneron, J. P., and Papatheodorou, V. (1982). Progressive sensitization of circulating basophils against *Ixodes ricinus* L. antigens during repeated infestations of rabbits. *Parasite Immunology* **4**: 335–61.

Brown, K. A., O'Donaghue, K., and Brown, T. A. (1995). DNA in cremated bones from an early Bronze Age cemetery cairn. *International Journal of Osteoarchaeology* **5**: 181–7.

Brown, S. J. (1982). Antibody- and cell-mediated immune resistance by guinea pigs to adult *Amblyomma americanum* ticks. *American Journal of Tropical Medicine and Hygiene* **31**: 1285–90.

Broza, M. and Halpern, M. (2001). Pathogen reservoirs. Chironomid egg masses and *Vibrio cholerae*. *Nature* **412**: 40.

Bruintjes, T. D. and Panhuysen, R. G. (1995). The paleopathological diagnosis of seronegative spondyloarthropathies. Proceedings of the Ninth European Meeting of the Paleopathology Association, Museu D'Arqueologica de Catalunya: 73–7.

Buikstra, J. (ed.) (1981). *Prehistoric tuberculosis in the Americas*. Northwestern University Archeological Program, Evanston, IL.

Buikstra, J. E. and Ubelaker, D. H. (ed.) (1994). *Standards for data collection from human skeletal remains—proceedings of a seminar at the Field Museum of Natural History*. Arkansas Archeological Survey Research Series No. 44, Arkansas Archeological Survey, Fayetteville, AR.

Buikstra, J. E., Konigsberg, L. W., and Bullington, J. (1986). Fertility and the development of agriculture in the prehistoric Midwest. *American Antiquity* **51**: 528–46.

Buka, S. L., Tsuang, M. T., Torrey, E. F., Klebanoff, M. A., Bernstein, D., and Yolken, R. H. (2001). Maternal infections and subsequent psychosis among offspring. *Archives of General Psychiatry* **58**: 1032–7.

Buonagurio, D., Nakada, S., Parvin, J., Krystal, M., Palese, P., and Fitch, W. (1986). Evolution of human influenza A viruses over 50 years: rapid, uniform rate of change in NS gene. *Science* **232**: 980–2.

Burgos, J. D., Correal-Urrego, G., and Arregoces, C. (1994). Treponematosis en restos oseos preceramicos de Columbia. *Revista de la Academia Colombiana de Ciencias Exactas, Fisicas y Naturales* **19**: 273–81.

Burnet, F. M. and White. D. O. (1972). *Natural history of infectious disease* (4th edn). Cambridge University Press, Cambridge.

Campillo, D. (1993). *Paleopathología: los primeros vestigios de la enfermedad* (Part I). Fundació Uriach, Barcelona.

Campillo, D. (1994). *Paleopathología: los primeros vestigios de la enfermedad* (Part II). Fundació Uriach, Barcelona.

Campillo, D. (1996). Paleopathology, predisposing causes, stress and pathocenosis. In: *Notes on populational significance of paleopathological conditions: health, illness and death in the past* (ed. A. Pérez-Pérez). Fundació Uriach, Barcelona: 67–75.

Cano, R. J. (1994). *Bacillus* DNA in amber—a window to an ancient symbiosis? *American Society for Microbiology News* **60**: 129–34.

Cano, R. J. (1997). DNA amplification techniques in fossilized samples. In: *Environmental applications of nucleic acid amplification techniques* (ed. G. A.Toranzos). Technomic Publishing Co., Lancaster, PA: 183–201.

Cano, R. J. (1998). The microbiology of amber. In: *Digging for pathogens: ancient emerging diseases—their evolutionary, anthropological and archaeological context* (ed. C. L. Greenblatt). Balaban Publishers, Rehovot, Israel: 71–95.

Cano, R. J. and Borucki, M. (1995). Revival and identification of bacterial spores in 25- to 40-million-year-old Dominican amber. *Science* **268**: 1060–4.

Cano, R. J., Poinar, H. N., Roubik, D. W., and Poinar, G. O., Jr (1992). Enzymatic amplification and nucleotide sequencing of portions of the 18s rRNA gene of the bee *Proplebeia dominicana* (Apidae: Hymenoptera) isolated from 25–40 million year old Dominican amber. *Medical Science Research* **20**: 619–22.

Cano, R. J., Poinar, H., Pieniazek, N., Acra, A., and Poinar, G. O., Jr (1993). Amplification and sequencing of DNA from a 120- to 143-million-year-old weevil. *Nature* **363**: 536–8.

Cano, R. J., Borucki, M., Higby-Schweitzer, M., Poinar, H., Poinar, G. O., Jr, and Pollard, K. (1994). *Bacillus* DNA in fossil bees: an ancient symbiosis? *Applied and Environmental Microbiology* **60**: 2164–7.

Cano, R. J., Tiefenbrunner, F., Ubaldi, M., Del Cueto, C., Luciani, S., Cox, T., *et al.* (2000). Sequence analysis of bacterial DNA in the colon and stomach of the Tyrolean Iceman. *American Journal of Physical Anthropology* **112**: 297–309.

Carlson, C. R., Caugant, D. A., and Kolsto A. (1994). Genotypic diversity among *Bacilllus cereus* and *Bacillus thuringiensis* strains. *Applied and Environmental Microbiology* **60**: 1719–25.

Castrucci, M. R., Donatelli, I., Sidoli, L., Barigazzi, G., Kawaoka, Y., and Webster, R. G. (1993). Genetic reassortment between avian and human influenza A viruses in Italian pigs. *Virology* **193**: 503–6.

Cattaneo, C., Gelsthorpe, K., Phillips, P., and Sokol, R. J. (1990). Blood in ancient human bone. *Nature* **347**: 339.

Cattaneo, C., DiMartino, S., Scali, S., Craig, O. E., Grandi, M., and Sokol, R. J. (1999). Determining the human origin of fragments of burnt bone: a comparative study of histological, immunological and DNA techniques. *Forensic Science International* **102**: 181–91.

Cerutti, N., Marin, A., Massa, E. R., and Savoia, D. (1999). Immunological investigation of malaria and new perspectives in paleopathological studies. *Bolletino Soceita Italiana di Biologia Sperimentale* **75**: 17–20.

Cha, Z. (1989). Deamidation of bovine calbindin. *Biochemistry* **28**: 8646–53.

Chakraborty, S., Mukhopadhyay, A. K., Bhadra, R. K., Ghosh, A. N., Mitra, R., Shimada, T., *et al.* (2000). Virulence genes in environmental strains of *Vibrio cholerae. Applied and Environmental Microbiology* **66**: 4022–8.

Champagne, D. E. and Ribeiro, J. M. (1994). Sialokinin I and II: vasodilatory tachykinins from the yellow fever mosquito *Aedes aegypti. Proceedings of the National Academy of Sciences, USA* **91**: 138–42.

Champagne, D. E., Nussenzveig, R. H., and Ribeiro, J. M. (1995a). Purification, partial characterization, and cloning of nitric oxide-carrying heme proteins (nitrophorins) from salivary glands of the blood-sucking insect *Rhodnius prolixus. Journal of Biological Chemistry* **270**: 8691–5.

Champagne, D. E., Smartt, C. T., Ribeiro, J. M., and James, A. A. (1995b). The salivary gland-specific apyrase of the mosquito *Aedes aegypti* is a member of the 5′-nucleotidase family. *Proceedings of the National Academy Sciences, USA* **92**: 694–8.

Chandrasekhar S. and Ratnam S. (1992). Studies on cell-wall deficient non-acid fast variants of *Mycobacterium tuberculosis. Tubercle and Lung Disease* **73**: 273–9.

Chang, Y., Cesarman, E., Pessin, M., Lee, F., Culpepper, J., Knowles, D., and Moore, P. (1994). Identification of herpes virus-like DNA sequences in AIDS-associated Kaposi's sarcoma. *Science* **266**: 1865–69.

Charlab, R. and Ribeiro, J. M. (1993). Cytostatic effect of *Lutzomyia longipalpis* salivary gland homogenates on *Leishmania* parasites. *American Journal of Tropical Medicine and Hygiene* **48**: 831–8.

Charlab, R., Rowton, E. D., and Ribeiro, J. M. (2000). The salivary adenosine deaminase from the sand fly *Lutzomyia longipalpis. Experimental Parasitology* **95**: 45–53.

Chen, C. C. and Laurence, B. R. (1985). The encapsulation of the sheaths of microfilariae of *Brugia pahangi* in the hemocoele of mosquitoes. *Journal of Parasitology* **71**: 834–6.

Chen, H.L, Jakes, K. A., and Foreman, D. W. (1998). Preservation of archaeological textiles through fibre mineralization. *Journal of Archaeological Science* **25**: 1015–21.

Chen, Z., Li, Y., and Krug, R. M. (1999). Influenza A virus NS1 protein targets poly(A)-binding protein II of the cellular 3′-end processing machinery. *EMBO Journal* **18**: 2273–83.

Child, A. M., Gillard, R. D., and Pollard, A. M. (1993). Microbially induced promotion of amino acid racemization in bone: isolation of the micro-organisms and detection of their enzymes. *Journal of Archaeological Science* **20**: 159–68.

Cholera Working Group Ic, Bangladesh (1993). Large epidemic of cholera-like disease in Bangladesh caused by *Vibrio cholerae* O139 synonym Bengal. *Lancet* **342**: 387–90.

Claas, E. C., Osterhaus, A. D., van Beek, R., De Jong, J. C., Rimmelzwaan, G. F., Senne, D. A., *et al.* (1998). Human influenza A H5N1 virus related to a highly pathogenic avian influenza virus. *Lancet* **351**: 472–7.

Clarkson, L. (1975). *Death, disease and famine in pre-industrial England.* Gill & Macmillan, Dublin.

Coatney, G. R., Collins, W. E., McWilson, W., and Contacos, P. G. (1971). *The primate malarias.* United States Government Printing Office, Washington, DC.

Cockburn, A. (1963). *The evolution and eradication of infectious diseases.* Johns Hopkins University Press, Baltimore, MD.

Cockburn, E. (1995). Forty years on: Are Aiden Cockburn's theories still valid. In: *L'Origin de la Syphilis en Europe Avant ou apres 1493* (ed. O. Dutour, G. Palfi, J. Berato, and J.-P. Brun). Centre Archeologique du Var, Toulon, France: 23–6.

Cockburn, T. A. (1963). The origin of treponematoses. *Bulletin of the World Health Organization* **24**: 221–8.

Cohen, M. N. (1989). *Health and the rise of civilization.* Yale University Press, New Haven, CT.

Cohen, M. N. (1996). Does paleopathology reflect prehistoric community health? In: *Notes on populational significance of paleopathological conditions: health, illness and death in the past* (ed. A. Pérez-Pérez). Fundació Uriach, Barcelona: 55–65.

Cohen, M. (1997). Does paleopathology measure community health? In: *Integrating archeological demography* (ed. R. Paine). Center for Archeological Investigation, Carbondale, IL: 242–62.

Cohen, M. N. and Armelagos, G. J. (1984). *Paleopathology at the origins of agriculture.* Academic Press, New York.

Collins, F. H., Sakai, R. K., Vernick, K. D., Paskewitz, S., Seeley, D. C., Miller, L. H., *et al.* (1986). Genetic selection of a *Plasmodium*-refractory strain of the malaria vector *Anopheles gambiae. Science* **234**: 607–10.

Collins, M. J., Westbroek, P., Muyzer, G., and de Leeuw, J. W. (1992). Experimental evidence for condensation reactions between sugars and proteins in carbonate skeletons. *Geochimica Cosmochimica Acta* **56**: 1539–44.

Collins, M. J., Waite, E. R., and van Duin, A. C. T. (1999). Predicting protein decomposition: the case of aspartic-acid racemization kinetics. *Philosophical Transactions of The Royal Society of London B, Biological Sciences* **354**: 51–64.

Collins, S. D. (1931). Age and sex incidence of influenza and pneumonia morbidity and mortality in the epidemic of 1928–1929 with comparative data for the epidemic of 1918–1919. *Public Health Reports* **46**: 1917–37.

Colman, P. M., Varghese, J. N., and Laver, W. G. (1983). Structure of the catalytic and antigenic sites in influenza virus neuraminidase. *Nature* **303**: 41–4.

Coluzzi, M. (1999). The clay feet of the malaria giant and its African roots: hypotheses and inferences about origin, spread and control of *Plasmodium falciparum*. *Parassitologia* **41**: 277–83.

Comstock, L. E., Maneval, D., Jr, Panigrahi, P., Joseph, A., Levine, M. M., Kaper, J. B., *et al.* (1995). The capsule and O antigen in *Vibrio cholerae* O139 Bengal are associated with a genetic region not present in *Vibrio cholerae* O1. *Infection and Immunity* **63**: 317–23.

Comstock, L. E., Johnson, J. A., Michalski, J. M., Morris, J. G., Jr, and Kaper, J. B. (1996). Cloning and sequence of a region encoding a surface polysaccharide of *Vibrio cholerae* O139 and characterization of the insertion site in the chromosome of *Vibrio cholerae* O1. *Molecular Microbiology* **19**: 815–26.

Conway, D. J. and Roper, C. (2000). Micro-evolution and emergence of pathogens. *International Journal for Parasitology* **30**: 1423–30.

Cooper, A. and Poinar, H. N. (2000). Ancient DNA: do it right or not at all. *Science* **289**: 530–1.

Copley, M. S., Rose, P. J., Clapman, A., Edwards, D. N., Horton, M. C., and Evershed, R. P. (2001). Detection of palm fruit lipids in archaeological pottery from Qasr Ibrim, Egyptian Nubia. *Proceedings of the Royal Society of London B, Biological Sciences* **268**: 593–7.

Covacci, A., Telford, J. L., Del Giudice, G., Parsonnet, J., and Rappuoli, R. (1999). *Helicobacter pylori* virulence and genetic geography. *Science* **284**: 1328–33.

Cox, M. (1991). A study of the sensitivity and specificity of four presumptive tests for blood. *Journal of Forensic Sciences* **36**:1503–11.

Cox, N. J. and Subbarao, K. (2000). Global epidemiology of influenza: past and present. *Annual Review of Medicine* **51**: 407–21.

Coy, J. P. (1975). Iron Age cookery. In: *Archaeozoological studies* (ed. A. T. Clason). North-Holland Publishing Company, Amsterdam: 426–30.

Cronje, G. (1984). Tuberculosis mortality decline in England and Wales, 1851–1910. In: *Urban disease and mortality in 19th century England* (ed. R. Woods and J. Woodward). Batsford, London: 79–101.

Crosby, A. W. (1989). *America's forgotten pandemic— the influenza of 1919*. Cambridge University Press, Cambridge.

Crubézy, E., Ludes, B., Poyeda, J. D., Clayton, J., Crouau-Roy, B., and Montagnon, D. (1998). Identification of *Mycobacterium bovis* DNA in an Egyptian Pott's

disease of 5,400 years old. *Comptes Rendus de l'Academie des Sciences, Sciences de la Vie* **321**: 941–51

Curry, G. B. (1987). Molecular palaeontology. *Geology Today* **3**: 12–16.

Daffe, M. and Draper, P. (1998). The envelope layers of mycobacteria with reference to their pathogenicity. *Advances in Microbial Physiology* **39**: 131–203.

Dales, G. F. (1964). The mythical massacre at Mohenjo Daro. *Expedition* **6**: 36–43.

Davies, P. D. O., Humpries, M. J., Byfield, S. P., Nunn, A. J., Darbyshire, J. H., Citron, I. C. M., *et al.* (1984). Bone and joint tuberculosis. A survey of notifications in England and Wales. *Journal of Bone and Joint Surgery British volume* **66**: 326–30.

Davis, S. J. M. (1987). *The archaeology of animals*. Yale University Press, New Haven, CT.

Day, J. F., Thornburg, R. W., Thorpe, S. R., and Baynes, J. W. (1979). Nonenzymatic glucosylation of rat albumin; studies *in vitro* and *in vivo*. *Journal of Biological Chemistry* **254**: 595–7.

Dean, M., Carrington, M., Winkler, C., Huttley, G. A., Smith, M. W., Allikmets, R., *et al.* (1996). Genetic restriction of HIV-1 infection and progression to AIDS by a deletion allele of the CKR5 structural gene. Hemophilia Growth and Development Study, Multicenter AIDS Cohort Study, Multicenter Hemophilia Cohort Study, San Francisco City Cohort, ALIVE Study. *Science* **273**: 1856–62.

DeBoer, S. H. and Sasser M. (1986). Differentiation of *Erwinia caratovora* ssp. *Caratovora* and *Erwinia caratovora* ssp. *Atroseptica* on the basis of cellular fatty acid composition. *Canadian Journal of Microbiology* **32**: 796–800.

Dees, S. B. and Moss, C. W. (1975). Cellular fatty acids of *Alcaligenes* and *Pseudomonas* species isolated from clinical specimens. *Journal of Clinical Microbiology* **1**: 414–19.

Den Boer, J. W., Yzerman, E. P. F., Schellekens, J., Lettinga, K. D., Boshuizen, H. C., Van Steenbergen, J. E., *et al.* (2002). A large outbreak of Legionnaires' disease at a flower show, the Netherlands, 1999. *Emerging Infectious Diseases* **8**: 37–43.

DeSalle, R., Gatesy, J., Wheeler, W., and Grimaldi, D. (1992). DNA sequences from a fossil termite in Oligo-Miocene amber and their phylogenetic implications. *Science* **257**: 1933–6.

Desowitz, R. S. (1991). *The malaria capers: more tales of parasites and people, research and reality*. W. W. Norton & Co., New York.

Desowitz, R. S. (1997). *Who gave pinta to the Santa Maria? Torrid diseases in a temperate world*. W. W. Norton & Co., New York.

Despres, L., Imbert-Establet, D., Combes, C., and Bonhomme, F. (1992). Molecular evidence linking hominid evolution to recent radiation of schistosomes (Platyhelminthes: Trematoda). *Molecular Phylogenetics and Evolution* **1**: 295–304.

Diamond, J. (1997). *Guns, germs and steel: a short history of everybody for the last 13,000 years.* Vintage, London

Distel, D.L, Lee, H. K.-W., and Cavanaugh, C. W. (1995). Intracellular coexistence of methano- and thioautotrophic bacteria in a hydrothermal vent mussel. *Proceedings of the National Academy of Sciences, USA* **92**: 9598–602.

Dixon R., Aveling, E., and Roberts, C. A. (1995). Detection of a mycobacterium tuberculosis specific insertion element from ancient human rib using the polymerase chain reaction. *Ancient DNA III International Conference. Oxford, July 20–22, 1995*: S19.

Dobzhansky, T. (1973). Nothing in biology makes sense except in the light of evolution. *American Biology Teacher* **35**: 125–9.

Donnenberg, M. S. and Whittam, T. S. (2001). Pathogenesis and evolution of virulence in enteropathogenic and enterohemorrhagic *Escherichia coli. Journal of Clinical Investigation* **107**: 539–48.

Donoghue, H. D., Spigelman, M., Zias, J., Gernaey-Child, A. M., and Minnikin, D. E. (1998). *Mycobacterium tuberculosis* complex in calcified pleura from remains 1,400 years old. *Letters of Applied Microbiology* **27**: 265–9.

Donoghue, H., Holton, J., and Spigelman, M. (2001). PCR primers that can detect low levels of *Mycobacterium leprae* DNA. *Journal of Medical Microbiology* **50**: 177–82.

Donoghue, H. D., Gladykowska-Ryyeczycka, J., Marscik, A., Holton, J., and Spigelman, M. (2002). *Mycobacterium leprae* DNA in archaeological specimens. In: *British archaeology reports: past and present of leprosy* (ed. C. Roberts and M. Lewis) **1054**: 271–85.

Doran, G. H., Dickel, D. N., Ballinger, W. E., Jr, Agee, O. F., Laipis, P. J., and Hauswirth, W. W. (1986). Anatomical, cellular and molecular analysis of 8,000-yr-old human brain tissue from the Windover archaeological site. *Nature* **323**: 803–6.

Dowdle, W. R. (1999). Influenza A virus recycling revisited. *Bulletin of the World Health Organization* **77**: 820–8.

Drancourt, M., Aboudharam, G., Signoli, M., Dutour, O., and Raoult, D. (1998). Detection of 400-year-old *Yersinia pestis* DNA in human dental pulp: an approach to the diagnosis of ancient septicemia. *Proceedings of the National Academy of Sciences, USA* **95**: 12637–40.

Dubos, R. (1965). *Man adapting.* Yale University Press, New Haven, CT.

Dutour, O. (1989). *Hommes fossiles du Sahara.* CNRS éditions, Paris.

Dutour, O., Pàlfi, G., Berato, J., and Brun, J. P. (ed.) (1994a). *L'Origin de la syphilis en Europe—avant ou apres 1493?* Centre Archeologique du Var, Toulon, France.

Dutour, O., Signoli, M., Georgeon, E., and Da Silva, J. (1994b). Le charnier de la grande peste de Marseille (rue Leca): données de la fouille de la partie centrale et premiers résultats anthropologiques. *Préhistoire et Anthropologie Méditerranéennes* **3**: 191–203.

Dutour O., Signoli, M., and Pálfi, G. (1998). How can we reconstruct the epidemiology of infectious diseases in the past? In: *Digging for pathogens: ancient emerging diseases—their evolutionary, anthropological, and archaeological context* (ed. C. L. Greenblatt). Balaban Publishers, Rehovot, Israel: 241–63.

Dutour, O., Pàlfi, G., Brun, J. P., Bérato, J., Panuel, M., Hass, C. J., *et al.* (1999). Morphological, paleoradiological and paleomicrobiological study of a French medieval case of tuberculosis spondylitis with cold abscess. In: *Tuberculosis past and present* (ed. G. Pálfi, O. Dutour, J. Deák, and I. Hutás). Golden Book Publishers and the Tuberculosis Foundation, Budapest–Szeged: 395–402.

Edwards, J. F., Higgs, S., and Beaty, B. J. (1998). Mosquito feeding-induced enhancement of Cache Valley virus (Bunyaviridae) infection in mice. *Journal of Medical Entomology* **35**: 261–5.

Eglinton, G. (1998). The archaeological and geological fate of biomolecules. In: *Digging for pathogens: ancient emerging diseases—their evolutionary, anthropological, and archaeological context* (ed. C. L. Greenblatt). Balaban Publishers, Rehovot, Israel: 299–327.

Eglinton, G. and Logan, G. A. (1991). Long term preservation of biomolecules. *Philosophical Transactions of the Royal Society of London B, Biological Sciences* **333**: 315–28.

Eisenach, K. D., Cave, M. D., Bates, J. H., and Crawford, J. T. (1990). Polymerase chain reaction amplification of a repetitive DNA sequence specific for *Mycobacterium tuberculosis. Journal of Infectious Diseases* **161**: 977–81.

El-Najjar, M., Desanti, M. V., and Ozebek, L. (1978). Prevalence and possible etiology of dental enamel hypoplasia. *American Journal of Physical Anthropology* **48**: 185–92.

Emslie, S. D. (1981). Prehistoric agricultural ecosystems: Avifauna from Pottery Mound, New Mexico. *American Antiquity* **46**: 853–61.

Engelthaler, D. M., Hinnebusch, B. J., Rittner, C. M., and Gage, K. L. (2000). Quantitative competitive PCR as a technique for exploring flea—*Yersinia pestis* dynamics. *American Journal Tropical Medicine and Hygiene* **62**: 552–60.

Evans, D. T., Jing, P., Allen, T. M., O'Connor, D. H., Horton, H., Venham, J. E., *et al.* (2000). Definition of five new simian immunodeficiency virus cytotoxic T-lymphocyte epitopes and their restricting major histocompatibility complex class I molecules: evidence for an influence on disease progression. *Journal of Virology* **74**: 7400–10.

Evershed, R P. (1990). Lipids from samples of skin from seven Dutch bog bodies: preliminary report. *Archaeometry* **32**: 139–53.

Evershed, R. P. (1993). Biomolecular archaeology and lipids. *World Archaeology* **25**: 74–93.

Evershed, R. P. and Tuross, N. (1996). Proteinaceous material from potsherds and associated soils. *Journal of Archaeological Science* **23**: 429–36.

Evershed, R. P., Arnot, K. I., Collister, J., Eglinton, G., and Charters, S. (1994). Application of isotope ratio monitoring gas chromotography-mass spectrometry to the analysis of organic residues of archaeological origin. *Analyst* **119**: 909–14.

Evershed, R. P., Bland, H. A., Van Bergen, P. F., Carter, J. F., Houghton, M. C., and Rowley-Conway, P. A. (1997). Volatile compounds in archaeological plant remains and the Maillard reaction during decay of organic matter. *Science* **278**: 432–3.

Ewald, P. W. (1983). Host–parasite relations, vectors, and the evolution of disease severity. *Annual Review of Ecology and Systematics* **14**: 465–85.

Ewald, P. (1991a). Waterborne transmission and the evolution of virulence among gastrointestinal bacteria. *Epidemiology and Infection* **106**: 83–119.

Ewald, P. W. (1991b). Transmission modes and the evolution of virulence, with special reference to cholera, influenza and AIDS. *Human Nature* **2**: 1–30.

Ewald, P. W. (1994a). *Evolution of infectious disease*. Oxford University Press, New York.

Ewald, P. W. (1994b). Evolution of mutation rate and virulence among human retroviruses. *Philosophical Transactions of the Royal Society of London, Series B, Biological Sciences* **346**: 333–43.

Ewald, P. W. (1998). Emerging diseases, ancient pathogens, and the evolution of virulence. In: *Digging for pathogens: ancient emerging diseases—their evolutionary, anthropological, and archaeological context* (ed. C. L. Greenblatt). Balaban Publishers, Rehovot, Israel: 47–69.

Ewald, P. W. (1999). Evolutionary control of HIV and other sexually transmitted viruses. In: *Evolutionary medicine* (ed. W. R. Trevathan, E. O. Smith, and J. J. McKenna). Oxford University Press, New York: 271–311.

Ewald, P. W. (2000). *Plague time: how stealth infections cause cancers, heart disease and other deadly ailments*. The Free Press, New York.

Ewald, P. W., Sussman, J. B., Distler, M. T., Libel, C., Chammas, W. P., Dirita, V. J., *et al.* (1998). Evolutionary control of infectious disease: prospects for vectorborne and waterborne pathogens. *Memórias do Instituto Oswaldo Cruz* **93**: 567–76.

Faerman, M., Jankauskas, R., Gorski, A., Bercovier, H., and Greenblatt, C. L. (1997). Prevalence of human tuberculosis in a medieval population of Lithuania studied by ancient DNA analysis. *Ancient Biomolecules* **1**: 205–14.

Faerman, M., Jankauskas, R., Gorski, A., Bercovier, H., and Greenblatt, C. L. (1999). Detecting *Mycobacterium tuberculosis* DNA in medieval skeletal remains from Lithuania. In: *Tuberculosis past and present* (ed. G. Pálfi, O. Dutour, J. Deák, and I. Hutás). Golden Book Publishers and the Tuberculosis Foundation, Budapest–Szeged: 371–6.

Fahrenholz, H. (1913). Ectoparasiten und Abstammungslehre. *Zoologie Analytique* **41**: 371–4.

Falkow, S., Isberg, R. R., and Portnoy, D. A. (1992). The interaction of bacteria with mammalian cells. *Annual Review of Cell Biology* **8**: 333–63.

Farmer, J. J. and Hickman-Brenner, F. W. (1992). The genera *Vibrio* and *Photobacterium*. In: *The prokaryotes* (2nd edn) (ed. A. Balows, H. G. Trüper, M. Dworkin, W. Harder, and K.-H. Schleifer). Springer Verlag, Berlin, Heidelberg, New York: 2938–3011.

Felbeck, H. and Distel, D. L. (1992). Prokaryotic symbionts of marine invertebrates. In: *The prokaryotes* (2nd edn) (ed. A. Balows, H. G. Trüper, M. Dworkin, W. Harder, and K.-H. Schleifer). Springer Verlag, Berlin, Heidelberg, New York: 3891–906.

Felsenstein, J. (1989). PHYLIP—Phylogeny Inference Package, version 3.2. *Cladistics* **5**: 164–6.

Ferigolo, J. (1988). Nonhuman vertebrate paleopathology of some Brazilian Pleistocene mammals. In *Paleopathologia paleoepidemiologia—estudos multidisciplinares* (ed. A. J. Goncalves de Araujo and L. F. Ferreira). Panorama, Brazil: 213–34.

Ferreira, B. R. and Silva, J. S. (1998). Saliva of *Rhipicephalus sanguineus* tick impairs T cell proliferation and IFN-gamma-induced macrophage microbicidal activity. *Veterinary Immunology and Immunopathology* **64**: 279–93.

Ferreira, L. F., Britto, C., Cardoso, M. A., Fernandes, O., Reinhard, K., and Araujo, A. (2000). Paleoparasitology of Chagas' disease revealed by infected tissues from Chilean mummies. *Acta Tropica* **75**: 79–84.

Fields, P. I. and Barnes, R. C. (1991). The genus *Chlamydia*. In: *The prokaryotes* (2nd edn) (ed. A. Balows, H. G. Trüper, M. Dworkin, W. Harder, and K.-H. Schleifer). Springer Verlag, Berlin, Heidelberg, New York: 3691–709.

Fields, P. I., Popovic, T., Wachsmuth, K., and Olsvik, O. (1992). Use of polymerase chain reaction for detection of toxigenic *Vibrio cholerae* O1 strains from the Latin American cholera epidemic. *Journal of Clinical Microbiology* **30**: 2118–21.

Fiennes, R. (1978). *Zoonoses and the origins and ecology of human disease*. Academic Press, New York.

Fine, P. (1984). Leprosy and tuberculosis—an epidemiological comparison. *Tubercle* **65**: 137–53.

Finlay, B. J. and Fenchel, T. (1992). Methanogens and other bacteria as symbionts of free-living anaerobic ciliates. *Symbiosis* **14**: 375–90.

Fischer, E. P. and Klose, S. (1996). *Infectious diseases.* Piper, Munich.

Fitch, W., Leiter, J., Li, X., and Palese, P. (1991). Positive Darwinian evolution in human influenza A viruses. *Proceedings of the National Academy of Sciences, USA* **88**: 4270–4.

Fletcher, H. A., Donoghue, H. D., Holton, J., Pap, I., and Spigelman, M. (2002). Widespread occurrence of *Mycobacterium tuberculosis* DNA from 18th–19th century Hungarians. *American Journal of Physical Anthropology* (in press).

Flint, J., Hill, A. V., Bowden, D. K., Oppenheimer, S. J., Sill, P. R., Serjeantson, S. W., *et al.* (1986). High frequencies of alpha-thalassaemia are the result of natural selection by malaria. *Nature* **321**: 744–50.

Fokin, S. I. and Skovorodkin, I. N. (1997). Experimental analysis of the resistance of *Paramecium caudatum* (Ciliophora) against infection by the bacterium *Holospora undulata. European Journal of Protistology* **33**: 214–18.

Fokin, S. I., Brigge, T., Brenner, J., and Görtz, H.-D. (1996). *Holospora* species infecting the nuclei of *Paramecium* appear to belong to two groups of bacteria. *European Journal of Protistology* **32** (Supplement 1): 19–24.

Folk, R. L. (1993). SEM imaging of bacteria and nanobacteria in carbonate sediments and rocks. *Journal of Sedimentary Petrology* **63**: 990–9.

Fosso, C. (1989). Alone with death on the tundra. In: *Alaska: reflections on land and spirit* (ed. R. Hedin and G. Holthaus). University of Arizona Press, Tucson, AR: 215–22.

Foster, P. L. (2000). Adaptive mutation: implications for evolution. *Bioessays* **22**: 1067–74.

Foulcault, M. (1972). *Naissance de la clinique. Une archeologie du regard medical 2ème edition.* Presses universitaires de France, Paris.

Francischetti, I. M., Valenzuela, J. G., and Ribeiro, J. M. (1999). Anophelin: kinetics and mechanism of thrombin inhibition. *Biochemistry* **38**: 16678–85.

Franco, E. L., Villa, L. L., Ruiz, A., and Costa, M. C. (1994). Transmission of cervical human papilloma virus infection by sexual activity: differences between low and high oncogenic risk types. *Journal of Infectious Diseases* **172**: 756–63.

Fraser, C. M. (1998). Complete genome sequence of *Treponema pallidum*, the syphilis spirochete. *Science* **281**: 375–8.

Frenkiel-Krispin, D., Levin-Zaidman, S., Shimoni, E., Wolf, S. G., Wachtel, E. J., *et al.* (2001). Regulated phase transitions of bacterial chromatin: a non-enzymatic pathway for generic DNA protection. *European Molecular biology Journal* **20**: 1184–91.

Fricker, C., Spigelman, M., and Fricker, E. (1997). The detection of *Escherichia coli* DNA in the ancient remains of Lindow Man using the polymerase chain reaction. *Letters in Applied Microbiology* **24**: 351–4.

Fritsche, T. R., Horn, M., Seyedirashti, S., Gautom, R. K., Schleifer, K.-H., and Wagner, M. (1999). *In situ* detection of novel bacterial endosymbionts of *Acanthamoeba* spp. phylogenetically related to members of the order Rickettsiales. *Applied and Environmental Microbiology* **65**: 206–11.

Froment, A. (1995). Les treponematoses: une perspective historique. In: *L'Origin de la syphilis en Europe—avant ou apres 1493?* (ed. O. Dutour, G. Pàlfi, J. Berato, and J.-P. Brun). Centre Archeologique du Var, Toulon, France: 260–8.

Frost, W. (1920). Statistics of influenza morbidity. *Public Health Reports* **35**: 584–97.

Frothingham, R., Strickland, P. L., Bretzel, G., Ramaswamy, S., Musser, J. M., and Williams, D. L. (1999). Phenotypic and genotypic characterization of *Mycobacterium africanum* isolates from West Africa. *Journal of Clinical Microbiology* **37**: 1921–6.

Fuchsberger, N., Kita, M., Hajnicka, V., Imanishi, J., Labuda, M., and Nuttall, P. A. (1995). Ixodid tick salivary gland extracts inhibit production of lipopolysaccharide-induced mRNA of several different human cytokines. *Experimental and Applied Acarology* **19**: 671–6.

Fujishima, M. and Heckmann, K. (1984). Intra- and interspecific transfer of endosymbionts in Euplotes. *Journal of Experimental Zoology* **230**: 339–45.

Fujishima, M., Nagahare, K., and Kojimand Y. (1990). Changes in morphology, buoyant density and protein composition in differentiation from the reproductive short form to the infectious long form of *Holospora obtusa*, a macronucleus-specific symbiont of the ciliate *Paramecium caudatum. Zoological Science* **7**: 849–60.

Gajdusek, D. C. (2001). Fantasy of a 'virus' from the inorganic world: molecular casting and atomic ghost replicas—amyloid enhancing factors are scrapie-like agents. In: *Conformational diseases—a compendium* (ed. B. Solomon, A. Taraboulos, and E. Katchalski-Katzir). The Center for the Study of Emerging Diseases, Jerusalem: 287–90.

Galil, B. S. and Hülsmann, N. (1997). Protist transport via ballast water—biological classification of ballast tanks by food web interactions. *European Journal of Protistology* **33**: 244–53.

Gammelin, M., Altmuller, A., Reinhardt, U., Mandler, J., Harley, V., Hudson, P., *et al.* (1990). Phylogenetic analysis of nucleoproteins suggests that human influenza A viruses emerged from a 19th-century avian ancestor. *Molecular Biology and Evolution* **7**: 194–200.

Ganapamo, F., Rutti, B., and Brossard, M. (1997). Identification of an *Ixodes ricinus* salivary gland fraction through its ability to stimulate CD4 T cells present in BALB/c mice lymph nodes draining the tick fixation site. *Parasitology* **115**: 91–6.

Gao, F., Yue, L., White, A. T., Pappas, P. G., Barchue, J., Hanson, A. P., *et al.* (1992). Human infection by genetically diverse SIVSM-related HIV-2 in West Africa. *Nature* **358**: 495–9.

Gao, F., Bailes, E., Robertson, D. L., Chen, Y., Rodenburg, C. M., Michael, S. F., *et al.* (1999). Origin of HIV-1 in the chimpanzee *Pan troglodytes*. *Nature* **397**: 436–41.

Garcia, M. M., Martin, D. M. J., Brooks, B. W., Fraser, A. D. E., and Lior, H. (1987). Development of a library for the rapid identification of campylobacters of veterinary significance by gas chromatography. The Fourth International Workshop on *Campylobacter* Infections, Goteborg, Sweden.

Garcia-Frias, J. (1994). La tuberculosis en los antiguos Peruanos. *Actualidad Médica Peruana* **5**: 274–91.

Garcia-Sastre A., Egorov, A., Matassov, D., Brandt, S., Levy, D. E., Durbin, J. E., *et al.* (1998). Influenza A virus lacking the NS1 gene replicates in interferon-deficient systems. *Virology* **252**: 324–30.

Garrett, L. (1994). *The coming plague: newly emerging diseases in a world out of balance.* Farrar, Straus and Giroux, New York.

Garrett, L. (2000). *Betrayal of trust.* Hyperion, New York.

Genton, B., al-Yaman, F., Mgone, C. S., Alexander, N., Paniu, M. M., Alpers, M. P., *et al.* (1995). Ovalocytosis and cerebral malaria. *Nature* **378**: 564–5.

Gerard, H. C., Wang, G. F., Balin, B. J., Schumacher, H. R., and Hudson, A. P. (1999). Frequency of apolipoprotein E (APOE) allele types in patients with *Chlamydia*-associated arthritis and other arthritides. *Microbial Pathogenesis* **26**: 35–43.

Gernaey, A. M., Minnikin, D. E., Copley, M. S., Ahmed, A. S., Roberston D. J., Nola, J., *et al.* (1999). Correlation of the occurrence of mycolic acids with tuberculosis in an archaeological population. In: *Tuberculosis past and present* (ed. G. Pálfi, O. Dutour, J. Deák, and I. Hutás). Golden Book Publishers and the Tuberculosis Foundation, Budapest–Szeged: 275–84.

Gernaey, A. M., Minnikin, D. E., Copley, M. S., Dixon, R. A., Middleton, J. C., and Roberts, C. A. (2001). Mycolic acids and ancient DNA confirm an osteological diagnosis of tuberculosis. *Tuberculosis* **81**: 259–65.

Gerston, K. F., Blumberg, L., and Gafoor, H. (1998). Viability of mycobacteria in formalin-fixed tissues. *International Journal of Tuberculosis and Lung Disease* **2**: 521–3.

Ghosh, A. (1982). Deurbanization of the Harappan civilization. In: *Harappan civilization* (ed. G. Possehl). Aris & Phillips, Warminster, UK: 321–3.

Gilbert, S. C., Plebanski, M., Gupta, S., Morris, J., Cox, M., Aidoo, M., *et al.* (1998). Association of malaria parasite population structure, HLA, and immunological antagonism. *Science* **279**: 1173–7.

Gilbert, T. and Cooper, A. (2001). Ancient DNA *Yersinia pestis*. Colloque international: La Peste: entre épidémies et sociétés. Session n° 4: Epidémiologie actuelle et microbiologie. Marseille, 23 au 26 juillet 2001: 72.

Gillespie, R. D., Mbow, M. L., and Titus, R. G. (2000). The immunomodulatory factors of bloodfeeding arthropod saliva. *Parasite Immunology* **22**: 319–31.

Gilliam, M. and Taber, E., III (1991). Diseases, pests, and normal microflora of honeybees, *Apis mellifera*, from feral colonies. *Journal of Invertebrate Pathology* **58**: 286–9.

Gilliam, M., Buchmann, S. L., Lorenz, B. J., and Schmalzel, R. J. (1990a). Bacteria belonging to the genus *Bacillus* associated with three species of solitary bees. *Apidologie* **21**: 99–105.

Gilliam, M., Roubik, D. W., and Lorenz, B. J. (1990b). Microorganisms associated with pollen, honey, and brood provisions in the nest of a stingless bee, *Melipona fasciata*. *Apidologie* **21**: 89–97.

Gilmcher, M. J. and Katz, E. P. (1965). The organization of collagen in bone: the role of noncovalent bonds in the relative insolubility of bone collagen. *Journal of Ultrastructure Research* **12**: 705–29.

Ginsburg, I. (1988). The biochemistry of bacteriolysis: paradoxes, facts and myths. *Microbiological Sciences* **5**: 137–42.

Gladykowska-Rzeczycka, J. (1976). A case of leprosy from a Medieval burial ground. *Folia Morphologica—Warsz.* **35**: 253–64.

Gofton, J. P., Bennett, P. H., Smythe, H. A., and Decker, J. L. (1972). Sacroilitis and ankylosing spondylitis in North American Indians. *Annals of the Rheumatic Diseases* **31**: 474–81.

Gokhale, B. G. (1959). *Ancient India* (4th edn). Asia Publishing House, New York.

Goldstein, L. (1995). Politics, law, pragmatics, and human burial excavations: an example from northern California. In: *Bodies of evidence: reconstructing history through skeletal analysis* (ed. A. L. Grauer). Wiley-Liss, New York: 3–17.

Goldstein, M. S., Arensburg, B., and Nathan, H. (1976). Pathology of Bedouin skeletal remains from two sites in Israel. *American Journal of Physical Anthropology* **45**: 621–40.

Golenberg, E. M., Giannasi, D. E., Clegg, M. T., Smiley, C. J., Durbin, M., Henderson, D., *et al.* (1990). Chloroplast DNA sequence from a miocene magnolia species. *Nature* **344**: 656–8.

Goma Epidemiology Group (1995). Public health impact of Rwandan refugee crisis: what happened in Goma, Zaire, in July 1994? *Lancet* **345**: 339–44.

Gooding, R. H. (1972). Digestive processes of hematophagous insects—II: Trypsin from the sheep ked *Melophagus ovinus* (L.) (Hippoboscidae, Diptera) and its inhibition by mammalian sera. *Comparative Biochemistry and Physiology B* **43**: 815–24.

Goodman, A. H. and Rose, J. C. (1996). Dental enamel hypoplasias as measures of developmental stress. In: *Notes on populational significance of paleopathological conditions: health, illness and death in the past* (ed. A. Pérez-Pérez). Fundació Uriach, Barcelona: 77–95.

Goodman, A. H., Brooke, T. R., Swedlund, A. C., and Armelagos, G. J. (1988). Biocultural perspectives on stress in prehistoric, historical and contemporary population research. *Yearbook of Physical Anthropology* **31**: 169–202.

Gordon, C. C. and Buikstra, J. E. (1981). Soil pH, bone preservation, and sampling bias at mortuary sites. *American Antiquity* **46**: 566–71.

Gordon, R. E., Haynes, W. C., and Pang, C. (1973). *The genus Bacillus*. Handbook No. 427, US Department of Agriculture, Washington, DC.

Görtz, H.-D. (1986). Endonucleobiosis in ciliates. *International Review of Cytology* **102**: 169–213.

Görtz, H.-D. and Wiemann, M. (1989). Route of infection of the bacteria *Holospora elegans* and *Holospora obtusa* into the nuclei of *Paramecium caudatum. European Journal of Protistology* **24**: 101–9.

Görtz, H.-D., Lellig, S., Miosga, O., and Wiemann, M. (1990). Changes in fine structure and polypeptide pattern during the development of *Holospora obtusa*, a bacterium infecting the macronucleus of *Paramecium caudatum. Journal of Bacteriology* **172**: 5664–9.

Götherström, A., Fischer, C., Lindén, K., and Lidén, K. (1995). X-raying ancient bone. A destructive method in connection with DNA analysis. *Laborativ Arkeologi* **8**: 26–8.

Goto, H. and Kawaoka, Y. (1998). A novel mechanism for the acquisition of virulence by a human influenza A virus. *Proceedings of the National Academy of Sciences, USA* **95**: 10224–8.

Goudsmit, J. and Lukashov, V. V. (1999). Dating the origin of HIV-1 subtypes. *Nature* **400**: 325–6.

Goudsmit, J., Dekker, J., Smit, L., Kuiken, C., Geelen, J., and Perizonius, R. (1993). Analysis of retrovirus sequences in c. 5300 years old Egyptian mummy DNA obtained via amplification and molecular cloning. In: *Biological anthropology and the study of ancient Egypt* (ed. W. V. Davies and R. Walter). British Museum Press, London: 91–7.

Grange, J. M. (1992). The mystery of the mycobacterial 'persister'. *Tubercle and Lung Disease* **73**: 249–51.

Grant, R. A., Filman, D. J., Finkel, S. E., Kolter, R., and Hogle, J. M. (1998). The crystal stucture of Dps, a ferratin homolog that binds and protects DNA. *Nature Structural Biology* **5**: 294–303.

Graw, M., Weisser, H.-J., and Lutz, S. (2000). DNA typing of human remains found in damp environments. *Forensic Science International* **113**: 91–5.

Greenblatt, C. L. (ed.) (1998). *Digging for pathogens: ancient emerging diseases — their evolutionary, anthropological, and archaeological context.* Balaban Publishers, Rehovot, Israel.

Greenblatt, C., Davis, A., Clement, B., Kitts, C., Cox, T., and Cano, R. (1999). Diversity of microorganisms isolated from amber. *Microbial Ecology* **38**: 58–68.

Greenwood, B. M. and Mutabingwa, T. (2002). Malaria in 2002. *Nature* **415**: 670–2.

Grossman, L. (1991). Repair of damaged DNA. In: *Encyclopedia of human biology.* Academic Press, San Diego: vol. 6, 547–53.

Grupe G. (1987) [Trace elements in buried human bones and their value. A review.] *Anthropologischer Anzeiger; Bericht uber die biologisch-anthropologische Literatur* **45**: 19–28.

Grupe, G. and Herrmann, B. (1988). *Trace elements in environmental history.* Springer, New York (in German).

Guhl, F., Jaramillo, C., Yockteng, R., Vallejo, G. A., and Cardenas-Arroya, F. (1997). *Trypanosoma cruzi* DNA in human mummies. *Lancet* **349**: 1370.

Guhl, F., Jaramillo, C., Vallejo, G. A., Yockteng, R., and Cárdenas-Arroyo, F. (1999). Isolation of *Trypanosoma cruzi* DNA in 4,000 year-old mummified human tissue from Northern Chile. *American Journal of Physical Anthropology* **108**: 401–7.

Haas, C. J., Zink, A., Molnar, E., Szeimies, U., Reischl, U., Marcsik, A., *et al.* (1999). Molecular evidence for tuberculosis in Hungarian tissue samples. In: *Tuberculosis past and present* (ed. G. Pálfi, O. Dutour, J. Deák, and I. Hutás). Golden Book Publishers and the Tuberculosis Foundation, Budapest–Szeged: 385–91.

Haas, C., Zink, A., Molnar, E., Szeimes, U., Reischl, U., Marcsik, A., *et al.* (2000a). Molecular evidence for different stages of tuberculosis in ancient bone samples from Hungary. *American Journal of Physical Anthropology* **113**: 293–304.

Haas, C. J., Zink, A., Pàlfi, G., Szeimies, U., and Nerlich, A. G. (2000b). Detection of leprosy in ancient human skeletal remains by molecular identification of *Mycobacterium leprae. American Journal of Clinical Pathology* **114**: 428–436.

Haas, F. and Haas, S. S. (1996). The origins of *Mycobacterium tuberculosis* and the notion of its contagiousness. In: *Tuberculosis* (ed. W. N. Rom and S. M. Garay). Little, Brown, and Co., Boston: 3–19.

Hacker, J. and Carniel, E. (2001). Ecological fitness, genomic islands and bacterial pathogenicity. A Darwinian view of the evolution of microbes. *EMBO Reports* **2**: 376–81.

Hackett, C. J. (1963). On the origin of the human treponematoses (pinta, yaws, endemic syphilis and venereal syphilis). *Bulletin of the World Health Organization* **29**: 7.

Hagelberg, E. and Clegg, J. B. (1991a). Isolation and characterization of DNA in archaeological bone. *Proceedings of the Royal Society of London B, Biological Sciences* **244**: 45–50.

Hagelberg, E., Bell, L. S., Allen, T., Boyde, A., Jones, S., and Clegg, J. B. (1991b). Analysis of ancient bone DNA: techniques and applications. *Philosophical Transactions of the Royal Society of London B, Biological Sciences* **333**: 399–407.

Haglund, W. D. and Sorg, M. H. (1997). Method and theory of forensic taphonomic research. In: *Forensic taphonomy: the postmortem fate of human remains* (ed. W. D. Haglund and M. H. Sorg). CRC Press, Boca Raton, FL: 13–26.

Hagnere, C. and Harf, C. (1993). Symbiotic interactions between free-living amoeba and harboured mercury-resistant bacteria. *European Journal of Protistology* **29**: 155–9.

Hahn, B. H., Shaw, G. M., De Cock, K. M., and Sharp, P. M. (2000). AIDS as a zoonosis: scientific and public health implications. *Science* **287**: 607–14.

Haldane, J. B. S. (1948). The rate of mutation of human genes. *Hereditas Supplement* **35**: 267–73.

Haldane, J. B. S. (1949). Disease and evolution. *La Ricerca Scientifica* **19**: S68–76.

Hall, R. H., Khambaty, F. M., Kothary, M. H., Keasler, S. P., and Tall, B. D. (1994). *Vibrio cholerae* non-O1 serogroup associated with cholera gravis genetically and physiologically resembles O1 El Tor cholera strains. *Infection and Immunity* **62**: 3859–63.

Hamblin, M.. and Di Rienzo, A. (2000). Detection of the signature of natural selection in humans: evidence from the Duffy blood group locus. *American Journal of Human Genetics* **66**: 1669–79.

Hamblin, M. T., Thompson, E. E., and Di Rienzo, A. (2002). Complex signatures of natural selection at the Duffy blood group locus. *American Journal of Human Genetics* **70**: 369–83.

Hamilton, A. (2000). Nanobacteria: gold mine or minefield of intellectual enquiry? *Microbiology Today* **27**: 182–4.

Han, Y. S., Thompson, J., Kafatos, F. C., and Barillas-Mury, C. (2000). Molecular interactions between *Anopheles stephensi* midgut cells and *Plasmodium berghei*: the time bomb theory of ookinete invasion of mosquitoes. *EMBO Journal* **19**: 6030–40.

Handt, O., Richards, M., Trommsdorff, M., Kilgar, C., Simanainen, J., Georgiev, O., et al. (1994). Molecular genetic analysis of the Tyrolean Ice Man. *Science* **264**: 1775–8.

Hänni, C., Laudet, V., Sakka, M., Begue, A., and Stehelin, D. (1990). Amplification of mitochondrial DNA fragments from ancient teeth and bones. *CR. Acad. Sci Paris Series III* **310**: 356–70.

Harb, O. S., Gao, L.-Y., and Kwaik, Y. A. (2000). From protozoa to mammalian cells: a new paradigm in the life cycle of intracellular bacterial pathogens. *Environmental Microbiology* **2**: 251–65.

Hare, P. E., Hoering, T. C., and King, K. (1980). *Biogeochemistry of amino acids*. John Wiley and Sons, New York.

Hartskeerl, R. A., De Wit, M. Y. L., and Klatser, P. R. (1989). Polymerase chain reaction for the detection of *Mycobacterium leprae*. *Journal of General Microbiology* **135**: 2355–64.

Heckmann, K. (1983). Endosymbionts of euplotes. In: *Intracellular symbiosis* (ed. K. W. Jeon). Academic Press, New York: 111–14.

Heckmann, K. and Görtz, H.-D. (1991). Prokaryotic symbionts of ciliates. In: *The prokaryotes* (2nd edn) (ed. A. Balows, H. G. Trüper, M. Dworkin, W. Harder, and K.-H. Schleifer). Springer Verlag, Berlin, Heidelberg, New York: 3865–90.

Heinz, H. J. (1961). Factors governing the survival of Bushman worm parasites in the Kalahari. *South African Journal of Science* **8**: 207–13.

Hemmer, H. (1990). *Domestication: the decline of environmental appreciation*. Cambridge University Press, Cambridge.

Herbert, W. J. and Parratt, D. (1979). Virulence of trypanosomes in the vertebrate host. In: *Biology of the kinetoplastida* (ed. W. H. R. Lumsden and D. A. Evans). Academic Press, New York: vol. 2, 481–521.

Herms, W. B. (1950). *Medical entomology*. Macmillan Company, New York: 70.

Herrmann, B. (1985). Parasitologisch-epidemiologische Auswertungen mittelalterlicher Kloaken. *Zeitschrift fuer Archaeologie des Mittelalters* **13**: 131–61.

Herrmann, B. and Hummel, S. (1998). Ancient DNA can identify disease elements. In: *Digging for pathogens: ancient emerging diseases—their evolutionary, anthropological, and archaeological context* (ed. C. L. Greenblatt). Balaban Publishers, Rehovot, Israel: 329–43.

Hershkovitz, I., Spiers, M., Katznelson, A., and Arensberg, B. (1992). Unusual pathological conditions in the lower extremities of a skeleton from ancient Israel. *American Journal of Physical Anthropology* **88**: 23–6.

Hershkovitz, I., Spiers, M., and Arensberg, B. (1993). Leprosy or Madura foot? The ambiguous nature of infectious disease in paleopathology: reply to Dr Manchester. *American Journal of Physical Anthropology* **91**: 251–3.

Hershkovitz, I., Rothschild, B. M., Wish-Baratz, S., and Rothschild, C. (1995). Natural variation and differential diagnosis of skeletal changes in Bejel (endemic syphilis). In: *L'Origin de la syphilis en Europe—avant ou apres 1493?* (ed. O. Dutour, G. Pàlfi, J. Berato, and J.-P. Brun). Centre Archeologique du Var, Toulon, France: 81–7.

Higgins, D. G., Bleasby, A. J., and Fuchs, R. (1992). CLUSTAL V: improved software for multiple sequence alignment. *Computer Applications in the Biosciences* **8**: 189–91.

Higuchi, P. and Wilson, A. C. (1984). Recovery of DNA from extinct species. *Federation Proceedings* **43**:1557.

Higuchi, R., Bowman, B., Freiberger, M., Ryder, O. A., and Wilson, A. C. (1984). DNA sequences from the quagga, an extinct member of the horse family. *Nature* **312**: 282–4.

Hill, A. V. S. (1996). Genetic of infectious disease resistance. *Current Opinion in Genetics and Development* **6**: 348–53.

Hill, A. V., Allsopp, C. E., Kwiatkowski, D., Anstey, N. M., Twumasi, P., Rowe, P. A., *et al.* (1991). Common West African HLA antigens are associated with protection from severe malaria. *Nature* **352**: 595–600.

Hinnebusch, B. J., Perry, R. D., and Schwan, T. G. (1996). Role of the *Yersinia pestis* hemin storage (*hms*) locus in the transmission of plague by fleas. *Science* **273**: 367–70.

Hirsch, A. (1860). *Handbuch der historisch-geographischen Pathologie.* Quoted from 2nd edition. Enke, Stuttgart, 1881.

Ho, T. B., Robertson, B. D., Taylor, G. M., Shaw, R. J., and Young, D. B. (2000). Comparison of *Mycobacterium tuberculosis* genomes reveals frequent deletions in a 20 kb variable region in clinical isolates. *Yeast* **4**: 272–82.

Hoberg, E. P., Alkire, N. L., de Queiroz, A., and Jones, A. (2000). Out of Africa: origins of the *Taenia* tapeworms in humans. *Proceedings of the Royal Society, London, Series B, Biological Sciences* **1579**: 781–7.

Hofreiter, M., Serre, D., Poinar, H., Kuch, M., and Pääbo, S. (2001). Ancient DNA. *Nature Reviews. Genetics* **2**: 353–9.

Holden T. (1990). Taphonomic and methodological problems in reconstructing diet from ancient human gut and faecal remains. Unpublished PhD thesis, University of London.

Holland, P., Abramson, R., Watson, R., and Gelfand, D. (1991). Detection of specific polymerase chain reaction product by utilizing the 5'-3' exonuclease activity of *Thermus aquaticus* DNA polymerase. *Proceedings of the National Academy of Sciences, USA* **88**: 7276–80.

Hooper, E. (1999). The river: a journey back to the source of HIV and AIDS. Little, Brown, & Co., Boston.

Hooton, E. A. (1930). *The Indians of Pecos Pueblo: a study of their skeletal remains.* Yale University Press, New Haven, CT.

Horn, M., Wagner, M., Müller, K.-D., Fritsche, T. R., Schleifer, K.-H., and Michel, R. (2000). *Neochlamydia hartmanellae* gen.nov., sp.nov. (Parachlamydiaceae), an endoparasite of the amoeba *Hartmanella vermiformis*. *Microbiology* **146**: 1231–9.

Horn, M., Harzenetter, M. D., Linner, T., Schmid, E. N., Müller, K.-D., Michel, R., *et al.* (2001). Members of the *Cytophaga-Flavobacterium-Bacteroides* phylum as intracellular bacteria of *Acanthamoeba*: proposal of 'Candidatus Amoebophilus asiaticus'. *Environmental Microbiology* **3**: 440–9.

Höss, M., Kohn, M., Pääbo, S., Knauer, F., and Schroder, W. (1992). Excrement analysis by PCR. *Nature* **359**: 199.

Höss, M., Dilling, A., Currant, A., and Pääbo, S. (1996a). Molecular phylogent of the extinct sloth *Mylodon darwinii*. *Proceedings of the National Academy of Sciences, USA* **93**: 181–5.

Höss, M., Jaruga, P., Zastawny, T. H., Dizdaroglu, M., and Pääbo, S. (1996b). DNA damage and DNA sequence retrieval from ancient tissue. *Nucleic Acids Research* **24**: 1304–7.

Hubbert, W. T., McCullogh, W. F., and Shurrenberger, D. V. M. (1975). *Diseases transmitted from animals to man.* Charles Thomas, Springfield, IL.

Hudson, E. H. (1958). *Non-venereal syphilis: a sociological and medical study of bejel.* Livingston, London.

Hummel, S. (2002). *Ancient DNA typing: methods, strategies and applications*, Springer, Heidelberg and New York (in press).

Hummel, S., Menzel, A., Uy, A., Hofferbert, S., and Herrmann, B. (1994). DNA-analysis of inherited and infectious diseases—perspectives for paleopathology. First reported at the Tenth European Meeting of the Palaeopathology Association, Goettingen, 29 August–3 September. *Homo* **45** (Supplement): S62.

Hurley, M. F., Scully, O. M., and McCutcheon, S. W. (1994). *Late Viking age and medieval Waterford excavations 1986–1992.* Waterford Corporation, Waterford, UK

Hutás, I. (1999). The history of tuberculosis in Hungary. In: *Tuberculosis past and present* (ed. G. Pálfi, O. Dutour, J. Deák, and I. Hutás). Golden Book Publishers and the Tuberculosis Foundation, Budapest–Szeged: 39–42.

Iwahashi, K., Watanabe, M., Nakamura, K., Suwaki, H., Nakaya, T., Nakamura, Y., *et al.* (1998). Positive and negative syndromes, and Borna disease virus infection in schizophrenia. *Neuropsychobiology* **37**: 59–64.

Jameson, K. R. (1993). *Touched with fire*. Free Press, New York.

Jamil, S., Wilson, S. M., Hacket, M., Hussain, R., and Stoker, N. G. (1994). A colorimetric PCR method for the detection of *M. leprae* in skin biopsies from leprosy patients. *International Journal of Leprosy and Other Mycobacterial Diseases* **62:** 512–20.

Janaway, R. C. (1985). Dust to dust: the preservation of textile materials in metal artifact corrosion products with reference to inhumation graves. *Science and Archaeology* **27:** 29–34.

Jelliffe, D. B., Woodbum, J. C., Bennett, F. J., and Jelliffe, K. (1962). The children of the Hadza hunters. *Tropical Paediatrics* **60:** 907–13.

Jeon, K. W. (1983). Integration of bacterial endosymbionts in amoebas. In: *Intracellular symbiosis* (ed. K. W. Jeon). Academic Press, New York: 29–47.

Jeon, K. W. (1991). Prokaryotic symbionts of amoebae and flagellates. In: *The prokaryotes* (2nd edn) (ed. A. Balows, H. G. Trüper, M. Dworkin, W. Harder, and K.-H. Schleifer). Springer Verlag, Berlin, Heidelberg, New York: 3855–64.

Johansson, S. R. and Horowitz, S. (ed.) (1986). Estimating mortality in skeletal populations: influence of the growth rate on the interpretation of levels and trends during the transition to agriculture. *American Journal of Physical Anthropology* **71:** 233–50.

Johnson, J. A., Salles, C. A., Panigrahi, P., Albert, M. J., Wright, A. C., Johnson, R. J., *et al.* (1994). *Vibrio cholera* O139 synonym Bengal is closely related to *Vibrio cholera* El Tor, but has important differences. *Infection and Immunity* **62:** 2108–10.

Joly, M. A. (1965). Physico-chemical approach to the denaturation of proteins. Academic Press, New York.

Jones, A. (1983). A coprolite from 6-8 pavement. In *Environment and living conditions at two Anglo-Scandinavian sites* (ed. A. R. Hall, H. K. Kenward, D. Williams and J. R. A. Greig), *The archaeology of York: the past environment of York*. Council for British Archaeology for York Archaeological Trust, London: Volume 14, 225–30.

Jones, M. (2001). *The molecule hunt: archaeology and the search for ancient DNA*. Allen Lane, Penguin Press, London.

Jordan, E. (1927). *Epidemic influenza: a survey*. American Medical Association, Chicago.

Kajander, E. O. and Çiftçioglu, N. (2001). Nanobacteria: an alternative mechanism for pathogenic intra- and extra-cellular calcification and stone formation. *Proceedings of the National Academy of Sciences, USA* **95:** 8274–9.

Kamerbeek, J., Schouls, L., Kolk, A., van Agterveld, M., van Soolingen, D., Kuijper, S., *et al.* (1997). Simultaneous detection and strain differentiation of *Mycobacterium*

tuberculosis for diagnosis and epidemiology. *Journal of Clinical Microbiology* **35:** 907–14.

Kaneda, T. (1977). Fatty acids of the genus *Bacillus*: an example of branched-chain preference. *Bacteriological Reviews* **41:** 391–418.

Kanegae, Y., Sugita, S., Sortridge, K., Yoshioka, Y., and Nerome, K. (1994). Origin and evolutionary pathways of the H1 hemagglutinin gene of avian, swine and human influenza viruses: cocirculation of two distinct lineages of swine viruses. *Archives of Virology* **134:** 17–28.

Kaper, J. B., Morris, J. G., Jr, and Levine, M. M. (1995). Cholera. *Clinical Microbiology Review* **8:** 48–86. Erratum in *Clinical Microbiology Review* **8:** 316.

Kaplan, C. W., Astaire, J., Sanders, M., and Kitts, C. L. (2001). 16S ribosomal DNA terminal restriction fragment pattern analysis of bacterial communities in feces of rats fed *Lactobacillus acidopholus* NCFM. *Applied and Environmental Microbiology* **67:** 1935–9.

Kapur, V., Whittam, T. S., and Musser, J. M. (1994). Is *Mycobacterium tuberculosis* 15,000 years old? *Journal of Infectious Diseases* **170:** 1348–9.

Karkanas, P., Kyparissi-Apostolika, N., Bar-Yosef, O., and Weiner, S. (1999). Mineral assemblages in Theopetra, Greece: a framework for understanding diagenesis in a prehistoric cave. *Journal of Archaeological Science* **26:** 1171–80.

Karlen, A. (1995). *Men and microbes*. Putnam & Sons, New York.

Karlsson, K. A. (2000). The human gastric colonizer *Helicobacter pylori*: a challenge for host–parasite glycobiology. *Glycobiology* **10:** 761–71.

Katz, B. J. and Man, E. H. (1980). The effects and implications of ultrasonic cleaning on the amino acid geochemistry of foraminifera. In: *Biogeochemistry of amino acids* (ed. P. E. Hare, T. C. Hoering, and K. King). John Wiley and Sons, New York: 215–22.

Katz, J., Lim, W., Bridges, C., Rowe, T., Hu-Primmer, J., Lu, X., *et al.* (1999). Antibody response in individuals infected with avian influenza A (H5N1) viruses and detection of anti-H5 antibody among household and social contacts. *Journal of Infectious Diseases* **180:** 1763–70.

Katzenberg, M. A. and Pfeiffer, S. (1995). Nitrogen isotope evidence for weaning age in a nineteenth century Canadian skeletal sample. In: *Bodies of evidence: reconstructing history through skeletal analysis* (ed. A. L. Grauer). Wiley-Liss, New York: 221–35.

Kauffman, S. A. (1993). *The origins of order: self-organization and selection in evolution*. Oxford University Press, New York.

Kawaoka, Y., Krauss, S., and Webster, R. G. (1989). Avian-to-human transmission of the PB1 gene of influenza A viruses in the 1957 and 1968 pandemics. *Journal of Virology* **63**: 4603–8.

Keilin, D. and Wang Y. L. (1947). Stability of haemoglobin and of certain endoerythrocytic enzymes *in vitro*. *Biochemistry* **41**: 491–500.

Kelley, M. A. and Micozzi, M. S. (1984). Rib lesions in chronic pulmonary tuberculosis. *American Journal of Physical Anthropology* **65**: 381–6.

Kennedy, M. J., Reader, S. L., and Swierczynski, L. M. (1994). Preservation records of micro-organisms: evidence of the tenacity of life. *Microbiology* **140**: 2513–29.

Kerley, E. R. and Bass, W. M. (1967). Paleopathology: meeting ground for many disciplines. *Science* **157**: 638–44.

Kidd, K. E. (1954). A note on the palaeopathology of Ontario. *American Journal of Physical Anthropology* **12**: 610.

Kilbourne, E. (1977). Influenza pandemics in perspective. *Journal of the American Medical Association* **237**: 1225–8.

Kilbourne, E. D. (1997). Perspectives on pandemics: a research agenda. *Journal of Infectious Diseases* **176** (Supplement 1): S29–31.

Kilgore, L. (1989). Possible case of rheumatoid arthritis from Sudanese Nubia. *American Journal of Physical Anthropology* **79**: 177–83.

Kiple, K. (ed.) (1993). *The Cambridge world history of human disease*. Cambridge University Press, Cambridge.

Kirkpatrick, B., Buchanan, R. W., Ross, D. E., and Carpenter, W. T., Jr (2001). A separate disease within the syndrome of schizophrenia. *Archives of General Psychiatry* **58**: 165–71.

Klein, J. (1987). Origin of major histocompatibility complex polymorphism: the trans-species hypothesis. *Human Immunology* **19**: 155–62.

Klompen, J. S., Black, W. C., IV, Keirans, J. E., and Oliver, J. H., Jr (1996). Evolution of ticks. *Annual Review of Entomology* **41**: 141–61.

Kobayashi, H., Yamamoto, M., and Aono, R. (1998). Appearance of a stress-response protein, phage-shock protein A, in *Escherichia coli* exposed to hydrophobic organic solvents. *Microbiology* **144**: 353–9.

Koch, R. (1884). An address on cholera and its bacillus. *British Medical Journal* **1884**: 403–7.

Koen, J. S. (1919). A practical method for field diagnoses of swine diseases. *American Journal of Veterinary Medicine* **14**: 468–70.

Kolata, G. (1999). *Flu, the story of the great influenza pandemic of 1918 and the search for the virus that caused it*. Simon & Schuster, New York.

Kolata, G. (2001). Kill all the bacteria. *New York Times*, January 7, Section 4, p. 1.

Kolman, C. J. and Tuross, N. (2000). Ancient DNA of human populations. *American Journal of Physical Anthropology* **111**: 5–23.

Kolman, C. J., Centurion-Lara, A., Lukeheart, S. A., Owsley, D. W., and Tuross, N. (1999). Identification of *Treponema pallidum* subspecies *pallidum* in a 200-year-old skeletal specimen. *Journal of Infectious Diseases* **180**: 2060–3.

Kolter, R., Siegele, D. A., and Tormo, A. (1993). The stationary phase of the bacterial life cycle. *Annual Review of Microbiology* **47**: 855–74.

Kopecky, J. and Kuthejlova, M. (1998). Suppressive effect of *Ixodes ricinus* salivary gland extract on mechanisms of natural immunity *in vitro*. *Parasite Immunology* **20**: 169–74.

Korber, B., Muldoon, M., Theiler, J., Gao, F., Gupta, R., Lapedes, A., *et al.* (2000). Timing the ancestor of the HIV-1 pandemic strains. *Science* **288**: 1789–96.

Koski, L. B., Morton, R. A., and Golding, G. B. (2001). Codon bias and base composition are poor indicators of horizontally transferred genes. *Molecular Biology and Evolution* **18**: 404–12.

Kraaijeveld, A. R., Van Alphen, J. J., and Godfray, H. C. (1998). The coevolution of host resistance and parasitoid virulence. *Parasitology* **116**: S29–45.

Krafft, A. E., Duncan, B. W., Bijwaard, K. E., Taubenberger, J. K., and Lichy, J. H. (1997). Optimization of the isolation and amplification of RNA from formalin-fixed, paraffin-embedded tissue: the Armed Forces Institute of Pathology Experience and Literature Review. *Molecular Diagnosis* **2**: 217–30.

Kramar, C., Lagier, R., and Baud, C. A. (1990). Rheumatic diseases in Neolithic and medieval populations of western Switzerland. *Zeitschrift fur Rheumatologie* **49**: 338–45.

Krause, R. M. (1994). Dynamics of emergence. *Journal of Infectious Diseases* **170**: 265–71.

Krause, R. M. (1998). Introduction to emerging infectious diseases. In: *Emerging infections* (ed. R. M. Krause). Academic Press, New York: 1–22.

Kreger, M. D. (1997). *Zoonotic diseases*. US Department of Agriculture, Beltsville, MD.

Krings, M., Stone, A., Schmitz, R. W., Krainitzki, H., and Pääbo, S. (1997). Neandertal DNA sequences and the origin of modern humans. *Cell* **90**: 19–30.

Krings, M., Geisert, H., Schmitz, R. W., Krainitzki, H., and Pääbo, S. (1999). DNA sequence of the mitochondrial hypervariable region II from the Neandertal type specimen. *Proceedings of the National Academy of Sciences, USA* **96**: 5581–5.

Krueger, H. W. and Sullivan, C. (1984). Models for carbon isotope fractionation between diet and bone. In: *Stable isotopes in nutrition* (ed. J. E. Turnland and P. E. Johnson). American Chemical Society Symposium Series 258, Washington, DC: 205–22.

Kubes, M., Fuchsberger, N., Labuda, M., Zuffova, E., and Nuttall, P. A. (1994). Salivary gland extracts of partially fed *Dermacentor reticulatus* ticks decrease natural killer cell activity *in vitro*. *Immunology* **82**: 113–16.

Kumar, S., Tamura, K., and Nei, M. (1993). MEGA: Molecular Evolutionary Genetic Analysis, Version 1.0. Pennsylvania State University, University Park, PA.

Kumar, S. S., Nasidze, I., Walimbe, S. R., and Stoneking, M. (2000). Brief communication: discouraging prospects for ancient DNA from India. *American Journal of Physical Anthropology* **113**: 129–33.

Lafond, E. M. (1958). An analysis of adult skeletal tuberculosis. *Journal of Bone and Joint Surgery* **40A**: 346–64.

Lambert, J. B., Simpson, S. V., Szpunar, C. B., and Buikstra, J. E. (1984). Copper and barium as dietary discriminants: the effects of diagenesis. *Archaeometry* **26**: 131–8.

Lambert, L. T., Cox, K., Mitchell, R., Rossello-Mora, del Cueto, C., Dodge, D., *et al.* (1998). Isolation and characterization of a novel organism, *Staphylococcus succinus*, sp. nov. from 25–35-million year old Dominican amber. *International Journal of Systematic Bacteriology* **48**: 511–15.

Landis, W. G. (1988). Ecology. In: *Paramecium* (ed. H.-D. Görtz). Springer Verlag, Berlin: 419–36.

Lange, R. and Hengge-Aronis, R. (1991). Identification of a central regulator of stationary-phase gene expression in *Escherichia coli*. *Molecular Microbiology* **5**: 49–59.

Lanphear, K. M. (1988). *Health and mortality in a nineteenth-century poorhouse skeletal sample*. PhD dissertation, State University of New York, Albany, NY.

Lanzaro, G. C., Lopes, A. H., Ribeiro, J. M., Shoemaker, C. B., Warburg, A., Soares, M., *et al.* (1999). Variation in the salivary peptide, maxadilan, from species in the *Lutzomyia longipalpis* complex. *Insect Molecular Biology* **8**: 267–75.

Larsen, C. S. (1995). Biological changes in human populations with agriculture. *Annual Review of Anthropology* **2**: 185–215.

Larsen, C. S. and Hutchinson, D. L. (1992). Dental evidence for physiological disruption: biocultural interpretations from the eastern Spanish borderlands, USA. In: *Recent contributions to the study of enamel developmental defects* (ed. A. H. Goodman and L. Capasso). Edigrafital, Teramo: 151–69.

Lassen, C., Hummel, S., and Herrmann, B. (1994). Comparison of DNA extraction and amplification from ancient human bone and mummified soft tissue. *International Journal of Legal Medicine* **107**: 152–5.

Latta, T. (1832). Relative to the treatment of cholera by the copious injection of aqueous and saline fluids into the vein. *Lancet* **2**: 274–7.

Lawrence, J. G. and Ochman, H. (1998). Molecular archaeology of the *Escherichia coli* genome. *Proceedings of the National Academy of Sciences, USA* **95**: 9413–17.

Lea, C. H., Hannan, R. S., and Greaves, R. I. N. (1950). The reaction between proteins and reducing sugars in the 'dry' state. *Biochemical Journal* **47**: 626–9.

LeCount, E. R. (1919). The pathologic anatomy of influenzal bronchopneumonia. *Journal of the American Medical Association* **72**: 650–2.

Lederberg, J. (1999). J. B. S. Haldane (1949) on infectious disease and evolution. *Genetics* **153**: 1–3.

Lee-Thorp, J. A. and van der Merwe, N. J. (1991). Aspects of the chemistry of modern and fossil biological apatites. *Journal of Archaeological Science* **18**: 343–54.

Lerner, E. A. and Shoemaker, C. B. (1992). Maxadilan-cloning and functional expression of the gene encoding this potent vasodilator peptide. *Journal of Biological Chemistry* **267**: 1062–6.

Lerner, E. A., Ribeiro, J. M., Nelson, R. J., and Lerner, M. R. (1991). Isolation of maxadilan, a potent vasodilatory peptide from the salivary glands of the sand fly *Lutzomyia longipalpis*. *Journal of Biological Chemistry* **266**: 11234–6.

Levin, B. R. (1993). The accessory genetic elements of bacterial existence conditions and (co)evolution. *Current Opinion in Genetics and Development* **3**: 849–54.

Levin, B. R. (1996). The evolution and maintenance of virulence in microparasites. *Emerging Infectious Diseases* **2**: 93–102.

Levin, E. J. (1970). Healing in congenital osseous syphilis. *American Journal of Roentgenology* **110**: 591–7.

Levine, M. M., DuPont, H. L., Khodabandelou, M., and Hornick, R. B. (1973). Long-term *Shigella* carrier state. *New England Journal of Medicine* **288**: 1169–71.

Levins, R. (1995). Towards an integrated epidemiology. *Trends in Ecology and Evolution* **10**: 304.

Levy, S. (1994). *The antibiotic paradox: how miracle drugs are destroying the miracle*. Plenum Press, New York.

Levy, S. B. (1998). The challenge of antibiotic resistance. *Scientific American* **278**: 46–53.

Lewin, P. K. (1967). Palaeo-electron microscopy of mummified tissue. *Nature* **213**: 416–17.

Li, H.-C., Fujiyoshi, T., Lou, H., Yashiki, S., Sonoda, S., Cartier, L., *et al.* (1999). The presence of ancient human T-cell lymphotropic virus type I provirus DNA in an Andean mummy. *Nature Medicine* **5**: 1428–32.

Li, S., Schulman, J., Itamura, S., and Palese, P. (1993). Glycosylation of neuraminidase determines the neurovirulence of influenza A/WSN/33 virus. *Journal of Virology* **67**: 6667–73.

Lima, H. C. and Titus, R. G. (1996). Effects of sand fly vector saliva on development of cutaneous lesions and the immune response to *Leishmania braziliensis* in BALB/c mice. *Infection and Immunity* **64**: 5442–5.

Lin, Y., Shaw, M., Gregory, V., Cameron, K., Lim, W., Klimov, A., *et al.* (2000). Avian-to-human transmission of H9N2 subtype influenza A viruses: relationship between H9N2 and H5N1 human isolates. *Proceedings of the National Academy of Sciences, USA* **97**: 9654–8.

Lindahl, T. (1993). Instability and decay of the primary structure of DNA. *Nature* **362**: 709–15.

Linder, F. and Grove, R. (1943). *Vital statistics rates in the United States: 1900–1940*. United States Government Printing Office, Washington, DC.

Linse, A. R. (1992). Is bone safe in a shell midden? In: *Deciphering a shell midden* (ed. J. K. Stein). Academic Press, San Diego: 327–45.

Lipshutz, R., Morris, D., Chee, M., Hubbell, E., Kozal, M. J., Shah, J., *et al.* (1995). Using oligonucleotide probe arrays to access genetic diversity. *Biotechniques* **19**: 442–7.

Lipsitch, M. (2001). The rise and fall of antimicrobial resistance. *Trends in Microbiology* **9**: 438–44.

Livingstone, F. B. (1958). Anthropological implications of sickle cell gene distribution in West Africa. *American Anthropologist* **60**: 533–62.

Livingstone, F. B. (1989). Who gave whom hemoglobin S: The use of restriction site haplotype variation for the interpretation of the evolution of the beta S-globin gene. *American Journal of Human Biology* **1**: 289–302.

Lockhart, A. B., Thrall, P. H., and Antonovics, J. (1996). Sexually transmitted diseases in animals: ecological and evolutionary implications. *Biological Reviews* **71**: 415–71.

Logan, A. N. (1994). *Bacterial systematics*. Blackwell Scientific Publications, Boston, MA.

Loreille, O., Roumat, E., Verneau, O., Bouchet, F., and Hanni, C. (2001). Ancient DNA from *Ascaris*: extraction, amplification and sequences from eggs collected in coprolites. *International Journal for Parasitology* **31**: 1101–6.

Loy, T. H. and Hardy, B. L. (1992). Blood residue analysis of 90,000-year-old stone tools from Tabun Cave, Israel. *Antiquity* **66**: 24–35.

Ludwig, G. V., Christensen, B. M., Yuill, T. M., and Schultz, K. T. (1989). Enzyme processing of La Crosse virus glycoprotein G1: a bunyavirus-vector infection model. *Virology* **171**: 108–13.

Ludwig, G. V., Israel, B. A., Christensen, B. M., Yuill, T. M., and Schultz, K. T. (1991). Role of La Crosse virus glycoproteins in attachment of virus to host cells. *Virology* **181**: 564–71.

Ludwig, G. V., Kondig, J. P., and Smith, J. F. (1996). A putative receptor for Venezuelan equine encephalitis virus from mosquito cells. *Journal of Virology* **70**: 5592–9.

Lukashov, V. V. and Goudsmit, J. (2001). Evolutionary relationships among parvoviruses: virus–host coevolution among autonomous primate parvoviruses and links between adeno-associated and avian parvoviruses. *Journal of Virology* **75**: 2729–40.

Ly, T. M. and Müller, H. E. (1990). Ingested *Listeria monocytogenes* survive and multiply in protozoa. *Journal of Medical Microbiology* **33**: 51–4.

Lyman, R. L. (1994). *Vertebrate taphonomy*. Cambridge University Press, Cambridge.

McDaniel, T. K., Jarvis, K. G.,Donnenberg, M. S., and Kaper, J. B. (1996). A genetic locus of enterocyte effacement conserved among diverse enterobacterial pathogens. *Proceedings of the National Academy of Sciences, USA* **92**: 1664–8.

McDonough, K. A., Barnes, A. M., Quan, T. J., Montenieri, J., and Falkow, S. (1993). Mutation in pla gene of *Yersinia pestis* alters the course of the plague bacillus–flea interaction. *Journal of Medical Entomology* **30**: 772–80.

McDonough, M. A. and Butterton, J. R. (1999). Spontaneous tandem amplification and deletion of the shiga toxin operon in *Shigella dysenteriae*. *Molecular Microbiology* **34**: 1058–69.

McGreevy, P. B., Bryan, J. H., Oothuman, P., and Kolstrup, N. (1978). The lethal effects of the cibarial and pharyngeal armatures of mosquitoes on microfilariae. *Transactions of the Royal Society of Tropical Medicine and Hygiene* **72**: 361–8.

Machado, J. O. (1971). Ecology of stingless bees. *Ciencia e Cultura* **23**: 625–33.

Mackay, E. (1931). Architecture and masonry. In: *Mohenjodaro and the Indus civilization* (ed. J. Marshall). Probsthain, London: 262–86.

McKenna, M. C. (1980). Eocene paleolatitude, climate and mammals of Ellesmere Island. *Palaeogeography, Palaeoclimatology, Palaeoecology* **24**: 349–62.

McLean, S. (1931). Roentgenographic and pathologic aspects of congenital osseous syphilis. *American Journal of Diseases of Children* **41**: 130–52, 411–18.

Madden, M., Salo, W. L., Streitz, J., Aufderheide, A. C., Fornaciari, G., Jaramillo, C., *et al.* (2001). Hybridization screening of very short PCR products for paleoepidemiological studies of Chagas' Disease. *Biotechniques* **30**: 102–9.

Maddrell, S. H. and O'Donnell, M. J. (1992). Insect malpighian tubules: V-ATPase action in ion and fluid transport. *Journal of Experimental Biology* **172**: 417–29.

Mallory, J. P. and Mair, V. H. (2000). *The Tarim mummies*. Thames & Hudson, London.

Manchester, K. (1993). Unusual pathological condition in the lower extremities of a skeleton from ancient Israel. *American Journal of Physical Anthropology* **91**: 249–50.

Margulis, L. (1981). *Symbiosis in cell evolution, life and its environment on the early earth*. W. H. Freeman & Co., San Francisco.

Margulis, L., Dolan, M. F., and Guerrero, R. (2000). The chimeric eukaryote: origin of the nucleus from the karyomastigont in amitochondriate protists. *Proceedings of the National Academy of Sciences, USA* **97**: 6954–9.

Marota, I. (1995). A search for treponemal DNA in a Renaissance arm dressing. First reported at the Ancient DNA III Conference, Oxford, 20–22 July, S20.

Marshall, J. (1931). The buildings. In: *Mohenjo-daro and the Indus civilization*. (ed. J. Marshall). Probsthain, London: 15–26.

Martin, L. D. (1994). *Cenozoic climatic history from a biological perspective*. Institute for Tertiary-Quaternary Studies Symposium Series 2: 39–56.

Martin L. D. and Hoffman, R. S. (1987). Pleistocene faunal provinces and Holocene biomes of the central Great Plains. In: *Quaternary environments of Kansas* (ed. W. C. Johnson). Kansas Geological Survey Series 5: 159–65.

Martin, L. D. and Neuner, N. M. (1978). The end of the Pleistocene in North America. *Nebraska Academy of Sciences Transactions* **6**: 117–26.

Martin L. D. and Rothschild, B. M. (1989). Paleopathology and diving mosasaurs. *American Scientist* **77**: 460–7.

Martin, L. D. and Rothschild, B. M. (1998). Earth history and the evolution of sickness. In: *Digging for pathogens: ancient emerging diseases—their evolutionary, anthropological, and archaeological context* (ed. C. L. Greenblatt). Balaban Publishers, Rehovot, Israel: 15–46.

Martin, W. and Müller, M. (1998). The hydrogen hypothesis for the first eukaryote. *Nature* **392**: 37–41.

Martinez, A. and Kolte, R. (1997). Protection of DNA during oxidative stress by the nonspecific DNA-binding protein Dps. *Journal of Bacteriology* **179**: 5188–94.

Masset, C. (1973). Influence du sexe et de l'âge sur la conservation des os humains. In: *L'Homme, hier et aujourd'hui. Recueil d'études en Hommage à André Leroi-Gourhan (Introduction by M. Saubert—a collective work)*. Cujas, Paris: 333–43.

Masurel, N. and Marine, W. M. (1973). Recycling of Asian and Hong Kong influenza A virus hemagglutinins in man. *American Journal of Epidemiology* **97**: 44–9.

Matheson, C., Donoghue, H. D., Fletcher, H., Holton, J., Thomas, M., Pap, I., *et al.* (2000). Tuberculosis in ancient populations: a linkage study. Presented at the 5th International Ancient DNA Conference, Manchester, UK, 12–14 July 2000 (Abstract 33).

Matrosovich, M., Gambaryan, A., Teneberg, S., Piskarev, V., Yamnikova, S., Lvov, D., *et al.* (1997). Avian influenza A viruses differ from human viruses by recognition of sialyloligosaccharides and gangliosides and by a higher conservation of the HA receptor-binding site. *Virology* **233**: 224–34.

Maurelli, A. T., Fernandez, R. E., Bloch, C. A., Rode, C. K., and Fasano, A. (1998). 'Black holes' and bacterial pathogenicity: a large genomic deletion that enhances the virulence of *Shigella* spp. and enteroinvasive *Escherichia coli*. *Proceedings of the National Academy of Sciences, USA* **95**: 3943–8.

May, R. M. and Anderson, R. M. (1990). Parasite–host coevolution. *Parasitology* **100** (Supplement): S89–101.

Mays, C. and Crane-Kramer, G. M. M. (2002). Two probable cases of treponemal disease of late medieval date in England. *American Journal of Physical Anthropology* **119** (in press).

Mays, S. (1992). Taphonomy factors in a human skeletal assemblage. *Circaea* **9**: 54–8.

Mays, S., Taylor, G. M., Legge, A. J., Young, D. B., and Turner-Walker, G. (2001). Paleopathological and biomolecular study of tuberculosis in a medieval skeletal collection from England. *American Journal of Physical Anthropology* **114**: 298–311.

Mbow, M. L., Rutti, B., and Brossard, M. (1994). IFN-gamma, IL-2, and IL-4 mRNA expression in the skin and draining lymph nodes of BALB/c mice repeatedly infested with nymphal *Ixodes ricinus* ticks. *Cellular Immunology* **156**: 254–61.

Mecsas, J. and Strauss, E. (1996). Molecular mechanisms of bacterial virulence: type III secretion and pathogenicity islands. *Emerging Infectious Diseases* **2**: 271–88.

Meindl, R. S. (1987). Hypothesis: a selective advantage for cystic fibrosis heterozygotes. *American Journal of Physical Anthropology* **74**: 39–45.

Ménard V. (1888). *Tuberculose vertébrale*. Asselin et Houzeau, Paris.

Menzel, A. (1994). Perspektiven fuer die Historische Humanoeoekologie durch den Einsatz PCR-gestuetzter Screeningverfahren zur Detektion erblicher und infektioeser Krankheiten an (prae-) historischem Quellenmaterial. Master's thesis, Faculty of Biology, University of Goettingen.

Mertens, P. P., Burroughs, J. N., Walton, A., Wellby, M. P., Fu, H., O'Hara, D., *et al.* (1996). Enhanced infectivity of modified blue tongue virus particles for two insect cell lines and for two Culicoides vector species. *Virology* **217**: 582–93.

Miller, M. F. and Wyckoff, R. W. G. (1963). Proteins in dinosaur bones. *Biochemistry* **60**: 176–8.

Miller, P. L., Nadkarni, P. M., and Carriero, N. M. (1991). Parallel computation and FASTA: confronting the problem of parallel database search for a fast sequence comparison algorithm. *Computer Applications in the Biosciences* **7**: 71–8.

Miller, R. L., Ikram, S., Armelagos, G. J., Walker, R., Harer, W. B., Shiff, C. J., *et al.* (1994). Diagnosis of *Plasmodium falciparum* infections in mummies using the rapid manual ParaSight-F test. *Transactions of the Royal Society of Tropical Medicine and Hygiene* **88**: 31–2.

Milner, G. R., Humpf, D. A., and Harpending, H. C. (1989). Pattern matching of age-at-death distributions in paleodemographic analysis. *American Journal of Physical Anthropology* 80: 49–58.

Mitchell-Olds, T. and Bergelson, J. (2000). Genomics and coevolution. *Current Opinion in Plant Biology* 3: 273–6.

Mitsherlich, E. and Marth, E. H. (1984). *Microbial survival in the environment.* Springer-Verlag, Berlin.

Modiano, D., Luoni, G., Sirima, B. S., Simpore, J., Verra, F., Konate, A., et al. (2001). Haemoglobin C protects against clinical *Plasmodium falciparum* malaria. *Nature* 414: 305–8.

Møller-Christensen, V. (1961). *Bone changes in leprosy.* Munksgaard, Copenhagen.

Molto, A. (1995). A treponematosis 'endemic' to the pre-contact population of the Cape Region of Baja California Sur. In: *L'Origin de la syphilis en Europe—avant ou apres 1493?* (ed. O. Dutour, G. Pàlfi, J. Berato, and J.-P. Brun). Centre Archaeologique du Var, Toulon, France: 176–84.

Montiel, R., Malgosa, A., and Subirà, E. (1991). Overcoming PCR inhibitors in ancient DNA extracts from teeth. *Ancient Biomolecules* 1: 221–5.

Moodie, R. L. (1923). *Paleopathology.* University of Illinois Press, Chicago.

Moran, N. A., Munson, M. A., Baumann, P., and Ishikawa, H. (1993). A molecular clock in endosymbiotic bacteria calibrated using the insect hosts. *Proceedings of the Royal Society of London B, Biological Sciences* 253: 167–71.

Morita, R. Y. (1999). Is H(2) the universal energy source for long-term survival? *Microbial Ecology* 38: 307–20.

Morral, N., Bertranpetit, J., Estivill, X., Nunes, V., Casals, T., Gimenez, J., et al. (1994). The origin of the major cystic fibrosis mutation (delta F508) in European populations. *Nature Genetics,* 7: 169–75.

Moshitch, S., Doll, L., Rubinfield, B. Z., Stocker, B. A., Schoolnik, G. K., Gafni, Y., et al. (1992). Mono- and bi-phasic *Salmonella typhi*: genetic homogeneity and distinguishing characteristics. *Molecular Microbiology* 6: 2589–97.

Moss, W. L. and Biegelow, G. H. (1922). Yaws: an analysis of 1046 cases in the Dominican Republic. *Bulletin of Johns Hopkins Hospital* 33: 43–7.

Mullis, K. B. and Faloona, F. (1987). Specific synthesis of DNA *in vitro* via a polymerase-catalysed chain reaction. *Methods in Enzymology* 155: 335–50.

Murphy, B. and Webster, R. (1996). Orthomyxoviruses. In: *Fields virology, vol. 1* (ed. B. Fields, D. Knipe, and P. Howley). Lippincott-Raven, Philadelphia: 1397–445.

Nagel, R. L. (1994). Origins and dispersion of the sickle gene. In: *Sickle cell disease: basic principles and clinical practice* (ed. S. H. Embury and R. P. Hebbel). Raven Press, New York: 381–94.

Nair, G. B., Ramamurthy, T., Bhattacharya, S. K., Mukhopadhyay, A. K., Garg, S., Battacharya, M. K., et al. (1994). Spread of *Vibrio cholerae* O139 Bengal in India. *Journal of Infectious Diseases* 169: 1029–34.

Nathanson, E. and Cohen, W. (1941). A statistical and roentgen analysis of two hundred cases of bone and joint tuberculosis. *Radiology* 36: 550–67.

Nataro, J. P. and Kaper, J. B. (1998). Diarrheagenic *Escherichia coli. Clinical Microbiology Reviews* 11: 142–201.

Nawrocki, S. P. (1995). Taphonomic processes in historic cemeteries. In: *Bodies of evidence: reconstructing history through skeletal analysis* (ed. A. L. Grauer). Wiley-Liss, New York: 49–66.

Nemeskeri, J. (1963). Die spätmittelalterliche Bevolkrung von Fonyod. *Anthropologia Hungarica* 6: no. 1–2.

Nerlich, A. G., Haas, C. J., Zink, A., Szeimies, U., and Hagedorn, H. G. (1997). Molecular evidence for tuberculosis in an ancient Egyptian mummy. *Lancet* 349: 1760–3.

Neumann, H. W. (1966). A preliminary survey of the paleopathology of an archaic American Indian population. *Bulletin of the Tulane University Medical Faculty* 25: 195–206.

Newman, M. E. and Julig, P. (1989). The identification of protein residues on lithic artefacts from stratified boreal forest site. *Canadian Journal of Archaeology* 13: 119–32.

Newton, B. M. and White, N. (1999). Malaria: new developments in treatment and prevention. *Annual Review of Medicine* 50: 179–92.

Nguyen Tang, A. (1988). *Epidémiologie tropicale.* Report of Agence de Cooperation Culturelle et Technique: 303–21.

Nielsen-Marsh, C. M. and Hedges, R. E. M. (2000). Patterns of diagenesis in bone—1: the effects of site environments. *Journal of Archaeological Science* 27: 1139–50.

Niemann, S., Richter, E., and Rüsch-Gerdes, S. (2000). Differentiation among members of the *Mycobacterium tuberculosis* complex by molecular and biochemical features: evidence for two pyrazinamide-susceptible subtypes of *M. bovis. Journal of Clinical Microbiology* 38: 152–7.

Noe-Nygaard, N. (1988). Taphonomy in archaeology with special emphasis on man as a biasing factor. *Journal of Danish Archaeology* 6: 7–52.

Noordhoek, G. T. A., Hermans, P. W. M., Paul, A. N., Schouls, L. M., van der Sluis, J. J., and van Embden, J. D. A. (1989). *Treponema pallidum* subspecies *pallidum* (Nichols) and *Treponema pallidum* subspecies *pertenue* (CDC 2575) differ in at least one nucleotide: comparison of two homologous antigens. *Microbiological Pathology* 156: 29–42.

Noordhoek, G. T. A., Cockayne, A., Schouls, L. M., Meleon, R. H., Stolz, E., and Van Embden, J. D. A. (1990a). A new attempt to distinguish serologically the subspecies of *Treponema pallidum* causing syphilis and yaws. *Journal of Clinical Microbiology* **28**: 1600–7.

Noordhoek, G. T. A., Wieles, B., Van der Sluis, J. J., and Van Embden, J. D. A. (1990b). Polymerase chain reaction and synthetic DNA probes: a means of distinguishing the causative agents of syphilis and yaws? *Infection and Immunity* **58**: 2011–13.

Noriega, F. G., Barillas-Mury, C., and Wells, M. A. (1994). Dietary control of late trypsin gene transcription in *Aedes aegypti*. *Insect Biochemistry and Molecular Biology* **24**: 627–31.

Noriega, F. G., Wang, X. Y., Pennington, J. E., Barillas-Mury, C. V., and Wells, M. A. (1996). Early trypsin, a female-specific midgut protease in *Aedes aegypti*: isolation, aminoterminal sequence determination, and cloning and sequencing of the gene. *Insect Biochemistry and Molecular Biology* **26**: 119–26

Noriega, F. G., Shah, D. K., and Wells, M. A. (1997). Juvenile hormone controls early trypsin gene transcription in the midgut of *Aedes aegypti*. *Insect Biochemistry and Molecular Biology* **27**: 63–6.

Noriega, F. G., Colonna, A. E., and Wells, M. A. (1999). Increase in the size of the amino acid pool is sufficient to activate translation of early trypsin mRNA in *Aedes aegypti* midgut. *Insect Biochemistry and Molecular Biology* **29**: 243–7.

Norris, S. J. and the *Treponema pallidum* Polypeptide Research Group (1993). Polypeptides of *Treponema pallidum*: progress toward understanding their structural, functional, and immunologic roles. *Microbiological Reviews* **57**: 750–79.

Nuorala, E. (1999). Tuberculosis on the 17th century man-of-war *Kronan*. *International Journal of Osteoarchaeology* **9**: 344–8.

O'Brien, S. J. (1991). Ghetto legacy. *Current Biology* **1**: 209–11.

Ohtani, S., Matsushima, Y., Ohhira, H., and Watanabe, A. (1995). Age-related changes in D-aspartic acid of rat teeth. *Growth, Development and Aging* **59**: 55–61.

Oliver, J. D. (1993). Formation of viable but non-culturable cells. In: *Starvation in bacteria* (ed. S. Kjelleberg). Plenum Press, New York: 239–72.

O'Neill, R. E., Talon, J., and Palese, P. (1998). The influenza virus NEP (NS2 protein) mediates the nuclear export of viral ribonucleoproteins. *EMBO Journal* **17**: 288–96.

O'Rourke, D. H., Hayes, M. G., and Carlyle, S. W. (2000). Ancient DNA studies in physical anthropology. *Annual Review of Anthropology* **29**: 217–42.

Ortner, D. (1992). Skeletal paleopathology: probabilities, possibilities and impossibilities. In: *Disease and demography in the Americas* (ed. J. W. Verano and D. Ubelaker). Smithsonian Institution Press, Washington, DC: 5–14.

Ortner, D. J. (1999). Paleopathology: implications for the history and evolution of tuberculosis. In: *Tuberculosis past and present* (ed. G. Pálfi, O. Dutour, J. Deák, and I. Hutás). Golden Book Publishers and the Tuberculosis Foundation, Budapest–Szeged: 255–62.

Ortner, D. J. and Putschar, W. G. J. (1985). Identification of pathological conditions in human skeletal remains. *Smithsonian Contributions to Anthropology* 28. Smithsonian Institution Press, Washington, DC.

Ortner, D. J., Tuross, N., and Stix, A. I. (1992). New approaches to the study of disease in archeological New World populations. *Human Biology* **64**: 337–60.

O'Shaunessy, W. B. (1832). Report on the chemical pathology of the malignant cholera. *Lancet* **1**: 929–36.

Pääbo, S. (1989). Ancient DNA: extraction, characterization, molecular cloning, and enzymatic amplification. *Proceedings of the National Academy of Sciences, USA* **86**: 1939–43.

Pääbo, S., Gifford, J. A., and Wilson, A. C. (1988). Mitochondrial DNA sequences from a 7000-year-old brain. *Nucleic Acids Research* **16**: 9775–87.

Pääbo, S., Irwin, D. M., and Wilson, A. C. (1990). DNA damage promotes jumping between templates during enzymatic amplification. *Journal of Biological Chemistry* **265**: 4718–21.

Padberg, B. (1992). Empirische Zugaenge zu einer Epidemiologie des Mittelalters. *Sudhoffs Archiv*, **76**: 164–78.

Palese, P. and Compans, R. W. (1976). Inhibition of influenza virus replication in tissue culture by 2-deoxy-2,3-dehydro-N-trifluoroacetylneuraminic acid (FANA): mechanism of action. *Journal of General Virology* **33**: 159–63.

Pàlfi, G. (1991). The first osteoarchaeological evidence of leprosy in Hungary. *International Journal of Osteoarchaeology* **1**: 99–102.

Pàlfi, G. (1993). Maladies, activités et environnements des populations anciennes en Europe centrale et occidentale: approche de paléopathologie comparée. *Acta Biologica Szeged* **39**: 19–24.

Pàlfi, G. and Marcsik, A. (1997). Palaeoepidemiological data of tuberculosis in Hungary. Abstract for the International Congress on the Evolution of Palaeoepidemiology of Tuberculosis, Szeged, Hungary.

Pàlfi, G., Ardagna, Y., Molnar, E., Dutour, O., Panuel, M., Haas, C.-J., *et al.* (1999). Coexistence of tuberculosis and ankylosing spondylitis in a 7–8th century specimen. In: *Tuberculosis past and present* (ed. G. Pálfi, O. Dutour, J. Deák, and I. Hutás). Golden Book Publishers and the Tuberculosis Foundation, Budapest–Szeged: 403–13.

Palkovich, A. M. (1981). Tuberculosis epidemiology in two Arikara skeletal samples: a study of disease impact. In: *Prehistoric tuberculosis in the Americas* (ed. J. E. Buikstra). Northwestern University Archeological Program, Evanston, IL: 161–75.

Palumbi, S. R. (2001). Humans as the world's greatest evolutionary force. *Science* 293: 1786–90.

Pap, I., Józsa, L., Repa, I., Bajzik, G., Lakhani, S. R., Donoghue, H. D., and Spigelman, M. (1999). 18–19th-century tuberculosis in naturally mummified individuals (Vác, Hungary). In: *Tuberculosis past and present* (ed. G. Pálfy, O. Dutour, J. Deák, and I. Hutás). Golden Book Publishers and the Tuberculosis Foundation, Budapest–Szeged: 421–8.

Park, P. (1991). Ice age bacteria return from the dead. *New Scientist* 130: 12.

Park, S.-J., Taton, T. A., and Mirkin, C. A. (2002). Array-based electrical detection of DNA with nanoparticle probes. *Science* 295: 1503–6.

Pasculle, A. W. (1992). The legionellas. In: *The prokaryotes* (2nd edn) (ed. A. Balows, H. G. Trüper, M. Dworkin, W. Harder, and K.-H. Schleifer). Springer Verlag, Berlin, Heidelberg, New York: 3281–303.

Passi, S., Rothschild-Boros, M. C., Fasella, P., Nazzaro-Porro, M., and Whitehouse, D. (1981). An analysis of high performance liquid chromotography to analysis of lipids in archaeological samples. *Journal of Lipid Research* 22: 778–84.

Pate, D. F., Hutton, J. T., and Norrish, K. (1989). Ionic exchange between soil solution and bone: toward a predictive model. *Applied Geochemistry* 4: 303–16.

Patterson, K. D. and Pyle, G. F. (1991). The geography and mortality of the 1918 influenza pandemic. *Bulletin of the History of Medicine* 65: 4–21.

Payne, D. (1988). Did medicated salt hasten the spread of chloroquine resistance in *Plasmodium falciparum*? *Parasitology Today* 4: 112–15.

Pelto, G. H. and Pelto, P. J. (1983). Diet and delocalization. In: *Hunger in history* (ed. R. I. Rotberg). Cambridge University Press, Cambridge: 309–30.

Pennisi, E. (1998). Genome reveals wiles and weak points of syphilis. *Science* 281: 324–5.

Pérez-Pérez, A. (ed.) (1996). *Notes on populational significance of paleopathological conditions: health, illness and death in the past.* Fundació Uriach, Barcelona.

Perna, N. T., Plunkett, G., Burland, V., Mau, B., Glasner, J. D., Rose, D. J., et al. (2001). Genome sequence of enterohaemorrhagic *Escherichia coli* O157:H7. *Nature* 409: 529–33.

Perry, W. L., III, Bass, W. M., Riggsby, W. S., and Sirotkin, K. (1988). The autodegradation of deoxyribonucleic acid (DNA) in human rib bone and its relationship to the time interval since death. *Journal of Forensic Science* 33: 144–53.

Petersen, G. M., Rotter, J. I., Cantor, R. M., Field, L. L., Greenwald, S., Lin, J. S., et al. (1993). The Tay–Sachs disease gene in North American Jewish populations: geographic variation and origin. *American Journal of Human Genetics* 35: 1258–69.

Petroni, G., Spring, S., Schleifer, K.-H., Verni, F., and Rosati, G. (2000). Defensive extrusive ectosymbionts of *Euplotidium* (Ciliophora) that contain microtubule-like structures are bacteria related to Verrucomicrobia. *Proceedings of the National Academy of Sciences, USA* 97: 1813–17.

Pfister, C. (1986). Grauzone des Lebens. Die aggregative Bevoelkerungsgeschichte des Kantons Bern vor dem Problem der totgeborenen und ungetauft verstorbenen Kinder. *Jb Schweizerische Gesellschaft fuer Familienforschung* 1986: 21–44.

Poinar, H. N. and Stankiewcz, B. A. (1999). Protein preservation and DNA retrieval from ancient tissues. *Proceedings of the National Academy of Sciences, USA* 96: 8426–31.

Poinar, H., Poinar, G. O., Jr, and Cano, R. J. (1993). Molecular phylogeny of an extinct legume *Hymenaea protera* from Dominican amber. *Nature* 363: 677.

Poinar, H. N., Hoss, M., Bada, J. L., and Pääbo, S. (1996). Amino acid racemization and the preservation of ancient DNA. *Science* 272: 864–6.

Poinar, H. N., Hofreiter, M., Spaulding, W. G., Martin, P. S., Stankiewicz, B. A., Bland, H., et al. (1998). Molecular coproscopy: dung and diet of the extinct ground sloth *Nothrotheriops shastensis*. *Science* 281: 402–6.

Poinar, H., Kuch, M., and Pääbo, S. (2001a). Molecular analyses of oral polio vaccine samples. *Science* 292: 743–4.

Poinar, H. N., Kuch, M., Sobolik, K. D., Barnes, I., Stankiewicz, A. B., Kuder, T., et al. (2001b). A molecular analysis of dietary diversity for three archaic Native Americans. *Proceedings of the National Academy of Sciences, USA* 98: 4317–22.

Pollitzer, R. (1959). *Cholera*. World Health Organization, Geneva.

Pond, F. R., Gibson, I., Lalucat, J., and Quackenbush, R. L. (1989). R-body producing bacteria. *Microbiological Reviews* 53: 25–67.

Potts, M. (1994). Dessication tolerance of prokaryotes. *Microbiology Reviews* 58: 755–805.

Powell, M. L. (1995). Why call it syphilis? Treponematosis before 1492 in the southeastern United States of America. In: *L'Origin de la syphilis en Europe—avant ou apres 1493?* (ed. O. Dutour, G. Pàlfi, J. Berato, and J.-P. Brun). Centre Archaeologique du Var, Toulon, France: 158–63.

Preer, J. R., Jr, Preer, L. B., and Jurand, A. (1974). Kappa and other endosymbionts in *Paramecium aurelia*. *Bacteriological Reviews* 38: 113–63.

Priest, F. G. (1993). Systematics and ecology of *Bacillus*. In: *Bacillus subtilis and other Gram positive bacteria: biochemistry, physiology, and molecular genetics* (ed. A. L. Sonenshein, J. A. Hoch, and R. Losick). American Society for Microbiology, Washington, DC: 3–16.

Pupo, G. M., Karaolis, D. K., Lan, R., and Reeves, P. R. (1997). Evolutionary relationships among pathogenic and nonpathogenic *Escherichia coli* strains inferred from multilocus enzyme electrophoresis and *mdh* sequence studies. *Infection and Immunity* 65: 2685–92.

Pupo, G. M., Lan, R., and Reeves, P. R. (2000). Multiple independent origins of *Shigella* clones of *Escherichia coli* and convergent evolution of many of their characteristics. *Proceedings of the National Academy of Sciences, USA* 97: 10567–72.

Purdy, B. A. (1988). *Wet site archaeology*. Telford Press, Caldwell, NJ.

Pusch, C. M., Nicholson, G. J., Bachmann, L., and Scholz, M. (2000). Degenerate oligonucleotide-primed preamplification of ancient DNA allows the retrieval of authentic DNA sequences. *Analytical Biochemistry* 279: 118–22.

Quetel, C. (1990). *History of syphilis*. Johns Hopkins University Press, Baltimore, MD.

Rader, A. E. and Murphy, J. R. (1988). Nucleotide sequences and comparison of the hemolysin determinants of *Vibrio cholerae* El Tor RV79(Hly⁺) and RV79(Hly⁻) and classical 569B(Hly⁻). *Infection and Immunity* 56: 1414–19.

Rafi, A., Spigelman, M., Stanford, J., Lemma, E., Donoghue, H., and Zias, J. (1994). *Mycobacterium leprae* DNA from ancient bone detected by PCR. *Lancet* 243: 1360–61.

Rafi, A., Spigelman, M., Stanford, J., Donoghue, H., Lemma, E., and Zias, J. (1994). DNA of *Mycobacterium leprae* detected by PCR in ancient bone. *International Journal of Osteoarchaeology* 4: 287–90.

Ramachandra, R. N. and Wikel, S. K. (1992). Modulation of host-immune responses by ticks (Acari: Ixodidae): effect of salivary gland extracts on host macrophages and lymphocyte cytokine production. *Journal of Medical Entomology* 29: 818–26.

Ramachandra, R. N. and Wikel, S. K. (1995). Effects of *Dermacentor andersoni* (Acari: Ixodidae) salivary gland extracts on *Bos indicus* and *B. taurus* lymphocytes and macrophages: *in vitro* cytokine elaboration and lymphocyte blastogenesis. *Journal of Medical Entomology* 32: 338–45.

Rameckers, J., Hummel, S., and Herrmann, B. (1997). How many cycles does a PCR need? Determinations of cycle numbers depend on the number of targets and the reaction efficiency factor. *Naturwissenschaften* 84: 259–62.

Rao, S. R. (1973). *Lothal and the Indus civilization*. Asia Publishing House, New York.

Raoult, D., Aboudharam, G., Crubézy, E., Larrouy, G., Ludes, B., and Drancourt, M. (2000). Molecular identification by 'suicide' PCR, of *Yersinia pestis* as the agent of Medieval Black Death. *Proceedings of the National Academy of Sciences, USA* 97: 12800–3.

Raoult, D., La Scola, B., Enea, M., Fournier, P. E., Roux, V., Fenollar, F., *et al.* (2001). A flea-associated Rickettsia pathogenic for humans. *Emerging and Infectious Diseases* 7: 73–81.

Razin, S., Yogev, D., and Naot, Y. (1998) Molecular biology and pathogenicity of mycoplasmas. *Microbiology and Molecular Biology Reviews* 62: 1094–156.

Redman, C. L. (1999). Human impact on ancient environments. University of Arizona Press, Tucson.

Reeve, C. A., Amy, P. S., and Matin, A. (1984). Role of protein synthesis in the survival of carbon-starved *Escherichia coli* K-12. *Journal of Bacteriology* 160: 1041–6.

Reichman, L. and Tanne, J. (2001). Timebomb: the global epidemic of multi-drug-resistant tuberculosis. McGraw-Hill, Washington, DC.

Reid, A. H. and Taubenberger, J. K. (1999). The 1918 flu and other influenza pandemics: 'over there, and back again'. *Laboratory Investigation* 79: 95–101.

Reid, A. H., Fanning, T. G., Hultin, J. V., and Taubenberger, J. K. (1999). Origin and evolution of the 1918 'Spanish' influenza virus hemagglutinin gene. *Proceedings of the National Academy of Sciences, USA* 96: 1651–6.

Reid, A. H., Fanning, T. G., Janczewski, T. A., and Taubenberger, J. K. (2000a). Characterization of the 1918 'Spanish' influenza virus neuraminidase gene. *Proceedings of the National Academy of Sciences, USA* 97: 6785–90.

Reid, S. D., Herbelin, C. J., Bumbaugh, A. C., Selander, R. K., and Whittam, T. S. (2000b). Parallel evolution of virulence in pathogenic *Escherichia coli*. *Nature* 406: 64–7.

Reitz, E. J. and Wing, E. S. (1999). *Zooarchaeology*. Cambridge University Press, Cambridge.

Rennie, J. (1992). Living together. *Scientific American* 266: 122–33.

Resnick, D. and Niwayama, G. (1988). Diagnosis of bone and joint disorders. Saunders, Philadelphia.

Reynaud, P., Moreau, N., Bouttevin, C., Rigaud, P., Castex, D., Dutour, O., and Signoli, M. (1996). *D.F.S. de l'opération archéologique effectuée sur le cimétière des Fédons, à Lambesc dans le cadre des fouille A.F.A.N. sur le tracé T.G.V. Vol. 5*. Report of Association pour les fouilles archéologiques nationales. Avignon-Marseille.

Rhine, J. A. and Taylor, R. K. (1994). TcpA pilin sequences and colonization requirements for O1 and O139 *Vibrio cholerae*. *Molecular Microbiology* 13: 1013–20.

Ribeiro, J. M. (1987a). *Ixodes dammini*: salivary anticomplement activity. *Experimental Parasitology* **64**: 347–53.

Ribeiro, J. M. (1987b). Vector salivation and parasite transmission. *Memorias do Instituto Oswaldo Cruz* **82**: 1–3.

Ribeiro, J. M. (1992). Characterization of a vasodilator from the salivary glands of the yellow fever mosquito *Aedes aegypti*. *Journal of Experimental Biology* **165**: 61–71.

Ribeiro, J. M. C. (1996). Common problems of arthropod vectors of disease. In: *Biology of disease vectors* (ed. W. C. Marquardt and B. J. Beaty). University Press of Colorado, Niwot, CO: 393–416.

Ribeiro, J. M. (2000). Blood-feeding in mosquitoes: probing time and salivary gland anti-haemostatic activities in representatives of three genera (*Aedes, Anopheles, Culex*). *Medical and Veterinary Entomology* **14**: 142–8.

Ribeiro, J. M. and Nussenzveig, R. H. (1993). Nitric oxide synthase activity from a hematophagous insect salivary gland. *FEBS Letters* **330**: 165–8.

Ribeiro, J. M., Makoul, G. T., Levine, J., Robinson, D. R., and Spielman, A. (1985). Antihemostatic, anti-inflammatory, and immunosuppressive properties of the saliva of a tick, *Ixodes dammini*. *Journal of Experimental Medicine* **161**: 332–44.

Ribeiro, J. M., Rossignol, P. A., and Spielman, A. (1986). Blood-finding strategy of a capillary-feeding sandfly, *Lutzomyia longipalpis*. *Comparative Biochemistry and Physiology A* **83**: 683–6.

Ribeiro, J. M., Vachereau, A., Modi, G. B., and Tesh, R. B. (1989). A novel vasodilatory peptide from the salivary glands of the sand fly *Lutzomyia longipalpis*. *Science* **243**: 212–14.

Ribeiro, J. M., Weis, J. J., and Telford, S. R., III (1990). Saliva of the tick *Ixodes dammini* inhibits neutrophil function. *Experimental Parasitology* **70**: 382–8.

Ribeiro, J. M., Hazzard, J. M., Nussenzveig, R. H., Champagne, D. E., and Walker, F. A. (1993). Reversible binding of nitric oxide by a salivary heme protein from a bloodsucking insect. *Science* **260**: 539–41.

Ribeiro, J. M., Katz, O., Pannell, L. K., Waitumbi, J., and Warburg, A. (1999). Salivary glands of the sand fly *Phlebotomus papatasi* contain pharmacologically active amounts of adenosine and 5'-AMP. *Journal of Experimental Biology* **202**: 1551–9.

Ribeiro, J. M., Charlab, R., Rowton, E. D., and Cupp, E. W. (2000). *Simulium vittatum* (Diptera: Simulidae) and *Lutzomyia longipalpis* (Diptera: Psychodidae) salivary gland hyaluronidase activity. *Journal of Medical Entomology* **37**: 743–7.

Ribeiro, J. M., Charlab, R., and Valenzuela, J. G. (2001). The salivary adenosine deaminase activity of the mosquitoes *Culex quinquefasciatus* and *Aedes aegypti*. *Journal of Experimental Biology* **204**: 2001–10.

Robbins, J., Rosteck, P., Haynes, J. R., Freyer, G., Cleary, M. L., Kalter, H. D., *et al.* (1979). The isolation and partial characterization of recombinant DNA containing genomic sequences from the goat. *Journal of Biological Chemistry* **254**: 6187–95.

Roberts, C. and Manchester, K. (1995). *The archaeology of disease*. Cornell University Press, Ithaca, NY.

Roberts, C., Lucy, D., and Manchester, K. (1994). Inflammatory lesions of ribs: an analysis of the Terry Collection. *American Journal of Physical Anthropology* **95**: 169–82.

Rogers, R. A. (1986). Language, human subspeciation, and Ice Age barriers in Northern Siberia. *Canadian Journal of Anthropology* **5**: 11–22.

Rogers, R. A., Martin, L. D., and Nicklas, T. D. (1990). Ice-age geography and the distribution of native North American languages. *Journal of Biogeography* **17**: 131–43.

Rogers, R. A., Rogers, L. A., Hoffmann, R. S., and Martin, L. D. (1991). Native American biological diversity and the biogeographic influence of Ice Age refugia. *Journal of Biogeography* **18**: 623–30.

Rogers, R. A., Rogers, L. A., and Martin, L. D. (1992). How the door opened: the peopling of the New World. *Human Biology* **64**: 281–302.

Rogers, R. A., Rogers, L. A., and Martin, L. D. (1994). The distribution of Nadene speakers and needle-leaf evergreen forests: a western North American parallel. *Nebraska Academy of Science TER-QUA Symposium Series 2*: 107–17.

Rollo, F. (1985). Characterisation by molecular hybridization of RNA fragments isolated from ancient (1400 BC) seeds. *Theoretical Applied Genetics* **71**: 330–3.

Rollo, F., Asci, W., Antonini, S., Marota, I., and Ubaldi, M. (1994). Molecular ecology of a neolithic meadow: the DNA of the grass remains from the archaeological site of the Tyrolean Iceman. *Experientia* **50**: 576–83.

Rosenberg, C. E. (1987). *The cholera years: the United States in 1832, 1849, and 1866*. University of Chicago Press, Chicago.

Rosenberg, R. (1985). Inability of *Plasmodium knowlesi* sporozoites to invade *Anopheles freeborni* salivary glands. *American Journal of Tropical Medicine and Hygiene* **34**: 687–91.

Rothschild, B. M. (1982). *Rheumatology: a primary care approach*. Yorke Medical Press, New York.

Rothschild, B. M. (2000). Rheumatoid arthritis at a time of passage. *Journal of Rheumatology* **28**: 245–50.

Rothschild, B. M. and Heathcote, G. M. (1993). Characterization of the skeletal manifestations of the treponemal disease yaws as a population phenomenon. *Clinical Infectious Diseases* **17**: 198–203.

Rothschild, B. M. and Heathcote, G. M. (1995). Characterization of gout in a skeletal population sample: presumptive diagnosis in a Micronesian population. *American Journal of Physical Anthropology* **98**: 519–25.

Rothschild, B. M. and Martin, L. D. (1987). Avascular necrosis in diving Cretaceous mosasaurs. *Science* **236**: 75–7.

Rothschild, B. M. and Martin, L. D. (1993). *Paleopathology: disease in the fossil record*. CRC Press, London.

Rothschild, B. M. and Rothschild, C. (1992). Inflammatory arthritis in the first century Negev. *Progress in Rheumatology* **5**: 112–15.

Rothschild, B. M. and Rothschild, C. (1993). Nineteenth century spondyloarthropathy independent of socioeconomic status: lack of skeletal collection bias. *Journal of Rheumatology* **20**: 314–19.

Rothschild, B. M. and Rothschild, C. (1994a). No laughing matter: spondyloarthropathy in Hyaenidae. *Journal of Zoo and Wildlife Medicine* **25**: 250–63.

Rothschild, B. M. and Rothschild, C. (1994b). Yaws, mine and ours: treponemal disease transitions in prehistory. *Journal of Comparative Human Biology* **45**: S115.

Rothschild, B. M. and Rothschild, C. (1995). Treponemal disease revisited: skeletal discriminators for yaws, bejel and venereal syphilis. *Clinical Infectious Diseases* **20**: 1402–8.

Rothschild, B. M. and Rothschild, C. (1996a). Treponemal disease in the New World. *Current Anthropology* **37**: 555–61.

Rothschild, B. M. and Rothschild, C. (1996b). Analysis of treponemal disease in North Africa: the case for bejel in the Sudan, but absence in West North Africa. *Human Evolution* **11**: 11–15.

Rothschild, B. M. and Rothschild, C. (1997). Congenital syphilis in the archeologic record: diagnostic insensitivity of osseous lesions. *International Journal of Osteoarchaeology* **7**: 39–42.

Rothschild, B. M. and Rothschild, C. (1999). Evolution of osseous/radiologic signs of tuberculosis. In: *Tuberculosis past and present* (ed. G. Pálfi, O. Dutour, J. Deák, and I. Hutás). Golden Book Publishers and the Tuberculosis Foundation, Budapest–Szeged: 293–8.

Rothschild, B. M. and Turnbull, W. (1987). Treponemal infection in a Pleistocene bear. *Nature* **329**: 61–2.

Rothschild, B. M. and Woods, R. J. (1989). Spondyloarthropathy in gorillas. *Seminars in Arthritis and Rheumatism* **18**: 267–76.

Rothschild, B. M. and Woods, R. J. (1991). Spondyloarthropathy: erosive arthritis in representative defleshed bones. *American Journal of Physical Anthropology* **85**: 25–34.

Rothschild, B. M. and Woods, R. J. (1992a). Spondyloarthropathy as an Old World phenomenon. *Seminars in Arthritis and Rheumatism* **21**: 306–16.

Rothschild, B. M. and Woods, R. J. (1992b). Character of pre-Columbian North American spondyloarthropathy. *Journal of Rheumatology* **19**: 1229–35.

Rothschild, B. M. and Woods, R. J. (1993). Arthritis in New World monkeys: osteoarthritis, calcium pyrophosphate deposition disease and spondyloarthropathy. *International Journal of Primatology* **14**: 61–78.

Rothschild, B. M., Wang, X., and Cifelli, R. (1993). Spondyloarthropathy in Ursidae: a sexually transmitted disease? *Research and Exploration* **9**: 382–4.

Rothschild, B. M., Wang, X.-M., and Shoshani, J. (1994). Spondyloarthropathy in proboscideans. *Journal of Zoo and Wildlife Medicine* **25**: 360–6.

Rothschild, B. M., Rothschild, C., and Hill, M. C. (1995a). Origin and transition of varieties of treponemal disease in the New World. *American Journal of Physical Anthropology* **20** (Supplement): 185.

Rothschild, B. M., Hershkovitz, I., and Rothschild, C. (1995b). Origin of yaws in Pleistocene East Africa: *Homo erectus* KNM-ER 1808. *Nature* **378**: 343–4.

Rothschild, B. M., Molnar, R. E., and Sebes, J. I. (1997). Mycobacteriosis in the Pliocene: imaging documentation in a fossil kangaroo. *Annals of Internal Medicine* **127**: 168–9.

Rothschild, B. M., Arriaza, B., Woods, R. J., and Dutour, O. (1999). Spondyloarthropathy identified as the etiology of Nubian erosive arthritis. *American Journal of Physical Anthropology* **109**: 259–67.

Rothschild, B. M., Calderon F. L., Coppa, A., and Rothschild, C. (2000). First European exposure to syphilis: the Dominican Republic at the time of Columbian contact. *Clinical Infectious Diseases* **31**: 936–41.

Rothschild, B. M., Martin, L. D., Lev, G., Bercovier, H., Kahila Bar-Gal, G., Greenblatt, C. L., et al. (2001). *Mycobacterium tuberculosis* complex DNA from an extinct bison dated 17,000 years before the present. *Clinical Infectious Diseases* **33**: 305–11.

Rothschild, C. and Rothschild, B. M. (1994). Syphilis, yaws and bejel: population distribution in North America. *American Journal of Physical Anthropology* **94**: 74–5.

Roubik, D. W. (1989). Ecology and natural history of tropical bees. Cambridge University Press, New York.

Rouquette, C., Ripio, M.T, Pellegrini, E., Bolla, J. M., Tascon, R. I., Vazquez-Boland, J. A., et al. (1996). Identification of a ClpC ATPase required for stress tolerance and *in vivo* survival of *Listeria monocytogenes*. *Molecular Microbiology* **5**: 977–87.

Rowbotham, T. J. (1980). Preliminary report on the pathogenicity of *Legionella pneumophila* for freshwater and soil amoebae. *Journal of Clinical Pathology* **33**: 1179–83.

Rowbotham, T. J. (1986). Current views on the relationships between amoebae, legionellae and man. *Israel Journal of Medical Sciences* **22**: 678–89.

Rowley, M. J., Rich, P. V., Rich, T. H., and Mackay, I. R. (1986). Immunoreactive collagen in avian and mammalian fossils. *Naturwissenschaften* **73**: 620–2.

Rudenko, G., Cross, M., and Borst, P. (1998). Changing the end: antigenic variation orchestrated at the telomeres of African trypanosomes. *Trends in Microbiology* **6**: 113–6.

Ruffer, M. A. (1921). *Studies in the paleopathology of Egypt.* University of Chicago Press, Chicago.

Ruffer, M. A. and Rietti, A. (1912). On osseous lesions in ancient Egyptians. *Journal of Pathological Bacteriology* **16**: 439–65.

Ruwende, C., Khoo, S. C., Snow, R. W., Yates, S. N., Kwiatkowski, D., Gupta, S., *et al.* (1995). Natural selection of hemi- and heterozygotes for G6PD deficiency in Africa by resistance to severe malaria. *Nature* **376**: 246–9.

Ryan, F. (1992). *Tuberculosis: the greatest story never told: the human story of the search for the cure of tuberculosis and the new global threat.* Swift Publishers, Sheffield, UK.

Saint-Hoyme, L. E. (1969). On the origins of New World paleopathology. *American Journal of Physical Anthropology* **31**: 295–302.

Sales, M. P. U., Taylor, G. M., Hughes, S., Yates, M., Hewinson, G., Young, D. B., *et al.* (2001). Genetic diversity among *Mycobacterium bovis* isolates: a preliminary study of strains from animal and human sources. *Journal of Clinical Microbiology* **39**: 4558–62.

Salo, W. L., Aufderheide, A. C., Buikstra, J., and Holcomb, T. A. (1994). Identification of *Mycobacterium tuberculosis* DNA in a pre-Columbian Peruvian mummy. *Proceedings of the National Academy of Sciences, USA* **91**: 2091–4.

Samuelson, J., Lerner, E., Tesh, R., and Titus, R. (1991). A mouse model of *Leishmania braziliensis braziliensis* infection produced by coinjection with sand fly saliva. *Journal of Experimental Medicine* **173**: 49–54.

Sanchez, J. L. and Taylor, D. N. (1997). Cholera. *Lancet* **349**: 1825–30.

Sandweiss, D. H. (1996). Environmental change and its consequences for human society on the central Andean coast: a malacological perspective. In: *Case studies in environmental archaeology* (ed. E. J. Reitz, L. A. Newsom, and S. J. Scudder). Plenum Press, New York: 127–47.

Santos, A. L. and Roberts, C. A. (2001). A picture of tuberculosis in young Portuguese people in the early 20th century: a multidisciplinary study of the skeletal and historical evidence. *American Journal of Physical Anthropology* **115**: 38–49.

Sattenspiel, L. and Harpending, H. C. (1983). Stable populations and skeletal age. *American Antiquity* **48**: 489–98.

Saunders, S. R., Herring, D. A., and Boyce, G. (1995). Can skeletal samples accurately represent the living population they come from? The St. Thomas, Cemetery site, Belleville, Ontario. In: *Bodies of evidence: reconstructing history through skeletal analysis* (ed. A. L. Grauer). Wiley-Liss, New York: 69–89.

Schafer, J. R., Kawaoka, Y., Bean, W. J., Suss, J., Senne, D., and Webster, R. G. (1993). Origin of the pandemic 1957 H2 influenza A virus and the persistence of its possible progenitors in the avian reservoir. *Virology* **194**: 781–8.

Schermer, S. J., Fisher, A. K., and Hodges, D. C. (1994). Endemic treponematosis in prehistoric western Iowa. In: *Skeletal biology in the Great Plains: migration, warfare, health, and subsistence* (ed. D. W. Owsley and R. L. Jantz). Smithsonian Institution Press, Washington, DC: 109–16.

Schlein, Y., Jacobson, R. L., and Shlomai, J. (1991). Chitinase secreted by *Leishmania* functions in the sandfly vector. *Proceedings of the Royal Society of London B, Biological Sciences* **245**: 121–6.

Schmid, E. N., Müller, K.-D., and Michel, R. (2001). Evidence for bacteriophages within *Neochlamydia hartmannellae*, an obligate endoparasitic bacterium of the free-living amoeba *Hartmannella vermiformis*. *International Journal of Endocytobiosis and Cell Research* **14**: 115–19.

Schmidt, H. J., Görtz, H.-D., Pond, F., and Quackenbush, R. L. (1988). Characterization of *Caedibacter* endonucleobionts from the macronucleus of *Paramecium caudatum* and the identification of a mutant with blocked R body synthesis. *Experimental Cell Research,* **174**: 49–57.

Schmidt, T., Hummel, S., and Herrmann, B. (1995). Evidence of contamination in PCR laboratory disposables. *Naturwissenschaften* **82**: 423–31.

Schneider, P., Rosat, J. P., Bouvier, J., Louis, J., and Bordier, C. (1992). *Leishmania major*: differential regulation of the surface metalloprotease in amastigote and promastigote stages. *Experimental Parasitology* **75**: 196–206.

Scholtissek, C., Rohde, W., Von Hoyningen, V., and Rott, R. (1978). On the origin of the human influenza virus subtypes H2N2 and H3N2. *Virology* **87**: 13–20.

Scholtissek, C., Ludwig, S., and Fitch, W. (1993). Analysis of influenza A virus nucleoproteins for the assessment of molecular genetic mechanisms leading to new phylogenetic virus lineages. *Archives of Virology* **131**: 237–50.

Schultz, M. (1999). The role of tuberculosis in infancy and childhood in prehistoric and historic population. In: *Tuberculosis past and present* (ed. G. Pálfi, O. Dutour, J. Deák, and I. Hutás). Golden Book Publishers and the Tuberculosis Foundation, Budapest–Szeged: 503–10. ·

Schwan, T. G. and Piesman, J. (2000). Temporal changes in outer surface proteins A and C of the Lyme disease-associated spirochete, *Borrelia burgdorferi*, during the chain of infection in ticks and mice. *Journal of Clinical Microbiology* **38**: 382–8.

Schwan, T. G., Piesman, J., Golde, W. T., Dolan, M. C., and Rosa, P. A. (1995). Induction of an outer surface protein on *Borrelia burgdorferi* during tick feeding. *Proceedings of the National Academy of Sciences, USA* **92**: 2909–13.

Schwartz, T. R., Schwartz, E. A., Mieszerski, L., McNally, L., and Kobilinsky, L. (1991). Characterization of deoxyribonucleic acid (DNA) obtained from teeth subjected to various environmental conditions. *Journal of Forensic Science* **36**: 979–90.

Scudder, S. J. (1993). Human influence on pedogenesis: midden soils on a southwest Florida Pliestocene dune island. Master's thesis, Department of Soil and Water Science, University of Florida, Gainsville, FL.

Selander, R. K., Beltran, P., Smith, N. H., Helmuth, R., Rubin, F. A., Kopecko, D. J., *et al.* (1990). Evolutionary genetic relationships of clones of *Salmonella* serovars that cause human typhoid and other enteric fevers. *Infection and Immunity* **58**: 2262–75.

Sensabaugh, G. F., Wilson, A. C., and Kirk, P. L. (1971). Protein stability in preserved biological remains. *International Journal of Biochemistry* **2**: 545–57.

Setlow P. (1994). Mechanisms which contribute to the long-term survival of spores of *Bacillus* species. *Journal of Applied Bacteriology Symposium Supplement* **76**: 49S–60S.

Setlow, P. (1995). Mechanisms for the preservation of damage to DNA in spores of *Bacillus* species. *Annual Review of Microbiology* **49**: 29–54.

Shadan, F. F. and Villarreal, L. P. (1993). Coevolution of persistently infecting small DNA viruses and their hosts linked to host-interactive regulatory domains. *Proceedings of the National Academy of Sciences, USA* **90**: 4117–21.

Shakespeare, M. (2001). *Zoonoses*. Pharmaceutical Press, London

Shimanuki, H. (1978). Bacteria. In: *Honey bee, pests, predators, and diseases* (ed. R. A. Morese). Cornell University Press, Ithaca, NY: 28–47.

Shipman, P. (1981). Application of scanning electron microscopy to taphonomic problems. *Annals of the New York Academy of Sciences* **376**: 357–85.

Shipman, P., Foster, G., and Schoeninger, M. (1984). Burnt bones and teeth: an experimental study of color, morphology, crystal structure and shrinkage. *Journal of Archaeological Science* **11**: 307–25.

Shirai, H., Nishibuchi, M., Ramamurthy, T., Bhattacharya, S. K., Pal, S. C., and Takeda, Y. (1991). Polymerase chain reaction for detection of the cholera enterotoxin operon of *Vibrio cholerae*. *Journal of Clinical Microbiology* **29**: 2517–21.

Shope, R. E. (1936). The incidence of neutralizing antibodies for swine influenza virus in the sera of human beings of different ages. *Journal of Experimental Medicine* **63**: 669–84.

Shope, R. E. and Lewis, P. A. (1931). Swine influenza, I, II, III. *Journal of Experimental Medicine* **54**: 349–85.

Signoli, M. (1998). Etude anthropologique des crises démographiques en contexte épidémique : Aspect paléo- et biodémographiques de la peste en Provence. Thèse d'Université. Université de la Méditerranée.

Signoli, M. and Dutour, O. (1997a). Etude anthropologique d'un charnier de la Grande Peste de Marseille (1720–1722): premiers résultats. *Anthropologie et Préhistoire* **108**: 147–58.

Signoli, M. and Dutour, O. (1997b). Le charnier du couvent de l'Observance: 1722. *Provence Historique* **189**: 469–88.

Signoli, M., Chausserie-Laprée, J., and Dutour, O. (1995). Etude anthropologique d'un charnier de la peste de 1720–1721 à Martigues. *Préhistoire et Anthropologie Méditerranéennes* **4**: 173–89.

Signoli, M., Leonetti, G., Astoul, P., Boetsch, G., and Dutour, O. (1997a). Mortality rate by phthisis in Marseilles from 1875 to 1892. Abstract for International Congress on the Evolution and Palaeoepidemiology of Tuberculosis, Szeged, Hungary.

Signoli, M., Léonetti, G., and Dutour, O. (1997b). The Great Plague of Marseilles (1720–1722): new anthropological data. *Acta Biologica* **42**: 123–33.

Signoli, M., Leonetti, G., Astoul, P., Boetsch, G., Dutour, O. (1997c). Mortality rate by phthisis in Marseilles from 1875 to 1892. *International congress on the Evolution and Palaeoepidemiolgy of Tuberculosis*. Szeges, Hungary. Abstract.

Signoli, M., Bello, S., and Dutour, O. (1998a). Marseille, 1722: le charnier de l'Observance et la rechute épidémique de la Grande Peste. *Médecine Tropicale* **58**: 7–13.

Signoli, M., Bello, S., Pàlfi, G., and Dutour, O. (1998b). Palaeoepidemiological study of the plague epidemics in Provence (XVIth–XVIIIth centuries): demographic aspects. Twelfth European Meeting of the Paleopathology Association, Prague-Pilsen, Czech Republic, August 26–29, 1998.

Signoli, M., Castex, D., Bizot, B., Duday, H., and Dutour, O. (2001). La gestion des cadavres en temps de peste. Colloque international: La Peste : entre épidémies et sociétés. Session n° 2. Apports des archives biologiques et de l'Archéologie. Marseille, 23 au 26 juillet 2001: 25.

Signoli, M., Chevé, D., Boëtsch, G., and Dutour, O. (2002). *Profil démographique et diffusion spatiale de la peste en*

Europe: le modèle de l,épidémie provençale de 1720–1723. Éditions du CTHS, Paris (in press).

Simonsen, L., Clarke, M. J., Schonberger, L. B., Arden, N. H., Cox, N. J., and Fukuda, K. (1998). Pandemic versus epidemic influenza mortality: a pattern of changing age distribution. *Journal of Infectious Diseases* **178**: 53–60.

Simonsen, L., Fukuda, K., Schonberger, L. B., and Cox, N. J. (2000). The impact of influenza epidemics on hospitalizations. *Journal of Infectious Diseases* **181**: 831–7.

Skibo, J. M. (1992). *Pottery function: a use-alteration perspective.* Plenum Press, New York.

Skinner, M. (1995). Osseous treponemal disease: limits on our understanding. In: *L'Origin de la syphilis en Europe — avant ou apres 1493?* (ed. O. Dutour, G. Pàlfi, J. Berato, and J.-P. Brun). Centre Archaeologique du Var, Toulon, France: 191–201.

Smith, P. and Kolska-Horowitz, L. R. (1998). Culture, environment and disease: Paleo-anthropological findings for the Southern Levant. In: *Digging for pathogens: ancient emerging diseases — their evolutionary, anthropological, and archaeological context* (ed. C. L. Greenblatt). Balaban, Rehovot, 201–40.

Sneath, P. H. A. (1986). Endospore-forming gram-positive rods and cocci. In: *Bergey's manual of systematic bacteriology* (ed. P. H. A. Sneath, N. S. Mair, M. E. Sharpe, and J. G. Holt). Williams and Wilkins, Baltimore, MD: vol. 2, 1104–207.

Snodgrass, R. E. (1943). The feeding apparatus of biting and disease-carrying flies: a wartime contribution to medical entomology. *Smithsonian Miscellaneous Collections* **104**: 1–51.

Snow, J. (1855). *On the mode of communication of cholera.* 2nd edn (1966 printing). Churchill, London.

Soeda, E., Maruyama, T., Arrand, J. R., and Griffin, B. E. (1980). Host-dependent evolution of three papova viruses. *Nature* **285**: 165–7.

Sola, C., Devallois, A., Horgen, L., Maisetti, J., Filliol, I., Legrand, E., and Rastogi, N. (1999). Tuberculosis in the Caribbean: using spacer oligonucleotide typing to understand strain origin and transmission. *Emerging Infectious Diseases* **5**: 404–14.

Sonoda, S., Li, H. C., Cartier, L., Nunez, L., and Tajima K. (2000). Ancient HTLV type 1 provirus DNA of Andean mummy. *AIDS Research and Human Retroviruses* **16**: 1753–6.

Sorrel, E. and Sorrel-Dejerine, C. (1932). *Tuberculose osseuse et ostéo-articulaire.* Masson, Paris.

Spence, K. D., Karlinsey, J. E., Kyriakides, T. R., Patil, C. S., and Minnick, M. F. (1992). Regulation and synthesis of selected bacteria-induced proteins in *Manduca sexta. Insect Biochemistry and Molecular Biology* **22**: 321–31.

Spigelman, M. (1996). The archeologist and ancient biomolecules: field sampling strategies to enhance recovery. *Papers of the Institute of Archeology* **7**: 69–74.

Spigelman, M. and Donoghue, H. D. (2001). Unusual pathological condition in the lower extremities of a skeleton from ancient Israel. *American Journal of Physical Anthropology* **114**: 92–3.

Spigelman, M. and Greenblatt, C. L. (1998). A guide to digging for pathogens. In: *Digging for pathogens: ancient emerging diseases — their evolutionary, anthropological, and archaeological context* (ed. C. L. Greenblatt). Balaban Publishers, Rehovot, Israel: 345–61.

Spigelman, M. and Lemma, E. (1993). The use of the polymerase chain reaction to detect *Mycobacterium tuberculosis* in ancient skeletons. *International Journal of Osteoarchaeology* **3**: 137–43.

Spigelman, M., Matheson, C., Lev, G., Greenblatt, C., and Donoghue, H. D. (2003). Confirmation of the presence of *Mycobacterium tuberculosis* complex-specific DNA in three archaeological specimens. *International Journal of Osteoarchaeology* (in press).

Spring, J. H. (1990). Endocrine regulation of diuresis in insects. *Journal of Insect Physiology* **36**: 13–22.

Springer, N., Ludwig, W., Drozanski, V., Amann, R., and Schleifer, K.-H. (1992). The phylogenetic status of *Sarcobium lyticum*, an obligate intracellular bacterial parasite of small amoeba. *FEMS Microbiology Letters* **96**: 199–202.

Springer, N., Ludwig, W., Amann, R., Schmidt, H. J., Görtz, H.-D., and Schleifer, K.-H. (1993). Occurrence of fragmented 16S rRNA in an obligate bacterial endosymbiont of *Paramecium caudatum. Proceedings of the National Academy of Sciences, USA* **90**: 9892–5.

Springer, N., Amann, R., Ludwig, W., Schleifer, K.-H., and Schmidt, H. (1996). *Polynucleobacter necessarius*, an obligate bacterial endosymbiont of the hypotrichous ciliate *Euplotes aediculatus*, is a member of the beta-subclass of Proteobacteria. *FEMS Microbiology Letters* **135**: 333–6.

Sreevatsan, S., Pan, X., Stockbauer, K. E., Connell, N. D., Kreiswirth, B. N. Whittam, T. S., *et al.* (1997). Restricted structural gene polymorphism in the *Mycobacterium tuberculosis* complex indicates evolutionarily recent global dissemination. *Proceedings of the National Academy of Sciences, USA* **94**: 9869–74.

Stankiewicz, B. A., Briggs, D. E. G., Evershed, R. P., Flannery, M. G., and Wuttke, M. (1997). Preservation of chitin in 25-million-year-old fossils. *Science* **276**: 1541–3.

Stankiewicz, B. A., Poinar, H. N., Briggs, D. E. G., Evershed, R. P., and Poinar, G. O., Jr (1998). Chemical preservation of plants and insects in natural resins. *Proceedings of the Royal Society B, Biological Sciences* **265**: 641–7.

Stark, K. R. and James, A. A. (1995). A factor Xa-directed anticoagulant from the salivary glands of the yellow fever mosquito *Aedes aegypti*. *Experimental Parasitology* **81**: 321–31.

Stark, K. R. and James, A. A. (1996). Salivary gland anticoagulants in culicine and anopheline mosquitoes (Diptera: Culicidae). *Journal of Medical Entomology* **33**: 645–50.

Stark, K. R. and James, A. A. (1998). Isolation and characterization of the gene encoding a novel factor Xa-directed anticoagulant from the yellow fever mosquito, *Aedes aegypti*. *Journal of Biological Chemistry* **273**: 20802–9.

Stast, P. (1974). HLA-A antigens in mummified pre-Columbian tissues. *Science* **183**: 864–6

Stead, W. W., Eisenach, K. D., Cave, M. D., Beggs, M. L., Templeton, G. L. Thoen, C. O., et al. (1995). When did *Mycobacterium tuberculosis* infection first occur in the New World? *American Journal of Respiratory and Critical Care Medicine* **151**: 1267–8.

Steckel, R. H. (ed.) (1996). Health index. *HEA Newsletter* **10**.

Steinbock, R. T. (1976). *Paleopathological diagnosis and interpretation*. Charles Thomas, Springfield, IL.

Steinert, M., Birkness, K., White, E., Fields, B., and Quinn, F. (1998). *Mycobacterium avium* bacilli grow saprozoically in coculture with *Acanthamoeba polyphaga* and survive within cyst walls. *Applied and Environmental Microbiology* **64**: 2256–61.

Stephens, J. C., Reich, D. E., Goldstein, D. B., Shin, H. D., Smith, M. W., Carrington, M., et al. (1998). Dating the origin of the CCR5-Delta32 AIDS-resistance allele by the coalescence of haplotypes. *American Journal of Human Genetics* **62**: 1507–15.

Sterling, T. R., Pope, D. S., Bishai, W. R., Harrington, S., Gershon, R. R., and Chaisson, R. E. (2000). Transmission of *Mycobacterium tuberculosis* from a cadaver to an embalmer. *New England Journal of Medicine* **342**: 246–8.

Stine, O. C. (2001). Utilizing microarrays of bacterial genes. *SIGBIO Newsletter* **21**: 23–5.

Stirland, A. (1995). Evidence for pre-Columbian treponematosis in medieval Europe. In: *L'Origin de la syphilis en Europe—avant ou apres 1493?* (ed. O. Dutour, G. Pàlfi, J. Berato, and J.-P. Brun). Centre Archaeologique du Var, Toulon, France: 109–15.

Stirland, A. and Waldron, T. (1990). The earliest cases of tuberculosis in Britain. *Journal of Archaeological Science* **17**: 221–30.

Stone, M. H. (1997). *Healing the mind*. Norton, New York.

Storey, R. (1992). Patterns of susceptibility to dental enamel defects in the deciduous dentition of a Precolumbian skeletal population. In: *Recent contributions to the study of enamel developmental defects* (ed. A. H. Goodman and L. Capasso). Edigrafital, Teramo: 171–83.

Stuart-Macadam, P. (1985). Porotic hyperostosis: representative of a childhood condition. *American Journal of Physical Anthropology* **66**: 391–8.

Stuart-Macadam, P. (1987a). A radiographic study of porotic hyperostosis. *American Journal of Physical Anthropology* **74**: 511–20.

Stuart-Macadam, P. (1987b). New evidence to support the anemia theory. *American Journal of Physical Anthropology* **74**: 521–6.

Stuart-Macadam, P. and Kent, S. (ed.) (1992). *Diet, demography, and disease: changing perspectives on anemia*. Aldine de Gruyter, New York.

Subbarao, K., Klimov, A., Katz, J., Regnery, H., Lim, W., Hall, H., et al. (1998). Characterization of an avian influenza A (H5N1) virus isolated from a child with a fatal respiratory illness. *Science* **279**: 393–6.

Sunagawa, K., Sirirungsi, W., Nakazato, I., Hirayasu, T., and Iwamasa, T. (1995). Pathologic studies and comparison of the virulence of herpes simplex virus type 2 from Okinawa, Japan and Chiang Mai, Thailand. *International Journal of Experimental Pathology* **76**: 255–62.

Swerdlow, D. L. and Ries, A. A. (1993). *Vibrio cholerae* non-O1—the eighth pandemic? *Lancet* **342**: 382–3.

Swofford, D. L. (1990). Phylogenetic analysis using PARSIMONY (computer program), Version 3.0. Illinois Natural History Survey, Champaign, IL.

Taramelli, D., Monti, D., Basilico, N., Parapini, S., Omodeo-Salé, F., and Olliaro, P. (1999). A fine balance between oxidised and reduced haem controls the survival of intererythrocytic plasmodia. In: *Parassitologia* (ed. Coluzzi, M. and Bradley, D.). Lombardo Editore, Rome: 205–8.

Taubenberger, J., Reid, A. H., Krafft, A. E., Bijward, K. E., and Fanning, T. G. (1997). Initial genetic characterization of the 1918 'Spanish' influenza virus. *Science* **275**: 1793–6.

Taubenberger, J. (1998). Influenza virus hemagglutinin cleavage into HA1, HA2: no laughing matter. *Proceedings of the National Academy of Sciences, USA* **95**: 9713–15.

Taubenberger, J.K., Reid, A. H., and Fanning, T. G. (2000). The 1918 influenza virus: a killer comes into view. *Virology* **274**: 241–5.

Taylor, G. M., Crossey, M., Saldanha, J., Waldron, T. (1996). DNA from *Mycobacterium tuberculosis* identified in medieval human skeleton remains using polymerase chain reaction. *Journal of Archaeological Science* **23**: 789–98.

Taylor, G. M., Rutland, P., and Molleson, T. (1997). A sensative polymerase chain reaction for the detection of *Plasmodium* species DNA in ancient human remains. *Ancient Biomolecules* **1**: 193–203.

Taylor, G., Goyal, M., Legge, A., Shaw, R. J., and Young, D. (1999). Genotyping analysis of *Mycobacterium tuberculosis* from medieval human remains. *Microbiology* **145**: 899–904.

Taylor, G. M., Widdison, S., Brown, I. N., and Young, D. (2000). A medieval case of lepromatous leprosy from 13–14th century Orkney, Scotland. *Journal of Archaeological Science* **27**: 1133–8.

Taylor, G. M., Mays, S., Legge, A. J., and Young, D. B. (2001). Genetic analysis of tuberculosis in human remains. *Ancient Biomolecules* **3**: 267–80.

Terra, W. R. (1990). Evolution of digestive systems of insects. *Annual Review of Entomology* **35**: 181–200.

Thanassi, D. G. and Hultgren, S. J. (2000). Multiple pathways allow protein secretion across the bacterial outer membrane. *Current Opinions in Cell Biology* **12**: 420–30.

Theodos, C. M. and Titus, R. G. (1993). Salivary gland material from the sand fly *Lutzomyia longipalpis* has an inhibitory effect on macrophage function *in vitro*. *Parasite Immunology* **15**: 481–7.

Theodos, C. M., Ribeiro, J. M., and Titus, R. G. (1991). Analysis of enhancing effect of sand fly saliva on *Leishmania* infection in mice. *Infection and Immunity* **59**: 1592–8.

Thom, S., Warhurst, D., Drasar, B. S. (1992). Association of *Vibrio cholerae* with fresh water amoebae. *Journal of Medical Microbiology* **36**: 303–6.

Thomas, R. H., Schaffner, W., Wilson, A. C., and Pääbo, S. (1989). DNA phylogeny of the extinct marsupial wolf. *Nature* **340**: 465–7.

Thompson, J. N. and Burdon, T. J. (1992). Gene-for-gene coevolution between plants and parasites. *Nature* **360**: 121–5.

Tishkoff, S. A., Varkonyi, R., Cahinhinan, N., Abbes, S., Argyropoulos, G., Destro-Bisol, G., *et al.* (2001). Haplotype diversity and linkage disequilibrium at human G6PD: recent origin of alleles that confer malarial resistance. *Science* **293**: 455–62.

Titus, R. G. and Ribeiro, J. M. (1988). Salivary gland lysates from the sand fly *Lutzomyia longipalpis* enhance *Leishmania* infectivity. *Science* **239**: 1306–8.

Titus, R. G., Theodos, C. M., Shankar, A. H., and Hall, L. R. (1994). Interactions between *Leishmania major* and macrophages. *Immunology Series* **60**: 437–59.

Torrey, E. F. (1980). *Schizophrenia and civilization*. J. Aronson, New York.

Torrey, E. F. (1999). Epidemiological comparison of schizophrenia and bipolar disorder. *Schizophrenia Research* **39**: 101–6.

Torrey, E. F., Miller, J., Rawlings, R., and Yolken, R. H. (1997). Seasonality of births in schizophrenia and bipolar disorder: a review of the literature. *Schizophrenia Research* **28**: 1–38.

Truswell, A. S. and Hansen, J. D.L (1976). Medical research among the !Kung. In: *Kalahari hunter gatherers* (ed. R. B. Lee and I. Devore). Harvard University Press, Cambridge, MA: 166–95.

True, H. and Lindquist, S. L. (2000). A yeast prion provides a mechanism for genetic variation and phenotypic diversity. *Nature* **407**: 477–83.

Turner, C. G. (1986). The first Americans: the dental evidence. *National Geographic Research* **2**: 37–46.

Tuross, J. D. (1993). The other molecules in ancient bone: noncollagenous protiens and DNA. In: *Molecular archaeology of prehistoric human bone* (ed. J. Lambert and G. Grupe). Springer, Berlin: 275–94.

Tuross, J. D. (1994). The biochemistry of ancient DNA in bone. *Experientia* **50**: 530–5.

Tuross, N., Eyre, D. R., Holtrop, M. E., Glimcher, M. J., and Hare, P. E. (1980). Collagen in fossil bones. In: *Biogeochemistry of amino acids* (ed. P. E. Hare, T. C. Hoering, and K. King). John Wiley & Sons, New York: 53–63.

Ubaldi, M., Luciani, S., Marota, I., Fornaciari, G., Cano, R. J., and Rollo, F. (1998). Sequence analysis of bacterial DNA in the colon of an Andean mummy. *American Journal of Physical Anthropology* **107**: 285–95.

Ubelaker, D. H. (1980). Human skeletal remains from OGSE-80, a Preceramic site on the Sta. Elena Peninsula, coastal Ecuador. *Journal of the Washington Academy of Sciences* **70**: 3–24.

Ubelaker, D. H. (1988) [1990]. Restos de esqueletos humanos del Sitio OGSE-80. In: *La prehistoria temprana de la península de Santa Elena, Ecuador: Cultura Las Vegas* (ed. Karen E. Stothert). Museos del Banco Central del Ecuador, Guayaquil, Ecuador: 105–32.

Ubelaker, D. H. (1992). Porotic hyperostosis in prehistoric Ecuador. In: *Diet, demography, and disease: changing perspectives on anemia* (ed. P. Stuart-Macadam and S. Kent). Aldine de Gruyter, New York: 201–17.

Ubelaker, D. H. (1995). Osteological and archival evidence for disease in historic Quito, Ecuador. In: *Grave reflections, portraying the past through cemetery studies* (ed. S. R. Saunders and A. Herring). Canadian Scholars Press Inc., Toronto, Canada: 223–39.

Ubelaker, D. H. (1996a). Skeletons testify: anthropology in forensic science. AAPA Luncheon Address, April 12, 1996. *Yearbook of Physical Anthropology* **39**: 229–44.

Ubelaker, D. H. (1996b). The remains of Dr Carl Austin Weiss: anthropological analysis. *Journal of Forensic Sciences* **41**: 60–79.

Ubelaker, D. H. (1996c). The population approach in paleopathology: a case study from Ecuador. In: *Notes on populational significance of paleopathological conditions: health, illness and death in the past* (ed. A. Pérez-Pérez). Fundació Uriach, Barcelona: 37–54.

Ubelaker, D. H. (1997). *Skeletal biology of human remains from La Tolita, Esmeraldas Province, Ecuador*. Smithsonian Contributions to Anthropology 41. Smithsonian Institution Press, Washington, DC.

Ubelaker, D. H. (2000). *Human skeletal remains from La Florida, Quito, Ecuador*. Smithsonian Contributions to Anthropology 43. Smithsonian Institution Press, Washington, DC.

Ubelaker, D. H. and Grant, L. G. (1989). Human skeletal remains: preservation or reburial? *Yearbook of Physical Anthropology* 32: 249–87.

Ubelaker, D. H. and Pap, I. (1997). Health profiles of a Bronze Age population from northeastern Hungary. *Annales Historico-Naturales Musei Nationalis Hungarici* 88: 271–96.

Ubelaker, D. H. and Pap, I. (1998). Skeletal evidence for health and disease in the Iron Age of northeastern Hungary. *International Journal of Osteoarchaeology* 8: 231–51.

Ubelaker, D. H., Berryman, H. E., Sutton, T. P., and Ray, C. E. (1991). Differentiation of hydrocephalic calf and human calvariae. *Journal of Forensic Sciences* 36: 801–12.

Ubelaker, D. H., Katzenberg, M. A., and Doyon, L. G. (1995). Status and diet in precontact highland Ecuador. *American Journal of Physical Anthropology* 97: 403–11.

Ubelaker, D. H., Jones, E. B., Donoghue, H. D., and Spigelman, M. (2000). Skeletal and molecular evidence for tuberculosis in a forensic case. *Anthropologie* 38: 193–200.

Ulrich, M. M., Perizonius, W. R., Spoor, C. F., Sandberg, P., and Vermeer, C. (1987). Extraction of osteocalcin from fossil bones and teeth. *Biochemical and Biophysical Research Communications* 149: 712–19.

Urioste, S., Hall, L. R., Telford, S. R., III, and Titus, R. G. (1994). Saliva of the Lyme disease vector, *Ixodes dammini*, blocks cell activation by a nonprostaglandin E2-dependent mechanism. *Journal of Experimental Medicine* 180: 1077–85.

Valenzuela, J. G., Walker, F. A., and Ribeiro, J. M. (1995). A salivary nitrophorin (nitric-oxide-carrying hemoprotein) in the bedbug *Cimex lectularius*. *Journal of Experimental Biology* 198: 1519–26.

Valenzuela, J. G., Charlab, R., Galperin, M. Y., and Ribeiro, J. M. (1998). Purification, cloning, and expression of an apyrase from the bed bug *Cimex lectularius*. A new type of nucleotide-binding enzyme. *Journal of Biological Chemistry* 273: 30583–90.

Valenzuela, J. G., Francischetti, I. M., and Ribeiro, J. M. (1999). Purification, cloning, and synthesis of a novel salivary anti-thrombin from the mosquito *Anopheles albimanus*. *Biochemistry* 38: 11209–15.

van der Zanden, A. G., Hoentjen, A. H., Heilmann, F. G., Weltevreden, E. F., Schouls, L. M., and Van Embden, J. D. A. (1998). Simultaneous detection and strain differentiation of *Mycobacterium tuberculosis* complex in paraffin wax embedded tissues and in stained microscopic preparations. *Journal of Clinical Pathology and Molecular Pathology* 51: 209–14.

van Valen, L. (1971). Group selection and the evolution of dispersal. *Evolution* 25: 591–8.

Vasan, S., Zhang, X., Zhang, X., Kapurniotu, A., Bernhagen, J., Teichberg, S., *et al.* (1996). An agent cleaving glucose-derived protien crosslinks *in vitro* and *in vivo*. *Nature* 382: 275–8.

Verano, J. W. and Ubelaker, D. H. (1992). *Disease and demography in the Americas*. Smithsonian Institution Press, Washington, DC.

Vinetz, J. M., Valenzuela, J. G., Specht, C. A., Aravind, L., Langer, R. C., Ribeiro, J. M., *et al.* (2000). Chitinases of the avian malaria parasite *Plasmodium gallinaceum*, a class of enzymes necessary for parasite invasion of the mosquito midgut. *Journal of Biological Chemistry* 275: 10331–41.

Visvesvara, G. S. (1995). Pathogenic and opportunistic free-living amoebae. In: *Manual of clinical microbiology* (6th edn) (ed. P. R. Murray). American Society for Microbiology Press, Washington, DC: 1196–203.

Vogel, F. and Motulsky, A. G. (1997). *Human genetics, problems and approaches*. Springer, Berlin.

Vogel, J. P., Andrews, H. L., Wong, S. K., and Isberg, R. R. (1998). Conjugative transfer by the virulence system of *Legionella pneumophila*. *Science* 279: 873–6.

Volkman, S. K., Barry, A. E., Lyons, E. J., Nielsen, K. M., Thomas, S. M., Choi, M., *et al.* (2001). Recent origin of *Plasmodium falciparum* from a single progenitor. *Science* 293: 482–4.

Voong, K., White, W., Titball, R. W., and Prentice, M. B. (2001). Attempted PCR of *Yersinia pestis* from dental pulp of Black Death victims. Colloque international: La Peste: entre épidémies et sociétés. Session n° 2: Apports des archives biologiques et de l'Archéologie. Marseille, 23 au 26 juillet 2001: 40.

Waldor, M. K. and Mekalanos, J. J. (1994). Emergence of a new cholera pandemic: molecular analysis of virulence determinants in *Vibrio cholerae* O139 and development of a live vaccine prototype. *Journal of Infectious Disease* 170: 278–83.

Waldron, T. (1985). DISH at Merton Priory: evidence for a new occupational disease? *British Medical Journal* 291: 1762–3.

Waldron, T. (1987). The relative survival of the human skeleton: implications for palaeopathology. In: *Approaches to archaeology and forensic science* (ed. A. Boddington,

A. N. Garland and R. C. Janaway). Manchester University Press, Manchester: 43–54.

Waldron, T. (1994). *Counting the dead: the epidemiology of skeletal populations*. Wiley, Chichester, UK

Waldron, T. (1999). The paleoepidemiology of tuberculosis: some problems considered. In: *Tuberculosis past and present* (ed. G. Pálfi, O. Dutour, J. Deák, and I. Hutás). Golden Book Publishers and the Tuberculosis Foundation, Budapest–Szeged: 471–8.

Walker, E. M., Arnett, J. K., Heath, J. D., and Norris, S. J. (1991). *Treponema pallidum* subsp. *pallidum* has a single, circular chromosome with a size of ~900 kilobase pairs. *Infection and Immunity* 60: 1568–76.

Walker, E. M., Howell. J. K, You, Y., Hoffmaster, A. R., Heath, J. D. Weinstock, G. M., *et al.* (1995). Physical map of the genome of *Treponema pallidum* subsp. *pallidum* (Nichols). *Journal of Bacteriology* 177: 1797–8.

Walker, P. L. (1996). Integrative approaches to the study of ancient health: an example from the Santa Barbara Channel area of southern California. In: *Notes on populational significance of paleopathological conditions: health, illness and death in the past* (ed. A. Pérez-Pérez). Fundació Uriach, Barcelona: 97–105.

Wang, G. C. and Wang, Y. (1997). Frequency of formation of chimeric molecules as a consequence of PCR coamplification of 16S rRNA genes from mixed bacterial genomes. *Applied and Environmental Microbiology* 63: 4645–50.

Wang, H. and Nuttall, P. A. (1994). Excretion of host immunoglobulin in tick saliva and detection of IgG-binding proteins in tick hemolymph and salivary glands. *Parasitology* 109: 525–30.

Warburg, A. and Schlein, Y. (1986). The effect of postbloodmeal nutrition of *Phlebotomus papatasi* on the transmission of *Leishmania major*. *American Journal of Tropical Medicine and Hygiene* 35: 926–30.

Warburg, A., Saraiva, E., Lanzaro, G. C., Titus, R. G., and Neva, F. (1994). Saliva of *Lutzomyia longipalpis* sibling species differs in its composition and capacity to enhance leishmaniasis. *Philosophical Transactions of the Royal Society of London B, Biological Sciences* 345: 223–30.

Waters, A. P., Higgins, D. G., and McCutchan, T. F. (1991). *Plasmodium falciparum* appears to have arisen as a result of lateral transfer between avian and human hosts. *Proceedings of the National Academy of Sciences, USA* 88: 3140–4.

Wayne, L. G., Brenner, D. J., Colwell, R. R., Grimont, P. A. D., and Kandler, O. (1987). Report of the *ad hoc* committee on reconciliation of approaches to bacterial systematics. *International Journal of Systematic Bacteriology* 37: 463–4.

Webb, S. (1995). *Palaeopathology of aboriginal Australians: health and disease across a hunter-gatherer continent.* Cambridge University Press, Cambridge.

Webster, R. G. and Rott, R. (1987). Influenza virus A pathogenicity: the pivotal role of hemagglutinin. *Cell* 50: 665–6.

Webster, R. G., Bean, W. J., Gorman, O. T., Chambers, T. M., and Kawaoka, Y. (1992). Evolution and ecology of influenza A viruses. *Microbiological Reviews* 56: 152–79.

Weed, L. A. and Baggenstoss, A. H. (1951). The isolation of pathogens from tissues of embalmed human bodies. *American Journal of Clinical Pathology* 2: 1114–20.

Weiner, S. and Lowenstein, H. A. (1980). Well-preserved fossil mollusc shells: characterization of mild diagenetic processes. In: *Biogeochemistry of amino acids* (ed. P. E. Hare, T. C. Hoering, and K. King). John Wiley and Sons, New York: 94–114.

Weiner, S., Goldberg, P., and Bar-Yosef, O. (1993). Bone preservation in Kebara Cave, Israel, using on-site Fourier transform infrared spectrometry. *Journal of Archaeological Science* 20: 613–27.

Weiss, E. (1992). Rickettsias. *Encyclopedia Microbiologica* 3: 585–610.

Weiss, R. A. (2000). Certain promise and uncertain peril: the debate on xenotransplantation. *EMBO Reports* 1: 2–4.

Westbrook, P., van der Meide, P. H., van der Weykloppers, J. S., van der Sluis, R. J., de Leeuw, J. W., and de Jong, E. W. (1979). Fossil macromolecules and cephalopod shells: characterization, immunological response and diagnosis. *Paleobiology* 5: 151–67.

Wheeler, C. H. and Coast, G. M. (1990). Assay and characterization of diuretic factors in insects. *Journal of Insect Physiology* 36: 23–34.

Wheeler, C. M., Coleman, J. L., Habicht, G. S., and Benach, J. L. (1989). Adult *Ixodes dammini* on rabbits: development of acute inflammation in the skin and immune responses to salivary gland, midgut, and spirochetal components. *Journal of Infectious Diseases* 159: 265–73.

Wheeler, M. (1966). *Civilizations of the Indus Valley and beyond.* Thames & Hudson, London.

Wheeler, M. (1968). *The Indus civilization.* Cambridge University Press, Cambridge.

Whelen, A. C. and Wikel, S. K. (1993). Acquired resistance of guinea pigs to *Dermacentor andersoni* mediated by humoral factors. *Journal of Parasitology* 79: 908–12.

WHO (2000). *WHO Expert Committee on Malaria.* World Health Organization Technical Report Series 892: i–v, 1–74.

Wiesenfeld, S. L. (1967). Sickle-cell trait in human biological and cultural evolution—development of agriculture causing increased malaria is bound to gene-pool changes causing malaria reduction. *Science* 157: 1134–40.

Wikel, S. K. (1982a). Immune responses to arthropods and their products. *Annual Review of Entomology* **27**: 21–48.

Wikel, S. K. (1982b). Influence of *Dermacentor andersoni* infestation on lymphocyte responsiveness to mitogens. *Annals of Tropical Medicine and Parasitology* **76**: 627–32.

Wikel, S. K. (1984). Immunomodulation of host responses to ectoparasite infestation—an overview. *Veterinary Parasitology* **14**: 321–39.

Wikel, S. K. and Osburn, R. L. (1982). Immune responsiveness of the bovine host to repeated low-level infestations with *Dermacentor andersoni*. *Annals of Tropical Medicine and Parasitology* **76**: 405–14.

Wikel, S. K., Graham, J. E., and Allen, J. R. (1978). Acquired resistance to ticks. IV. Skin reactivity and *in vitro* lymphocyte responsiveness to salivary gland antigen. *Immunology* **34**: 257–63.

Williamson, B. S. (2000). Direct testing of rock painting pigments for traces of haemoglobin at Rose Cottage Cave, South Africa. *Journal of Archaeological Science* **27**: 755–62.

Willman, M. R. (2001). Studying the historic migrations of the Irish potato famine pathogen using ancient DNA. *Trends in Plant Science* **6**: 450.

Wills, C. (1996). *Yellow fever, black goddess: the coevolution of people and plagues*. Addison-Wesley, New York.

Wilson, M. (1996). The emergence of new diseases: learning from the past and preparing for the future. In: *Infectious diseases* (ed. E. P. Fischer and S. Klose). Piper, Munich: 11–69.

Wilson, M. E. (1994). Disease in evolution: introduction. *Annals of the New York Academy of Sciences* **740**: 1–12.

Woese, C. R. (2000). Interpreting the universal phylogenetic tree. *Proceedings of the National Academy of Sciences, USA* **97**: 8392–6.

Wolf, S. G., Frenkiel, D., Arad, T., Finkel, S. E., Kolter, R., and Minsky, A. (1999). DNA protection by stress-induced biocrystallization. *Nature* **400**: 83–6.

Wood, J. W., Milner, G.R, Harpending, H. C., and Weiss, K. M. (1992). The osteological paradox: problems of inferring prehistoric health from skeletal samples. *Current Anthropology* **33**: 343–70.

Wright, D. J. (1971). Syphilis and Neanderthal man. *Nature* **229**: 409.

Wyss, A. (2001). Paleontology—digging up fresh clues about the origin of mammals. *Science* **292**: 1496–7.

Xu, G., Wilson, W., Mecham, J., Murphy, K., Zhou, E. M., and Tabachnick, W. (1997). VP7: an attachment protein of bluetongue virus for cellular receptors in *Culicoides variipennis*. *Journal of General Virology* **78**: 1617–23.

Yamada, H., Muramatsu, S., and Mizuno, T. (1990). An *Escherichia coli* protein that preferentially binds to sharply curved DNA. *Journal of Biochemistry* **108**: 420–5.

Yohe, R. M., II, Newman, M. E., and Schneider, J. S. (1991). Immunological identification of small-mammal proteins on aboriginal milling equipment. *American Antiquity* **56**: 659–66.

Yolken, R. H., Bachmann, S., Rouslanova, I., Lillehoj, E., Ford, G., Torrey, E. F., *et al.* (2001). Antibodies to *Toxoplasma gondii* in individuals with first-episode schizophrenia. *Clinical Infectious Diseases* **32**: 842–4.

Yoon, K.-H., Cho, S.-N., Lee, M.-K., Abalos, R. M., Cellona, R. V., Fajardo, T. T., Jr, *et al.* (1993). Evaluation of polymerase chain reaction amplification of *Mycobacterium leprae*-specific repetitive sequence in biopsy specimens from leprosy patients. *Journal of Clinical Microbiology* **31**: 895–9.

Zanotto, P. M., Gould, E. A., Gao, G. F., Harvey, P. H., and Holmes, E. C. (1996). Population dynamics of flaviviruses revealed by molecular phylogenies. *Proceedings of the National Academy of Sciences, USA* **93**: 548–53.

Zhou, N. N., Senne, D. A., Landgraf, J. S., Swenson, S. L., Erickson, G., Rossow, K., *et al.* (1999). Genetic reassortment of avian, swine, and human influenza A viruses in American pigs. *Journal of Virology* **73**: 8851–6.

Zhu, K., Bowman, A. S., Brigham, D. L., Essenberg, R. C., Dillwith, J. W., and Sauer, J. R. (1997). Isolation and characterization of americanin, a specific inhibitor of thrombin, from the salivary glands of the lone star tick *Amblyomma americanum* (L.). *Experimental Parasitology* **87**: 30–8.

Zias, J. (1998). Tuberculosis and the Jews in the ancient Near East: the biocultural interaction. In: *Digging for pathogens: ancient emerging diseases—their evolutionary, anthropological, and archaeological context* (ed. C. L. Greenblatt). Balaban Press, Rehovot, Israel: 277–97.

Zimmerman, M. R. (1979). Pulmonary and osseous tuberculosis in Egyptian mummy. *Bulletin of the New York Academy of Medicine* **55**: 604–8.

Zink, A., Haas, C. J., Reischl, U., Szeimies, U., and Nerlich, A. G. (2001). Molecular analysis of skeletal tuberculosis in an ancient Egyptian population. *Journal of Medical Microbiology* **50**: 355–66.

Zinsser, H. (1949). *Ratten, Laeuse und die Weltgeschichte.* Stuttgart [translated from *Rats, lice, and history*, (1934) Little, Brown, Boston].

Zohary, D., Tchernov, E., and Kolska Horwitz, L. (1998). The role of unconscious selection in the domestication of sheep and goats. *Journal of Zoology* **245**: 129–35.

Zorab, P. A. (1961). The historical and prehistorical background of ankylosing spondylitis. *Proceedings of the Royal Society of Medicine* **54**: 415–20.

Zumárraga, M. Bigi. F. Alito, A. Romano, M. I., and Cataldi, A. (1999). A 12.7 kb fragment of the *Mycobacterium tuberculosis* genome is not present in *Mycobacterium bovis*. *Microbiology* **145**: 893–7.

Index

Printed in the United States
86504LV00003B/25/A

9 780198 509011